The Formation
of the
American Scientific Community

Sally Gregory Kohlstedt

THE FORMATION OF THE AMERICAN SCIENTIFIC COMMUNITY

The American Association
for the Advancement of Science
1848-60

University of Illinois Press

URBANA   CHICAGO   LONDON

Publication of this work has been supported by a grant from the Oliver M. Dickerson Fund. The Fund has been established by Mr. Dickerson (Ph.D., Illinois, 1906) to enable the University of Illinois Press to publish selected works in American history, designated by the executive committee of the Department of History. Publication has also been supported by a grant from the American Association for the Advancement of Science.

LIBRARY OF CONGRESS CATALOGING IN PUBLICATION DATA

Kohlstedt, Sally Gregory, 1943–
  The formation of the American scientific community.

  Bibliography: p.
  Includes index.
  1. American Association for the Advancement of Science.
2. Science—History—United States. I. Title.
Q11.A53K63   506'.273   75-37748
ISBN 0-252-00419-1

TO MY MOTHER
AND THE MEMORY OF MY FATHER

Contents

Preface

Aside from a few cursory sketches, there has been no analysis of the scientific organization whose role was fundamental in creating a viable scientific community in the United States. Retrospective presidents and scientific popularizers eulogized the broad-based inclusive character of the American Association for the Advancement of Science. Biographers of eminent scientists prefer to hint at its limitations as promoter of basic scientific research in the nineteenth century. This study proposes to analyze both the nature and the function of the association during its formative years, 1848 to 1860, assuming only that the scientific organization was unique in its broadly based, national membership and activities and noting that it was expansive in its initial expectations of the role it would play in developing American science.

The term "institutional history" is eschewed today because it connotes a dry, superficial chronicle of persons and events. Yet the historian of an institution, like the biographer, finds that after thorough research the subject assumes a distinct personality and even a life style. Like a child, an institution is born into an environment replete with expectations, but it inevitably develops a personality and distinctive character which are more than the simple sum of the individual persons who form it. Traditional institutional histories too often slip into vague and impressionistic surveys revealing little about individuals and the setting from which they emerged. I hope that this investigation will join the growing rank of exceptions to this general rule, which are demonstrating that the study of institutions is in itself important and relevant to both social and intellectual history.[1]

1. While in the process of revising this material, the author attended a conference on Early American Learned Societies, the first in a series of three sponsored by the American Academy of Arts and Sciences, which suggests that important new research may provide both data and insight into the nature of early scientific institutions.

Practicing scientists who sought formal and informal affiliation with like-minded persons established the AAAS. Correspondence networks, less methodical than that of Henry Oldenburg of the Royal Society or Marie Mersenne of Paris, had been a stimulus and reinforcement for early American scientists. The organization met first the needs of individual scientists; ultimately it provided a vehicle for promotion and diffusion of scientific information. But correspondence could not satisfy the need for group discussion, the settlement of disputes over priority in a scientific discovery, and additional publication outlets. Local scientific societies developed in response to such needs but proved only a first step in the institutional development of science. Debate over the means and ends of association has tended to cloud the basic agreement of mid-nineteenth-century scientists on the exigency of a national organization. In today's centralized, bureaucratic society, both the unrealistic anticipations as well as the subconscious fears of scientists contemplating a national organization are difficult to comprehend. Because they lacked prior experience and clear professional definitions, tension among the active scientists often ran high. The effort to establish a national organ for science indicates the new concerns as well as the initial resolutions formulated by those scientists.

The American Association for the Advancement of Science was created in the ebb tide of reform movements, the period when gains of earlier "enthusiasms" were either consolidated or lost entirely. For science the 1840s were progressive years, during which it was established as a distinct discipline. The Smithsonian Institution, the two scientific schools at Harvard and Yale, increasing federal funding for science-related agencies, and finally the founding of the AAAS signaled the "take-off" period for American science.

The American Association, voted into existence as an expansion of the more specialized Association of American Geologists and Naturalists in 1847, provided a forum for discussion of the problems facing scientists. It was born in a decade ready for national voluntary organization, at a time when local scientific and philosophical societies had proved inadequate to meet the needs of specialized researchers and to serve as spokesmen for science.

While the British Association for the Advancement of Science provided an apparently clear model, the American experiment was to have its own distinctive characteristics. From the outset, because the aggressive Americans did not implicitly follow the deferential societal patterns of Europe, leadership would not easily emerge based on a "natural" hierarchical order. Because membership in the American Association was essentially open-ended, the disagreements over organizational form and function were largely internal—few outsiders were interested enough to offer specific criticisms. The history of the association during its first thirteen years of existence evidences all the tensions implicit in being an agent for the transition from amateur to professional. With expectations not clearly defined, the AAAS became the center for discussion of the structure and function pertinent to a scientific organization. It would be fallacious to suggest that the AAAS should be viewed as the precursor for modern scientific institutions. In concrete and explicit ways the organization was close to other antebellum institutions. Yet it did confront issues that required the formulation of new policy, especially in the area of professionalization and specialization. Recurring throughout the period were questions of ethics, critical review of presentations and papers, and the need to maintain open discussion between disciplines.

The purpose of this study is to trace the internal development of the association, which served as the experimental laboratory for testing modes of organization and professional aspirations. Until it began special sectional meetings, distinctions among scientists (Englishman William Whewell coined the word "scientist" in 1840), amateurs, and interested onlookers were blurred in most fields of science. By 1860 these categories formed a commonly accepted hierarchy, and popular interest turned away from sessions "too scientific." Recognizing the importance of a generally informed public, some scientific leaders urged the association to maintain its commitment to the popularization of science. The organization has never fully resolved the tension implicit in trying to fulfill both professional and popular needs. A major theme of this study is the nature and outcome of the struggle of the scientific community to establish a cohesive self-identity in the middle of the nineteenth century and the effect of that effort on the AAAS.

The other basic concern of this book is the collective character of the American scientific community at mid-century. Because nearly every prominent and minor American scientist was affiliated with the association during this period, it becomes an index to the general community of scientists. The manuscripts studied reveal much about the organization and about attitudes of its members, and local newspapers reflect public response to association meetings, but these are individual and impressionistic. The tools and insights of the social scientists can help evaluate and expand the initial conclusions. Chapter VIII summarizes the findings of a collective biographical study of the association and its leadership. In general, these results reinforce other contemporary and historical findings on the nature of American science and the characteristics of American scientists.

Writing this history of the early years of the American Association for the Advancement of Science was similar to piecing together a shattered mosaic; the organization has no formal archival materials covering the period under consideration. While some pieces of information were never recovered, much detail and a distinct outline emerged from scattered manuscript sources. Accordingly, grateful thanks go to numerous libraries and their helpful staffs. The association headquarters in Washington, D. C., was very helpful, despite its minimal records for the early years. In particular, Hans Nussbaum remained interested, gracious, and informative about the more recent history of the association. Among the libraries and rare-book rooms at which I spent fruitful days of research are those of the Smithsonian Institution, the Library of Congress, the American Philosophical Society, the Pennsylvania Historical Society, the New York Botanical Garden, the New York State Library at Albany, Yale University, Harvard University, and the Henry L. Huntington Library. Funds from the National Science Foundation and a predoctoral fellowship with the Smithsonian Institution made this research possible. A grant from the University of Illinois provided for travel costs and computer time, and a faculty fund grant from Simmons College helped complete the manuscript.

During the period I worked on this study, first as a dissertation and then as a book, numerous persons guided and encouraged my venture into a relatively unexplored area, the history of American science. Among these was the late Raymond P. Stearns of the University of Illinois, whose enthusiasm for American science and personal concern supported my first efforts in the field. During my affiliation with the Smithsonian Institution, the staff and visiting scholars at the Museum for the History of Science and Technology condensed into one delightful year insights and knowledge that greatly broadened my horizons on developments in modern science and scientific applications. Most central to my learning experience at the Smithsonian was Nathan Reingold; his own research on Joseph Henry, Alexander Dallas Bache, and their peers made him an invaluable commentator on my early work. Throughout the entire research and writing, Winton U. Solberg maintained the needed balance of patience, encouragement, and candid criticism. His perceptive work on nineteenth-century intellectual life in America clarified my own thinking, and his careful and precise comments greatly improved an early draft of this manuscript. Other colleagues and friends, especially Margaret Rossiter, Michele Aldrich, Barbara Rosenkrantz, and Henry Shapiro, offered thoughtful criticism of some detail and support for the enterprise.

My husband, David, has lived with this project from its inception, listening to reports of undigested data, commenting on vague but enthusiastic generalizations, and consulting on computer problems. His support and faith were critical factors in its completion. My parents' immeasurable guidance is given far too brief acknowledgement in the dedication of the book.

SALLY GREGORY KOHLSTEDT

Boston
21 February 1975

I

Image and Aspirations
for Science, 1815-48

It was perhaps fortuitous that the deliberate and rapid expansion of science coincided with an enthusiastic self-culture movement in the maturing United States. Science drew on the nationalistic zeal and optimism as well as on the tendency to organize that propelled social reform movements and inspired literary efforts in the second quarter of the nineteenth century.[1] Americans considered science a special assistant for the progress and prosperity of their rising republic. Science coordinated well with the common-sense philosophy which formulated responses to the social, political, and economic issues of the day.

Although popular enthusiasm for useful knowledge reinforced the rising aspirations of men of science, it ultimately engendered a rejection by the scientists of facile assertions about the automatic application of basic science to technological problems. Research scientists working together perceived that their role was changing and aspired to be professionals. The self-culture movement and the more introspective, exclusive professional concerns existed side by side by the 1840s. Recognizing the ambiguous, sometimes unhappy, but often mutually invigorating relationship between the two phenomena is essential to understanding the pattern of scientific institutional growth in the second quarter of the nineteenth century.

1. Arthur A. Ekirch, *The Idea of Progress in America, 1815–1860* (New York, 1944), offers the best analysis of the pervasive belief in progress and its implications in nearly every aspect of American life. The best monograph on all major reform movements remains Alice Felt Tyler's *Freedom's Ferment: Phases of American Social History to 1860* (Minneapolis, Minn., 1944). Perry Miller's posthumously published *The Life of the Mind in America from the Revolution to the Civil War* (New York, 1965) is suggestive but incomplete.

SCIENCE IN SELF-CULTURE

Never fully agreed upon by contemporaries, definitions of science in this period were, by implication, based on philosophies ranging from a strictly orthodox religion to belief in a first principle, from including only natural and physical science to incorporating "mechanics" (technology) and invention, and from an inductive to a deductive principle of scientific discovery. In popular thought, science was linked to the self-evident. A steady stream of inventions in the new nation brought more applause for utilitarian applications of science than for less easily understood European advances in laboratory research.[2] Conscious of their stereotype as "ingenious Yankees" in Europe, Americans felt pride mixed with inadequacy; combined with approbation for the inventiveness of Americans was an underlying denigration of achievements that lacked "real" scientific investigation or posited no new brilliant theory. This criticism most acutely affected men of science themselves, for whom the critique held personal as well as national implications. It is therefore not surprising that American scientists imbibed the nationalism of their era but were diffident about practical results which gained local acclaim.

Nationalism was a hallmark of the nineteenth century, and in America it peaked in the generation reaching maturity after the War of 1812. Coincident with the general rise of science in the century, it held far-reaching implications with reference to American science. The philosophical idea of progress was not new, but Americans reinterpreted it in light of their own interests and experience. Based on the evidence of material and cultural advancement, Americans made progress a law of their history and a mark of divine providence.[3] National achievement fostered the belief that man could conquer and control the immense forces of nature. Individual activism—Emerson's Man Thinking—established self-advancement as the essential key to national progress. Public orators borrowed language from the sciences when they spoke of

2. Richard J. Storr, *The Beginnings of Graduate Education in America* (Chicago, 1953), p. 5. Storr analyzes contemporary references to the "spirit of the age" and concludes that although the phrase was loosely used, it usually connoted democratic and utilitarian principles. The term "technology" was not commonly used; rather, men spoke of "practical information," "the arts," "useful knowledge," or simply "science."

3. Ekirch, *Idea of Progress*, p. 267. In 1848 Charles Sumner delivered a Phi Beta Kappa address at Union College entitled "The Law of Human Progress." See *ibid.,* pp. 258–259. Ralph H. Gabriel's *The Course of American Democratic Thought: An Intellectual History Since 1815*, part I (New York, 1940), pp. 3–102, is still a useful starting point for discovering the attitudes and ideas which motivated Americans in this period.

America as a "great laboratory" of the world, conducting a "grand, noble experiment."[4] The choice of image was significant.

The return of Revolutionary War hero Lafayette to the United States in 1824 ushered in an exuberant age. Numerous retrospective addresses recalled with pride the American advances during the fifty years since the Revolution.[5] Lafayette responded to praise and adulation by assuring his audiences across the nation that "at every step of my visit through the twenty-four United States . . . I have had to admire wonders of creation and improvement."[6] Americans could look with pride at their material progress: prosperous farms, growing industry, and a rapidly expanding network of canals, roads, and bridges. Fostered by independence, freedom, and a republican spirit, such improvement was assumed to be continuous and at an unprecedented pace. Looking at the products of human invention, Jacob Bigelow, a prominent spokesman for what he termed "technology," concluded in 1829:

> It would be natural to suppose that after the ingenuity of mankind had been devoted for so many centuries to the combination and application of materials, the field of new experiment would at length become exhausted, and that improvements would at length cease to appear. But experience has proved that the opposite state of events continually occurs. Since about the beginning of the present century and within the lives of many who are now upon the stage, some of the most important revolutions have taken place in the customs of society, derived entirely from innovations in the arts.[7]

Less effusive than Lafayette, Bigelow credited the ongoing improvement in the physical condition of society to the advanced state of the natural

4. Charles L. Sanford, *The Quest for Paradise: Europe and the American Moral Imagination* (Urbana, Ill., 1961), p. 173.

5. Fred Somkin, *The Unquiet Eagle: Memory and Desire in the Idea of American Freedom, 1815–1860* (Ithaca, N.Y., 1967), pp. 131–174. Somkin records in detail the response to the "National Guest," the Marquis de Lafayette, who was invited by President James Monroe and the Congress to return to the United States and participate in fiftieth-anniversary celebrations of the American Revolution. Also see Richard W. Van Alstyne, *Genesis of American Nationalism* (Waltham, Mass., 1970), pp. 169–180.

6. Quoted from the *New-York American*, 6 July 1825, in Somkin, *Unquiet Eagle*, p. 155.

7. Jacob Bigelow, *The Useful Arts, Considered in Connexion with the Applications of Science* (New York, 1863), p. 80. Bigelow, a Boston physician, held the chair of "application of science to the useful arts" at Harvard from 1816 to 1827. The volume cited above was first published in 1829 under the title *Elements of Technology*, and Bigelow is often credited with popularizing the new term "technology." See Miller, *Life of the Mind*, p. 289.

sciences, a more general diffusion of knowledge, order, and morality, and the state of general peace following the War of 1812.[8]

Prosperity provided another critical component. Expanding geographically and economically after the war, the nation easily absorbed new immigrants into the work force. Many of the arrivals found employment in manufacturing, construction, and mining as well as in agriculture, helping accelerate the rate of economic growth. The industries, in turn, required new skills and helped establish a group of experts that included chemical analysts and civil engineers. By the early 1840s the United States was ready to "take off" into the self-sustained growth necessary for a modern industrial nation.[9] The capital investment was usually private, but state governments provided substantial support for certain activities. Even Jacksonians outspoken in their opposition to federal sponsorship acknowledged that it might be appropriate for the state to be involved in public transportation, geological surveys, and education.[10] Technology and sometimes science benefited from the fiscal arrangements. The nation had not only an expansive optimism but also the economic base on which to develop.

Publication of Alexis de Tocqueville's *Democracy in America* in the 1830s heightened self-consciousness about American science and technology.[11] Americans received Tocqueville's assessment favorably, despite its undertone of criticism, because it coincided so well with their own self-analysis. Conscious of a supposed colonialism in literature, art, and letters, they realized that they were taking a technological lead and were proud of Yankee ingenuity expressed through invention. Although theoretical science was making longer strides during this period than is generally recognized, the inventors were the pride of the nation.[12]

---

8. Bigelow, *The Useful Arts*, p. 81.

9. The classic formulation is Walt W. Rostow, "The Take-Off into Self-Sustained Growth," *Economic Journal*, LXVI (Mar. 1956), 25–48. The precise dates Rostow suggests and his dramatic emphasis on a relatively abrupt take-off have provoked debate among economic historians. Most scholars do agree, however, that the foundation for industrial growth was established by the 1850s.

10. See, for example, Carter Goodrich, *Government Promotion of American Canals and Railroads* (New York, 1960).

11. Edited by Phillips Bradley, 2 vols. (New York, 1945). Vol. I was first published in 1835 and vol. II in 1840. One fact which continuously impressed Tocqueville was the technical advancement and mechanical ingenuity of the Americans, which, however, he felt limited their interest in less directly applicable basic research in the sciences.

12. A Hunter Dupree, "Science and Technology," in David T. Gilchrist and W. David Lewis, eds., *Economic Change in the Civil War Era: Proceedings of a Con-*

Tocqueville everywhere found men interested in applying science to the useful arts. Nor could his conclusion that there was a lack of interest in theoretical physical science in America be readily disproven, for there was no institution which embodied the aspirations of individual researchers such as Joseph Henry and Alexander Dallas Bache, who experimented after regular working hours in small, private laboratories, although some institutions did exist for advancing natural science. A full decade after Tocqueville's publication, Henry himself commented to an English scientist that "our forte is the application of science to art."[13] This theme echoed again and again, as in the *North American Review*'s critique of David Wells's *Annual of Scientific Discovery*; the *Review* observed that not only did inventive genius abound in America but "the Yankees are unparalleled in this respect among all the nations of the earth, both past and present."[14]

Not all Americans assumed that technological progress was positive and pervasive. More skeptical observers increasingly noted, as did James DeKay in 1826, that progress occurred "in those departments most immediately connected with the wants of society."[15] Moral philosophers in the liberal arts colleges, along with the New England Transcendentalists, denounced the resulting utilitarian enthusiasm and its consort materialism. Undoubtedly they concurred with a writer for the *North American Review*, who in 1847 summed up "The Intellectual Aspect of the Age" as

> eminently utilitarian, in the lowest sense of the word. The powers of outward nature are so fully developed, and in such vigorous exercise, as to claim a disproportionate share of men's time and energies. Such great mechanical inventions are yet recent, such marvellous applications of science to art are among the wonders, or rather, have ceased to be the wonders, of the day, that most persons talk, reason, and feel, as if the rapid creation of value were the test of genius, the supreme end

ference on *American Economic and Institutional Change, 1850–1873, and the Impact of the War* (Greenville, Del., 1965), p. 117.

13. Joseph Henry to Charles Babbage, Smithsonian Institution, 27 Apr. 1850, Babbage MSS on microfilm at the American Philosophical Society Library, Philadelphia (hereafter APS).

14. *North American Review*, LXXIV (1852), 517.

15. James DeKay, *Anniversary Address on the Progress of the Natural Sciences of the United States, Delivered before the Lyceum of Natural History of New York, February, 1826* (New York, 1826), p. 7. For a study of this theme, see Leo Marx, *Machine in the Garden* (New York, 1964). Also see Marvin Fisher, "The Iconology of Industrialism, 1830–1860," in Hennig Cohen, ed., *The American Culture: Approaches to the Study of the United States* (Boston. 1968), pp. 228–245.

of being, and the crowning purpose of life. Man is looked upon as a mechanical power, and men educate themselves for the same uses to which they consecrate spinning jennies and steam-boilers.[16]

The writer concluded that man stood at the "summit of lower creation," where he seemed content to bask in self-congratulation.[17] The *Review* typically praised advances in technology while strongly denouncing the simultaneous lack of spiritual and philosophical development.[18] Such skeptics were a minority, however.

In general, the scientific enthusiasm, with its technological overtones, introduced little social or religious unease. There were no philosophers to reconcile possible incompatibilities between revealed religion and science, for that enterprise seemed unnecessary in the 1830s. Americans were basically theists and deeply committed to traditional Protestant Christianity. Professors of moral philosophy, such as Francis Wayland at Brown, were the chief spokesmen for these religious tenets. Sweeping and unimaginative, their philosophy depended too long on the imported works of Englishman William Paley. When they finally turned to natural theology, they found that doctrinal and apologetic theology had no adequate language through which to answer modern men's inquiries about new scientific advances.[19]

For the time, however, religion and science existed without overt conflict. Assertive rhetoric on the unity of the two in disclosing God's plan for the universe masked the increasing anxiety felt about a breach beween them. As science became the source of knowledge for natural events, most natural theologians felt that there was a lag in understanding implications of the new sciences rather than a conflict. This, they believed, could be anticipated simply because scientific advances were occurring more rapidly than philosophical explanations of new phenomena. Science itself was not provocative and was often even conciliatory; in turn, religious men acknowledged science as a social good. Religious revivalism and enthusiasm for science could coexist in the popular mind. When scientists of the stature of James Dwight Dana, Louis Agassiz, and Edward Hitchcock took pains to point out their own religious beliefs and

16. *North American Review*, LXIV (1847), 277.
17. *Ibid.*, p. 276.
18. *Ibid.*, LXIII (1846), 357.
19. The best discussion of the collegiate moral philosophers is Donald H. Meyer, "The American Moralists: Academic Moral Philosophy in the United States, 1835–1880" (Ph.D. dissertation, University of California at Berkeley, 1967). Also see Paul R. Anderson and Max H. Fisch, eds., *Philosophy in America from the Puritans to James with Representative Selections* (New York, 1939), p. 327.

even to argue specific correlations between scientific findings and religious beliefs, public doubts were assuaged.[20] For others a growing secularization modified the need to justify scientific with religious truths.

Encouragement for science took several forms: popular lecture series and textbooks, self-help societies, and more collegiate training. Historians differ over the validity of Tocqueville's conclusions about American indifference to basic research but concur with his general obselvations on American social tendencies.[21] Among Tocqueville's most perceptive comments were those on the number of voluntary associations which pervaded all levels of society. He argued that there was an inherent connection between democracy and voluntary associationism, explaining the relationship as one of necessity: "democratic laws and habits" made individuals weak and caused them to band together to effect change.[22] Extending Tocqueville's argument, Stow Persons suggested that voluntary associations fulfilled two essential functions, simultaneously acting as a means of educating and organizing public opinion and as a pressure group to secure legislation.[23] The tendency manifested itself in numerous political reform movements as well as in organizations for self-help. Taken to its logical extreme, the movement culminated in Fourier's associationism or in the "Society of societies" sponsored by Horace Greeley in his *New York Daily Tribune*.[24] Coordination of local groups was successful primarily when the groups themselves were well established.

The most notable of the self-culture associations was the lyceum system which flourished in the 1830s and 1840s. It originated with the

20. Chapter V discusses the stance of scientific leaders toward religion. Stanley M. Guralnick, "Geology and Religion before Darwin: The Case of Edward Hitchcock, Theologian and Geologist (1793–1864)," *Isis*, LXIII (Dec. 1972), 519–543, demonstrates Hitchcock's dedication to reconciling geology and Protestant orthodoxy.

21. Perhaps the first to articulate this theme was Richard Shryock, "American Indifference to Basic Science during the Nineteenth Century," *Journal of World History*, II (1948–49), 50–65. Nathan Reingold in "American Indifference to Basic Research: A Reappraisal," in George Daniels, ed., *Nineteenth-Century American Science: A Reappraisal* (Evanston, Ill., 1972) pronounces the debate over the question of American indifference to basic research an essentially sterile one, especially in view of inadequate comparative data. Whether the proposition is true or not, Americans believed it to be true, and this belief had an effect on their institution building.

22. Alexis de Tocqueville, *Journey to America*, ed. J. P. Mayer (New Haven, Conn., 1960), p. 252. This volume contains Tocqueville's notebook observations, which were later elaborated in his *Democracy in America*.

23. Stow Persons, *American Minds: A History of Ideas* (New York, 1058), p. 163.

24. "The Doctrine of Association—Its Aims and Its Character," *New York Daily Tribune*, 2 May 1845.

goal of raising the level of general education of voluntary participants, but by the late 1840s and 1850s it degenerated to providing general entertainment. At its peak the lyceum included 800 town and country lyceums with various organizational structures; all predicted personal enlightment (self-culture) for everyone who paid the fee and listened attentively to circuit speakers.[25] An 1831 pamphlet by the lyceum leaders outlined the scope and goals of the national organization, stressing that progress was not inevitable but derived from human endeavor: "Every rational man . . . is endowed with capabilities for improvement—wherever he is placed he is surrounded with materials for his improvement; that intellectual, moral and *social* faculties are confined to no favored few of our race; that science is confined to no favored spot under heaven; that intellects and affections are coexistent with the race of man, and that science is as boundless as the earth and the heaven."[26] Elaborating on such egalitarian principles, the pamphlet suggested that local lyceums should be established for the dual purpose of advancing personal improvement and developing national potential.[27]

When Josiah Holbrook suggested projects for the lyceum movement, he specifically included agricultural and geological surveys and state collections of minerals; these were to promote science itself as well as to contribute to agricultural productivity and internal improvements.[28] Holbrook himself offered "lyceum apparatus" for sale, from a set of "mechanical powers" demonstrating principles of the lever, pulley, wheel and axle, screw, and inclined plane to a geological cabinet.[29] The entire lyceum movement coincided with a period of cosmological optimism, of belief in the goodness of human nature, and of a lively and inquiring spirit.[30] It also partook of the enthusiasm for science wihout, at least during the early years, necessarily emphasizing technology.

25. *American Lyceum, or Society for the Improvement of Schools and Diffusion of Useful Knowledge* (Boston, 1831), p. 4. The most comprehensive analysis of the movement is Carl Bode, *The American Lyceum: Town Meeting of the Mind* (New York, 1956).

26. *American Lyceum of Useful Knowledge*, p. 4.

27. *Ibid., passim.*

28. Holbrook at the same time urged support for surveys and collections, arguing that they would aid agriculture and internal improvements by "placing before legislators and others specimens of their own productions and a knowledge of their own resources in the mineral kingdom, by which industry would be encouraged and individual and public wealth and prosperity increased." Quoted in Bode, *American Lyceum*, pp. 25–26. A similar point is made in Dirk J. Struik, *Yankee Science in the Making* (New York, 1962), pp. 215–216.

29. Bode, *American Lyceum*, pp. 186–187.

30. *Ibid.*, pp. 27–37.

An irrefutable result of popular lectures given on the lyceum circuit or under other auspices was increased general consciousness of science and access to elementary scientific understanding. Among the most popular lectures were those on chemistry, geology, and astronomy, sciences which lent themselves readily to graphic representation and elaborate display.[31] Initially, leading men in the scientific world presented the popular lectures, although not often on the lyceum circuit. But they represented a generation of men whose scientific education had been piecemeal and eclectic. Most notably Amos Eaton of upper New York wrote numerous popular textbooks and gave public lectures as well as maintaining his busy schedule of geological and botanical field-work and teaching at Rensselaer Institute. Largely self-taught himself, his enthusiasm precipitated out of the engineering and student body and summer visitors a generation of natural scientists.[32] Benjamin Silliman of Yale lectured not only to New Haven but also to large Boston audiences. As did their collegiate contemporaries, they used the income from public lectures to supplement inadequate faculty salaries, but they were also responsive to intellectual curiosity and the desire to share their findings with an enthusiastic audience.

Some scientists had different aspirations to further by popular lectures. The example of the potential power of public support cited by contemporaries was the lecture series on astronomy presented by Ormsby MacKnight Mitchel. The lecture series, concluding with a statement of the need for an observatory in the West, resulted in a truly popular subscription by a broad cross-section of Cincinnati citizens for the construction of their own observatory complete with the then largest telescope in the United States.[33] Diffusion of science would, it seemed

31. Wyndham D. Miles, "Public Lectures on Chemistry in the United States," *Ambix: The Journal of the Society for the Study of Alchemy and Early Chemistry,* XV (Oct. 1968), 129–153.

32. Eaton, himself inspired to study science after hearing Benjamin Silliman lecture, was eager to publicize his teaching methods of student participation and demonstration. In 1826 he took twenty students on a trip, making a geological survey of central New York and gathering specimens. Both he and the students delivered lectures on botany, chemistry, and geology to interested townspeople along the route. His students included James Hall, Ebenezer Emmons, Asa Fitch, Eben Horsford, and Douglass Houghton. See the account in Samuel Rezneck's "A Traveling School of Science on the Erie Canal in 1826," *New York History,* XL (1959), 255–269. The best summary of Eaton's career is Ethel McAllister, *Amos Eaton: Scientist and Educator* (Philadelphia, 1941).

33. Stephen Goldfarb, "Science and Democracy: A History of the Cincinnati Observatory, 1842–1872," *Ohio History,* LXXVIII (Summer 1969), 172–178. The support of 300 subscriptions of $25 each from Cincinnati citizens was touted by contemporaries as evidence of what popularization might accomplish. But subse-

in the 1830s, facilitate American material and political aspirations.

Basically, the lyceum professed to diffuse and advance human knowledge, assuming that once men knew of the unresolved problems in science they would, in Baconian fashion, contribute to the data gathering necessary for deducing first principles. Close observers soon recognized that the only accomplishment—certainly not a minor one in a society without a comprehensive educational system—was that of spreading information and ideas. The new enthusiasm for the scientific lectures probably aided men of science indirectly in gaining funding for "useful" scientific projects such as the state geological surveys and the U.S. Coast Survey and Depot of Charts and Instruments. Practical achievement most impressed the public, which rarely recognized that results came from "a mind practically cognizant of the scientific truth it embodies, or even simultaneously by several minds in a department, the scientific bearings of which have recently become public property."[34] The lyceums coordinated with a popular textbook movement.

Fulfilling the need for more popular materials were several women, most of whom were also educators. The fact of their interest in promoting natural sciences at a time when opportunities for women were limited is indicative of the diffusion of the subject. Several women naturalists collected or drew sketches of specimens, as Maria Martin did for Audubon. Others published popular textbooks. Almira Lincoln Phelps, a pioneer in women's education, had studied under Amos Eaton and published a very popular study, *Familiar Lectures on Botany* (1829), as well as texts on chemistry, geology, and natural philosophy.[35] By mid-century, natural history had a place in most women's seminaries and colleges, and a few women were granted admission into local museums.[36] But the contributions of women remained peripheral. Caroline Wells Dall, abolitionist and women's rights reformer, said flatly at a lecture in Boston in the 1860s, "Scientific pursuits cannot be said to be fairly opened to women here."[37] The women were not on a professional career track

quent demands to open the observatory to the general public during prime observation times limited its importance as a research facility.

34. *North American Review*, LXXIX (1854), 229.

35. Almira Phelps and her sister Emma Willard figured prominently in the establishment of several schools for girls and were innovative in curriculum design. See Emma Lydia Bolzan, *Almira Hart Lincoln Phelps: Her Life and Work* (New York, 1956).

36. Thomas Woody, *A History of Women's Education in the United States*, II (New York, 1929), 474–480.

37. Caroline Wells Dall, *The College, the Market, and the Court; or, Woman's Relation to Education, Labor and the Law* (Boston, 1967).

despite serious, active interest. Through their local activities and publications they were an important factor in the popularization of science.

Lectures on the circuit ranged from rank amateurs to persons of proven scientific capability. The growth of a specialized vocabulary and techniques compelled scientists to adapt their arguments and demonstrations to ensure public understanding. Benjamin Silliman, who drew large crowds to the Lowell Institute in Boston as well as to his appearances in New Haven, New York, and Baltimore, took his lectures as a serious responsibility. A Yale professor, he thought that his success had come from making his topics "intelligible and attractive, without diminishing the dignity of science."[38] Such reconciliation of popular understanding and scientific accuracy became increasingly difficult in the late 1830s. As a result scientists began to insist on giving several lectures rather than a single talk, arguing that only a full series could adequately introduce the general public to any complex aspect of scientific inquiry.[39] Too few scientists, however, had the capacity or time to make their subject intelligible to a general audience by eliminating technical terminology and compressing theory into a simple outline.

A resurgent interest in Francis Bacon and his logic reinforced the progressive tenets of the early nineteenth century and seemed especially applicable to observational science. Although Bacon's work was more reverenced than read in America, his inductive system invited participation. Bacon's elaborately defined Solomon's House offered a level of participation to all seekers after truth.[40] A local lyceum, the Boston Society for the Diffusion of Useful Knowledge, republished the Englishman Henry Brougham's edition of Bacon's *Novum Organum Scien-*

---

38. Benjamin Silliman, "Reminiscences," VIII, 55, unbound MSS, Beinecke Library, Yale University (hereafter BLY). In 1845 he had written to a friend abroad, "I reaped the virgin harvest of the country in popular lectures and my success incited so many to flow that they have run the thing down...." Silliman to Mantell, New Haven, 18 July 1845, Silliman MSS, BLY. A more accurate explanation is that general enthusiasm for self-culture was itself on the wane by the mid-1840s. Also see Margaret W. Rossiter, "Benjamin Silliman and the Lowell Institute: The Popularization of Science in Nineteenth Century America," *New England Quarterly*, XLIV (Dec. 1971), 602–626.

39. Struik, *Yankee Science*, p. 203.

40. Benjamin Farrington, *Francis Bacon: Philosopher of Industrial Science* (New York, 1949). This provocative biographer understands Bacon not unlike nineteenth-century Americans did. He suggests that Bacon included a place for "counters, measurers, and weighers" in the scientific Solomon's House because he realized that the taking of a vast inventory of natural phenomena required the organized teamwork of many men. Appendix A, pp. 179–191, is an excerpt from the *New Atlantis*, describing Solomon's House.

tarum (1620) in 1831.[41] If men did not understand its rationalistic arguments, they were nevertheless struck by Bacon's eloquent appeal for the study of facts over idle speculation. Vague in definition, Baconianism assumed three related meanings—empiricism, anti-theorizing, and identification of all science with taxonomy.[42] The latter was a key conception for the interested scientific amateur. Scientists were also "Baconian" in their antipathy for wild hypothesizing and were deeply committed to establishing classifications for new national findings in geology, zoology, botany, and meteorology. But they simultaneously realized that fact gathering must be expertly done and generalizations kept cautious.

The inherent problem of active but undertrained amateurs was over-confidence and a naive sense of what discovery meant. When orator William E. Channing considered "The Present Age" in 1841, he recognized this possibility in an age of expansion, diffusion, and universality. Since science had "left her retreats, her shades, her selected company of votaries," she had become an inexhaustible mechanic.[43] Showing no deference toward scientific knowledge, Channing urged a spirit of free inquiry:

> In truth, nothing is more characteristic of our age than the vast range of inquiry which is opening more and more to the multitude of men. Thought frees the old bounds to which men used to confine themselves. It holds nothing too sacred for investigation. It calls the past to account; and treats hoary opinions as if they were of yesterday's growth.... Undoubtedly this is a perilous tendency. Men forget the limits of their powers. They shock the pious and revering minds, and rush into an extravagance of doubt more unphilosophical and foolish than the weakest credulity.[44]

But Channing went on blithely to dismiss such disadvantages, stressing instead the potential advantages which accrued to mankind through undisciplined creativity and unanticipated discovery. Ironically, Channing announced that science was public property just as her most ardent practitioners moved to establish limiting, professional criteria for participation.

41. Ekirch, *Idea of Progress*, p. 106.
42. For an excellent summary of Baconian thought in America, see George Daniels, "The Reign of Bacon," in *American Science in the Age of Jackson* (New York, 1968), pp. 63–85.
43. "The Present Age," in *Works of William Ellery Channing*, 6th ed., VI (Boston, 1848), 152.
44. *Ibid.*, p. 154.

Denison Olmsted's assumption that science fostered political equality operated among reformers, as demonstrated in the utopian community of New Harmony. Robert Owen in 1826 founded the New Harmony settlement in Indiana explicitly to develop a more rational, scientific community. With his "boatload of knowledge" that included three prominent members of the Academy of Natural Sciences of Philadelphia —Thomas Say, Gerald Troost, and William Maclure—Owen hoped not only to create a new society but also to bring notable scientific advance. This grand social experiment failed within two years, with Troost leaving for Tennessee and William Maclure establishing his residence in Mexico. But the town of New Harmony retained its scientific character chiefly because David Dale Owen resided there while he worked with his brother Richard on the state and federal geological surveys in the Midwest. New Harmony temporarily became a training ground for a new generation of geologists. Isolated from the East, the community of scholars in the wilderness had no real sustaining power and by the 1850s had lost both its social and scientific fervor.[45]

Higher education also reflected the new consciousness of science. Linked to the reformist outlook of the age, science and education were closely allied during this quarter-century, with historians of both disciplines giving credit to the pioneering efforts of Alexander Dallas Bache, Joseph Henry, and William Barton Rogers. Progress-minded educators in the old liberal arts colleges sought to modernize the curriculum, which usually meant more science and modern language training, among other changes.[46] A lack of training relevant to subsequent careers became evident in declining collegiate enrollments, and leading educators such as Francis Wayland responded to the challenge with limited success.[47] Science had been a part of the liberal arts curriculum

45. N. Gary Lane, "New Harmony and Pioneer Geology," *Geotimes*, LX (Sept. 1966), 18–22. Recently the August M. Kelly Publishers of New York reprinted several volumes relating to the Owenite experiment, including George B. Lockwood's pioneering *The New Harmony Communities*, William Maclure's *Opinions on Various Subjects* in three volumes, and Robert Dale Owen's autobiographical *Threading My Way*. This author knows no study which analyzes the motivation for the scientific grouping or the rapid disintegration of the community.

46. Winton U. Solberg, *The University of Illinois, 1867–1894: An Intellectual and Cultural History* (Urbana, Ill., 1968), pp. 22–36, has a lucid summary of trends in higher education for the period.

47. Wayland's "Report to the Corporation of Brown University on Changes in the System of Collegiate Education" was one of the most outspoken critiques. Also see Donald H. Fleming, *Science and Technology in Providence, 1760–1914: An Essay in the History of Brown University in the Metropolitan Community* (Providence, R.I., 1952), p. 75, which quotes from Benjamin Peirce's "Plan of a School

already in the eighteenth century,[48] and by the 1830s the leading colleges established professorships in mathematics and the sciences. The textbooks and teaching techniques utilized suggest, however, that most colleges followed rather than led the progress of science in the United States. The most significant changes down to the Civil War were qualitative: better textbooks, more well-trained faculty, and more laboratory instruction.[49] The last had its most impressive beginnings at the Lawrence Scientific School at Harvard and the Sheffield Scientific School at Yale. That colleges were part of a movement to institutionalize science is clear from the proposals for a national university in the late 1840s and early 1850s. Most faculty, however, remained basically tied to their teaching functions. While this responsibility absorbed time and energy, it also offered greater psychological and social rewards.[50] While still a student, Oliver Wolcott Gibbs observed to his mother, "What Dr. [John] Torrey says about the professor in America having too much work to do is very true. Most of them are worked to death & . . . in the course of their whole lives publish one single original paper [or] contribute one single new fact to science."[51] Only after the Civil War did Charles W. Eliot at Harvard and Daniel Coit Gilman at the new Johns Hopkins University encourage scientific research as coequal with teaching responsibility.[52]

The striking exception to the general level of scientific teaching was in medical schools and in the few polytechnic schools; there the best professors of natural and physical sciences inspired their students to explore the world of science as well as to read textbooks, sometimes

---

of Practical and Theoretical Science" (1846), proposing to "bring the university into more immediate and intimate connection with the community to which it belongs, by supplying the public demand for education of various kinds."

48. Theodore Hornberger, *Scientific Thought in American Colleges, 1638–1800* (Austin, Tex., 1945).

49. Stanley M. Guralnick, "Science and the American Colleges, 1828–1860" (Ph.D. dissertation, University of Pennsylvania, 1969), pp. 212–221. Guralnick succumbs a bit to the progressive rhetoric of the men he studied. His survey of fifteen colleges indicates that colleges were not initiators of scientific inquiry but were, in fact, responding to external advances. Because adaptation was slow, many students went to Europe to study, and men of science turned to independent scientific schools to implement their ideas on scientific education.

50. *Ibid.*

51. O. W. Gibbs to Laura Gibbs, Berlin, 25 July 1846, Gibbs Family MSS, Wisconsin Historical Society, Madison (hereafter WHS).

52. Richard Hofstadter and Walter P. Metzger, *The Development of Academic Freedom in the United States* (New York, 1955), p. 286.

distracting them from pursuing far more lucrative professions. The University of Pennsylvania Medical School, for example, had an exceptional record of producing scientists, for many men acquiring medical degrees in the period did not practice medicine.[53] Two technically oriented institutions in the Hudson River Valley also produced many mid-nineteenth-century scientific leaders, although their goal was to produce professional engineers. The U.S. Military Academy at West Point had been transformed from a military training school to a scientific center largely through the Military and Philosophical Society's efforts in the first decades of the century. At Troy, New York, Amos Eaton introduced laboratory techniques and field study into Rensselaer Polytechnic Institute, and his enthusiasm inspired a new generation of naturalists who moved beyond their mentor's limited theoretical base.[54] Men of science were often introduced to science in collegiate or other institutions for advanced training but relied on informal contacts to supplement their education.

In some instances scientific societies on the college campuses helped small enclaves of professors and students to study science outside the curriculum. Dependent on an interested student group, these local societies were often transitory, sometimes lasting no more than a four-year college generation. Proliferation of such groups, however, suggests student interest and indicates one way of meeting scientific curiosity even within a nonscientific curriculum.[55]

Professors of science, many of whom had independently developed their knowledge of a specific area of interest, worked to keep abreast of current research and to find funding for the collections and apparatus essential to teaching science adequately. Eager to train colleagues capable of contributing to American scientific advancement, they encouraged students to study science as both curricular and extracurricular activity. The American Association for the Advancement of Education, during its brief existence, most clearly enunciated the neces-

53. Data collated for Chapter VIII indicate that at least fifteen leaders of the AAAS attended the University of Pennsylvania to study medicine.

54. Frederick Rudolph, *The American College and University: A History* (New York, 1962), pp. 228–231.

55. *Ibid.*, p. 277. Guralnick and other historians of education in this period neglect what was evidently an important aspect of science in the colleges, especially in the 1830s. See, for example, the activity of C. B. Adams at Amherst and later Bowdoin Colleges. Adams to D. Humphrey Storer, 12 June 1847 and 10 Aug. 1842, Storer MSS, Museum of Science, Boston.

sity of developing men trained in science at the collegiate or even post-graduate level.[56] This association was a proponent of the attempt to found a national university at Albany, whose major goal for science was to provide a special science curriculum under leading national scientists. The failure of this plan in the New York legislature indicated public response to the enthusiastic but elitist proposals of the men of science.[57] Reform-minded educators and scientists continued to work toward a specific program in education to prepare men for careers in science.

Typically, however, the lyceum movement and general interest in science enhanced the prestige of scientists, many of whom were college professors. The scientists, in turn, consciously built a support group when they lectured to popular audiences. During the 1830s such reciprocity was evident on both a formal and an informal level, as suggested by the local societies considered in the following chapter. Science had a place in the American striving for self-culture and, as with many other components of the movement, it became institutionalized.

Yet popular involvement had negative results as well. Lectures were time-consuming and brought few direct results for science itself. The alternative of leaving the responsibility to less qualified men raised a fearsome specter—quackery.[58] The need to provide scientific spokesmen in a rampant democracy became an important factor in the shaping of national scientific organization.

Social reform movements, a leading feature of the age, demonstrated that while Americans were defensive about outside criticism, they nevertheless moved to rectify the most obvious discrepancies between reality and democratic ideals. Bolstered by the "winning" of the War of 1812 and by the strength of fifty years of political existence as a democracy, the second quarter of the nineteenth century was marked by conscious self-assessment within as well as outside the scientific community. In science these years indicated the shift from a disorganized group of amateurs without common goals or interests to one of scientists with vague but discernible professional goals. During the transition men of science concentrated on the prosaic tasks of justifying themselves to the public, organizing institutions, founding media of communication,

56. American Association for the Advancement of Education, *Proceedings*, I–V (1850–55), *passim*.

57. Robert Silverman and Mark Beach, "A National University for Upstate New York," *American Quarterly*, XXII (Fall 1970), 701–713.

58. Margaret W. Rossiter discusses the serious problem of quackery that confronted agricultural chemists in her study, *The Emergence of Agricultural Science: Justus Liebig and the Americans, 1840–1880* (New Haven, Conn., 1975).

and describing natural science objects which were arriving in the East almost more rapidly than scientific names could be assignel. Gradually, following the example of English reformist scientists, they worked to eliminate aristocratic tendencies in the older scientific societies and to gain more research support.[59] The Americans, however, lacked a common ideology and even a common foe. Rather, they hoped to see "the land of our birth and our affections rising as rapidly in scientific fame as in wealth and political power."[60] In order to establish their reputation at home and abroad, they needed regular funding and more rigorous standards.

Better science depended on financial support for individual research projects. Whether in a democracy or an autocracy, demonstration of the value of science was essential to securing support; as long as men of science lacked individual wealth or endowed institutions, the necessity of persuading the power with the purse remained. As the proposals for geological surveys indicate, scientists responded to and thus encouraged utilitarian expectations.[61] Because science in popular thinking was a discipline which produced material advancement, many scientists made deliberate efforts to acquaint the public with useful results. Seeking more support, one naturalist urged his fellows to demonstrate "the utility, the beneficial influence which natural history exercises on the common mind."[62] In reinforcing already prevalent assumptions, men of science built into public consciousness an expectation which the rising generation of professional scientists would fervently resent.

THE DILEMMA OF SCIENTISTS

The movement for self-culture created a dilemma for science because, ironically, it not only paralleled but gave rise to new professional aspirations. Self-culture sought to promote individual familiarity with complex literary, artistic, and scientific concepts. Undoubtedly it realized its goal of distribution or diffusion of general knowledge for per-

59. George A. Foote, "The Place of Science in the British Reform Movement, 1830–1850," *Isis*, XLII (Oct. 1951), 192–208.

60. William H. Allen, *An Address before the Cuvierian Society of the Wesleyan University, Middletown, Connecticut, July 31, 1838* (New York, 1838), p. 13.

61. C. W. Panghorn, Jr., "A History of the Popularization of Geology in America," Washington Academy of Science, *Journal*, XLIX (July 1959), 224–227.

62. Walter W. Ruschenberger, *A Notice of the Origin, Progress, and Present Condition of the Academy of Natural Sciences of Philadelphia* (Philadelphia, 1852), p. 5.

sonal edification but without an obvious increase in any given field of knowledge. The new professionalism, in contrast, sought carefully to delimit role and status so that specializations were more clearly defined and science might advance more rapidly. Because self-culture encouraged general and sometimes naive inquiry by local amateurs, the new professionals became more determined to guarantee sound research and publication. The readiness for professionalization gained by similar European activity and increasing self-confidence among American scientists received further impetus in response to the aggressive self-culture movement.

In their initially ambivalent response to the spirit of popular culture, men of science demonstrated their own uncertain definitions and goals, as well as a lack of coordination. Because the American Philosophical Society and the American Academy of Arts and Sciences had not exercised any real control in the loosely defined community of science, there was no clear distinction between scientist and amateur or between researcher and observer; nor did the scientific societies' periodicals exercise sufficient editorial authority. Their failure to respond on any significant scale to a rising professional spirit shifted responsibility for creating standards to the later organizations and intensified reaction to a policy whose implications were relatively rapidly implemented.

At the outset lyceum audiences and interested amateurs found ready scientific lecturers, often college professors who were eager to demonstrate their field of research to an interested and responsive public. Best known were Benjamin Silliman of Yale and Amos Eaton from upstate New York, who gave nearly 3,000 lectures outside the classroom. Such men believed in the possibilities of self-culture; books and travel, well used, might provide the necessary background for scientific research. Occasionally resentful of the time required by popular lecturing, they seemed not to have questioned its validity or importance. When, however, the growing number of self-confident amateurs presumed too much and as the difficulties of simplification increased, scientists grew restive about this function. As science became a complex of interrelated specializations, men of science began to recognize that they could provide little "for the scientific few which [would be] agreeable to the ignorant many." [63]

63. Robert Hare to Silliman, quoted in George P. Fisher, *The Life of Benjamin Silliman*, I (New York, 1866), 289. The younger Silliman later observed that his father's lectures were "like casting pearls before swine to take so much pain for those who for the most part appreciate it so little." Quoted in Edward Weeks, *The Lowells and Their Institute* (Boston, 1966), p. 43.

The limitations seemed clear to the next generation of men of science, often graduates of Yale and Rensselaer. In the midst of the movement for self-culture, they grew skeptical of the results of diffused knowledge and were less responsive to requests for public addresses. Their reasons were complex and often personal. Physical scientists such as Joseph Henry and Alexander Dallas Bache were not so dependent on field observation and collections as were the geologists and naturalists.[64] For them the advantages which drew peers in the natural sciences to lecturing seemed economic or egotistical, neither of which was a fit scientific motive.

In general, the generation of scientists just coming of age resisted the implied responsibility that they popularize accurate scientific knowledge and comment on the irrelevant or obtuse contributions amateurs were apt to present. They also resented having to supplement faculty salaries through public lectures, a need which Silliman and Eaton had taken for granted. John Torrey, a leading New York botanist, complained frequently about a lack of support for college professors of science, and especially about his own financial difficulties.[65] When, in addition, popular efforts proved the public fickle in support and surprisingly uncritical, reaction was negative and pronounced.

Public expectations could not be denied with impunity, however. James Dwight Dana warned an aspiring young scientist studying chemistry in Germany, "I have no doubt that your course [studying at Freyberg] is the right one, whatever the prospects before you. Science, pure & unadulterated, commands small pay in the world; there is little that satisfies the vulgar appetites of the people; but diluted and mixed with a sufficient amount of the *spirit of the age* [italics his], it stands high in favor. Some seem to think it well to deny with humbug [*sic*]; but this is poor economy in the end, for the pocket & character."[66] "Pure" science was not self-justifying, and Dana's political acumen made him recognize that it was inopportune to ignore popular demands. Most contemporaries, although grumbling, agreed. This did not prevent them from complaining of the loss of time and energy which might other-

64. These two physical scientists, heads of the Smithsonian Institution and the Coast Survey respectively, were to be the titans in American science for their generation, and they self-consciously embodied characteristics of the new professionalism.

65. Jacob W. Bailey to John Torrey, West Point, 18 Nov. 1836, Torrey MSS, New York Botanical Garden, New York (hereafter NYBG).

66. James Dwight Dana to George Brush, New Haven, 21 Sept. 1854, Brush Family MSS, Yale University Archives, Yale University (hereafter YUA).

wise be used for study and research, and they determined to establish more appropriate and consistent support than was derived from lectures. Colleagues who too frequently presented lectures were criticized as attempting "to make capital as the politicians say, by parading the results of [their] own labors . . . [and having them] pupped in the newspapers."[67] Scientists similarly resented what they considered an over-emphasis on applications of science. As William Brewer wrote to the same young scholar in Germany, "I am not surprised that you should be at work on applied sciences, rather than pure, when I consider the demands of American society, American learning and American Mind, where all must worship at the shrine of *Utility*. . . . My daily bread has been earned by my labor, and I have been *obliged* to cultivate such fields as yielded *that*, however pleasant it might have been to have wandered elsewhere, in search of pure gems."[68] Public expectations of entertainment and usable science created a very real resentment among some scientists.

Preoccupied with private research and a full teaching or work load, the new breed of scientist was more concerned with establishing legitimacy for himself and for scientific institutions—societies, publications, and research facilities—than with relating past accomplishments to the general public. This, however, required resources of time and money that brought scientists back full circle to the public. Yet even public enthusiasm proved incapable of sustained financial support. Conscious of European laboratories, the scientific community sought not simply approval of specific projects such as the Cincinnati observatory but continuing and consistent backing for research facilities. Contemporary enthusiasm for science ostensibly enhanced the potential for support, but it proved without dynamic force except when popular imagination was sparked to provide for an observable monument such as an observatory or a cabinet. Impatient and subject to changing economic currents, legislative support also suffered when results were slow, as demonstrated by the state surveys; the public was unaware of the time required for productive research.[69]

67. *Ibid.*
68. W. H. Brewer to George Brush, Ovid, N.Y., 22 Nov. 1854, Brush Family MSS, YUA.
69. See, for example, the state surveys mentioned in George P. Merrill, *The First One Hundred Years of American Geology* (New Haven, Conn., 1924), pp. 127–390. In a private communication Michele Aldrich pointed out that her analysis of voting patterns on the New York survey bills suggests that support came from both Whigs and Democrats and that party politics was not a critical predictor of scientific support of legislators.

The popular scientific lectures had quickly demonstrated the limitations of the assumption on which they rested, namely, that men were capable of understanding and profiting from theoretical knowledge. Scientists invited to participate were asked by lecture managers to speak "in a plan intelligible in manner, divested as far as practicable of technological phraseology."[70] Simplified presentations inevitably made even the uneducated listener believe he was capable of understanding and making decisions about science, as he assumed he was in social or political issues. Such science was subject to popular taste and public approval. Men with a facile pen or glib tongue might arouse the same, if not greater, attention than the researcher who systematically demonstrated his theory.

Publication of nonsense or "curiosities" was less frustrating to men of science than attempts to make authoritative statements on misleading information. Whether faulty identifications of theories originated with an unknown amateur or a well-known researcher, publication abroad inevitably resulted in embarrassment for American scientists. Thus fairly early there were attempts to monitor publications and to exercise peer pressure. Both Constantine Rafinesque and George W. Featherstonehaugh experienced restrictions imposed by fellow scientists. Although in retrospect both were seriously interested in geology, their tendency to generalize too readily and to assume a stature not granted them by their peers resulted in virtual ostracism by American scientists.[71] The control in scientific publications and societies was impossible on the lecture circuit.

One result of unsolicited and ill-informed participation was the derogatory connotations which accrued to the term "amateur." Its meaning during this period shifted gradually, related not only to chronology but also to the area of inquiry and the level of competence; nor was the transition ever total. The term might be used during the early nineteenth

70. Quoted in Edward Lurie, "Science in American Thought," *Journal of World History*, VIII (1964-65), 656.

71. Numerous private letters complain of the two scientists, and they were aware of the hostility. See, for example, Rafinesque to Torrey, Philadelphia, 9 June 1819, Lexington, Ky., Sept. 1820, and Philadelphia, 12 Apr. 1828; Lewis Caleb Beck to Torrey, Schenectady, 30 May 1819; and William Darlington to Torrey, West Chester, Pa., 3 May 1841, all in Torrey MSS, NYBG. Also see George W. Featherstonehaugh, *The Monthly American Journal of Geology and Natural Science*, intro. by George W. White, I (New York and London, 1969; facsimile of 1831-32 ed.), 82-90, 140-42. Charles Bouvé, "The Manuscripts of C. S. Rafinesque (1783-1840)," American Philosophical Society, *Proceedings*, CII (Oct. 1958), 590, also comments on Rafinesque's high productivity (939 pieces) and the mixed assessment given his work by contemporaries and historians alike.

century to apply to a person whose occupation was related to science but who made no contribution to his field, to a person avocationally interested in a specific area of research, or even to a gentleman who enjoyed science in general and helped sponsor popular lectures and participated in a local scientific or philosophical society. The amateurs' decline in status was proportional to rising professional aspirations.

Nationalism was less strident but no less real in science than in other cultural or social aspects of nineteenth-century America. It seems almost ironic that the desire for a national scientific association to serve as a symbol of and vehicle for scientific advance was inextricably linked to the same force that had given rise to popular self-culture and sustained belief in inevitable progress. Pulled at by the ambitions of individuals as well as of subgroups, and uncertain of overlapping goals in relation to government aid, questions of priority, matters of science, and professional standards, the men of science seemed unable to make firm and dynamic decisions concerning scientific organization itself. That they needed and wanted operative institutions, like those they saw functioning effectively in Europe, was abundantly evident. Yet they inevitably spotted inadequacies in proposed institutions and were critical of those already established; the resulting hesitation and delays certainly frustrated men requiring intercourse for their research and perhaps even inhibited the advance of science. Articulated doubts offered ample evidence that the topic of national organization was under discussion and had at least some supporters. Encouraged by the example of federalism and despite the sectional conflict clouding the horizon, men of science envisioned a time when science could be a truly cooperative enterprise.

Combining a concern for the demands of both self-culture and professional aspirations, William Allen, professor of chemistry and experimental philosophy at Dickinson College, responded to the implication of Tocqueville's remarks in 1838 by urging that every American support natural science out of a desire "to promote the reputation of his country."[72] Allen concurred with contemporary assertions that "the genius of our institutions" forbade any overt assistance from the general government for research not directly related to the ends of government.[73] In America the people—in contrast to the kings, princes, and nobles of Europe—constituted the government. This fact led Allen to argue for popularization of science: "When the popular mind shall have

72. Allen, *Address before the Cuvierian Society*, p. 13.
73. *Ibid.*

been instructed, and the popular taste rightly directed, the public voice will demand what it now would reject."[74] The effort required time, perseverance, and especially a combination of like-minded men, but it could produce what patronage did in Europe. Allen added that "the numerous associations which are springing up in every part of our land for the cultivation of natural history are scattering their influence over the whole face of the community."[75] Ideally, one general organization would bring together the efforts of "farmers, miners, seamen, and soldiers, and even the hunters of the Rocky Mountains, as well as the more liberally educated classes of engineers, travellers, and military and naval officers."[76] Well led, such a group would develop financial and political power; science in popular culture would be mutually satisfying. Time would prove that Allen's expectations could not be met by any single organization in his century or the next.

In summary, it seems that many scientists, like Allen, accepted the progressive tenets of popular culture. They early participated in the publication and lecture efforts of the movement to popularize science. Increasingly pressed for time as scientific advance accelerated and a new professional self-image demanded personal research, and also skeptical of the results of their efforts at popularization, by the late 1830s men of science reconsidered their role. At the same time the evident results of the radically democratic impulse in fact forced the community of science to continue to speak out publicly. It realized that collectively it must establish professional standards as well as direct public attention to scientific research which might not have utilitarian implications.

Understanding the history of science in America requires recognition of the curious ambivalence that existed in the minds of mid-nineteenth-century scientists. On the one hand, they shared the nationalistic zeal of their compatriots for things American. On the other hand, they were defensive about the general level of national achievement in original research and resentful of continuous emphasis on application. True, at the beginning of the century American science was at the level of most developing nations and, like art, literature, and theater, did not reflect the sophistication and maturity of the more advanced cultural centers in Europe. Despite the obvious cultural lag, however, the sense of inferiority persisted unreasonably long and ultimately inhibited more

74. *Ibid.*
75. *Ibid.*, p. 14.
76. *Ibid.*

rapid expansion of scientific research. As late as 1850 Louis Agassiz, having lived in the United States for only four years, announced that such deference was unnecessary even as he cautioned against the opposite evil of overconfidence: "[The] time has come where American scientific men should aim at establishing their respective standing without reference to the expression of opinion of Europeans respecting them, and at the same time to be cautious not to allow national feeling to exaggerate their value. I have been surprised to find American men of science value their correspondence with Europeans of no standing at home, and on the other hand seen things and characters praised beyond bonds, simply because they are American." [77]

The introspective aspect of the reform era had had a direct effect on science, as had the spirit of progress and nationalism. The economics of an expanding and a consciously assertive nation provided an essential support for cultural advance. Nor did any overt hostility confine the aspirations of science in general. These preconditions were not sufficient for promoting a comprehensive self-defined organization for science, however. It was necessities within what might be loosely called the national scientific community which propelled the people in science toward national association.

77. Agassiz to S. S. Haldeman, Cambridge, 2 May 1850, Joseph Leidy MSS, Academy of Natural Sciences of Philadelphia (hereafter ANSP).

II

To Create *Esprit de Corps*

Scientific or learned societies were a common component of cultural and intellectual life by the nineteenth century. Educated individuals banded into local, often short-lived organizations designed to advance their own knowledge of science and to diffuse their findings to the public. Having a scientific society became one measure of maturity for an aspiring urban area.[1] Yet despite the numerous scientific societies formed in the years following 1815, by 1840 American men of science still lacked national *esprit de corps*.[2] None of the existing associations provided a focus for national scientific activity, nor were they capable of representing American science on an international level. Their almost exclusively local activity was often as appropriate to the popular culture movement as it was to genuine scientific inquiry. External factors had established an environment capable of sustaining a national organization, but none was created. Before the transportation revolution of the early nineteenth century, which saw the rapid expansion of a public receptive to broad concepts of science, before the pervasive trend toward associationism, and before the inclusion of more basic science into the collegiate curriculum—before all of these, the possibilities of a national organization were severely limited. Given these, however, men of science could consider several European models as options for a national scientific institution, at the same time drawing on their own

1. Walter B. Hendrickson, "Science and Culture in the American Middle West," *Isis,* LXIV (Sept. 1973), 326–340.
2. The lack of *esprit de corps* was a recurring theme throughout the 1830s and 1840s, and Henry R. Schoolcraft repeated it in his unpublished address on "Scientific Associations Abroad," delivered before the New York Historical Society on 6 Nov. 1842, New York Historical Society Archives, New York.

practical experience in local American groups. In the latter those men who were to establish the first successful national organization came of age, and their experience armed them with caution and realistic (even pessimistic) expectations of what a national organization might accomplish.

Modern scientific organization developed in response to the needs of the seventeenth-century scientific revolution. A classic study of that phenomenon indicates that the major advantages of organization were soon realized by men of science of that century: that is, the concentration of scientists and information, the publication and translation of books and articles, and the involvement in problems of more general public concern.[3] As a result, the organized support which modern science required derived not from the universities but from its own corporate creation, the scientific society. By the nineteenth century some universities were assuming a new posture toward science, but only in specific locations such as Scotland and Germany were they supporting research.[4] Thus it is hardly surprising that American men of science should also look to societies to fulfill their similar needs for research and communication. Although the nineteenth century marked a turning point for the relationship between science and the universities or research foundations, at the close of the century William H. Brewer could legitimately conclude, "They [societies] have been, directly or indirectly, a most potent factor of progress in material advancement and in intellectual culture."[5]

As a leading sociologist of science suggests, science organization has always been loose.[6] Historically, few scientific societies have managed a central, ongoing source of authority, a clear hierarchical structure, or

3. Martha Ornstein, *The Role of Scientific Societies in the Seventeenth Century* (Chicago, 1928), pp. 260–261. Ornstein's pioneer work overemphasizes the experimental outlook of the major societies in England, Germany, and France but remains an important survey of early activity. For more detailed monographs on early organizations, see Harcourt Brown, *Scientific Organizations in Seventeenth Century France, 1620–1680* (Baltimore, 1934), and Margery Purver, *The Royal Society: Concept and Creation* (Cambridge, Mass., 1967). Another important study is Roger Hahn, *Anatomy of a Scientific Institution: The Paris Academy of Sciences, 1666–1803* (Berkeley, Calif., 1971).

4. For a lively and provocative assessment of the German universities, see Eric Ashby, *Technology and the Academics: An Essay on Universities and the Scientific Revolution* (New York, 1963), pp. 20–27.

5. William H. Brewer, "The Debt of This Century to Learned Societies," Connecticut Academy of Arts and Sciences, *Transactions*, XI, part 1 (1901–3), p. xlvii.

6. Bernard Barber, *Sociology of Science* (New York, 1962), pp. 207–208.

well-defined goals. Yet specialization and division of labor in scientific research brought a need for the coordination of special efforts as well as new requirements by government and the public for definition and justification. From within and from without, the older scientific organizations increasingly felt a demand that they take responsibility in educating the general public while simultaneously contributing to scientific advance. Closer scrutiny from outside required that the organizations define themselves more specifically and tighten structural patterns.

AMERICAN INSTITUTIONAL PATTERNS

In establishing their scientific and educational institutions, Americans clearly emulated basic British patterns. Throughout the colonial period North Americans served as field researchers for English and Continental scientists, with the most active becoming "corresponding members" of the Royal Society.[7] Increase and Cotton Mather's early attempt to found a Boston society in the 1680s and Benjamin Franklin's subsequent effort in Philadelphia in the 1740s reflected the influence of the Royal Society.[8] Parallels with British institutions were evident in the founding of the American Philosophical Society, various Linnaean societies, the Franklin Institute, and the American Association for the Advancement of Science. Americans stridently declared their independence from the mother country while emulating her social, intellectual, and cultural patterns.[9] Diffident about a cultural lag, the citizens of the young nation themselves most frequently manufactured a critical yardstick for making comparisons.[10] It seems that while a belief in progress led them into

7. The two best studies insist that Americans made substantive contributions, but the essential fact remains that, with limited texts and apparatus, most Americans were relegated to data collecting. See Brooke Hindle, *The Pursuit of Science in Revolutionary America, 1735–1789* (Chapel Hill, N.C., 1956), and Raymond P. Stearns, *Science in the British Colonies of North America* (Urbana, Ill., 1970). Also see Whitfield J. Bell, Jr., *Early American Science: Needs and Opportunities for Study* (Williamsburg, Va., 1955).

8. Michael G. Hall, "The Introduction of Modern Science into Seventeenth-Century New England: Increase Mather," *Ithaca: Proceedings of the Tenth International Congress of the History of Science . . .*, I (1962), 261–263; Hindle, *Pursuit of Science*, pp. 67–74.

9. Frank Thistlethwaite, *The Anglo-American Connection in the Early Nineteenth Century* (Philadelphia, 1959), pp. 76–110. Thistlethwaite does not comment on the scientific interplay across the Atlantic, which was as direct as that of the reformers: extensive correspondence, visits, exchange of materials, and unity of thinking on institutions.

10. An undated clipping in the Association of American Geologists and Natural-

the nineteenth century, they were also prodded by a sense of national inadequacy.

But the striking parallels belie significant differences in the development and operation of organizations in the two nations. In the first place, American scientists believed that the membership in the American societies often had a weaker scientific background than their European counterparts and that the level of papers presented was not as uniformly high.[11] In addition, American institutions demonstrated a distinctive style of public instruction. The British tended to group in rather exclusive circles for the "cultivation" of science and then on occasion offer didactic lectures to the public. Responding to republican rhetoric and widespread interest in the utilization of science, the Americans gradually developed a pattern of lecture sponsorship and open membership which, by comparison, seemed less exclusive (and perhaps less rigorous) than the British pattern. Still, the simple fact of similar society names does suggest that Americans retained an unusually long and unnecessary deference toward Europe, and especially toward England.[12]

In the second half of the eighteenth century two "philosophical" societies were founded in America. Devoted to all knowledge, including science, the arts, and the useful arts (technology), they provided a meeting place for intellectuals. High ideals and overexpectations led, rather quickly, to some disillusionment and a retreat to essentially local activity for both, although they continued to aspire to national recognition and maintained a cordial rivalry.[13] The American Philosophical

ists MSS, ANSP, from the *Albany Daily Advertiser* (1843) pointed out the quality of Silliman's *American Journal of Science and Arts*, self-consciously adding, "We hope that our intelligent readers need not be told of the high character which this Journal has maintained not only at home, but even in Aristocratic Europe, which is rapidly becoming aware that some benefit may be derived from reading an American book or an American Journal."

11. See Jeremiah Van Rensselaer, *Lecture on Geology: Being an Outline of the Science, Delivered in the New York Atheneum* (New York, 1825).

12. As late as 1847 Louis Agassiz, after less than a year in America, observed to a fellow European, "This deference toward England (unhappily, to them, Europe means almost exclusively England) is a curious fact in the life of the American people. . . . From England they receive their literature and the scientific work of central Europe reaches them through English channels. . . . Notwithstanding this kind of dependence upon England, in which American savants have voluntarily placed themselves, I have formed a high opinion of their acquirements, since I have learned to know them better, and I think we should render a real service to them and to science, by freeing them from this tutelage." Elizabeth C. Agassiz, ed., *Louis Agassiz: His Life and Correspondence*, I (Boston, 1885), 435.

13. *American Journal of Science and Arts* (usually referred to by contemporaries as Silliman's *Journal*; the name was shortened in 1879 and will hereafter be cited as *AJS*), XXXVII (1837), 311.

Society was founded in the 1760s, when Philadelphia led the nation politically, culturally, and intellectually, for American men of science.[14] Its hegemony was only slightly challenged by the second major learned society, the American Academy of Arts and Sciences, founded in Boston in 1779 by John Adams and others seeking to assert Boston's intellectual leadership. Although the latter's name deliberately implied emulation of the French Academy of Sciences, the American Academy's organizational structure and subsequent activity were clearly influenced by the Royal Society.[15] Both developed a network of "corresponding members" throughout the young nation and abroad, but their activity was chiefly local. In this they were not unlike the Royal Society of the early nineteenth century; a contemporary noted that it had become largely a London club, "most of whose members were united by a more or less vague interest in science."[16] In America, too, membership was often honorary, and local prestige maintained a sizable proportion of socially and politically eminent persons.

The only other significant attempt to establish a national scientific society was in Washington, the new national capital.[17] But the federally chartered Columbian Institute for the Promotion of Science (1816) lacked the nucleus of scientific men found in Boston or Philadelphia, and attempts to compensate with political membership proved unsuccessful. As a result the institute faded from view in the 1830s.[18] Only in the post–Civil War years did Washington establish a successful, ongoing local society based on scientists working in federal agencies.[19]

14. Whitfield J. Bell, Jr., is currently working on a history of the early American Philosophical Society and is particularly interested in the cultural life of Philadelphia. See especially his "The American Philosophical Society as a National Academy of Sciences, 1780–1846," *Ithaca: Proceedings of the Tenth International Congress of the History of Science* ..., I (1962), 165–177.

15. Hindle, *Pursuit of Science*, pp. 263–265. *Belles lettres* had a larger role in the Boston society than in the Philadelphia society, but still the "natural curiosities" dominated.

16. Quoted in L. Pearce Williams, "The Royal Society and the Founding of the British Association for the Advancement of Science," Royal Society of London, *Notes and Records*, XVI (Nov. 1961), 221–233.

17. Joel Barlow's often cited proposal for a national university in the capital, outlined in his *Prospectus of a National Institution to Be Established in the United States* (Washington, D.C., 1806), had the dual purpose of advancing knowledge by the association of scientific men and disseminating science among youth. The plan of the Jeffersonians never went beyond speculative stages.

18. See Richard Rathbun, *The Columbian Institute for the Promotion of Arts and Sciences: A Washington Society of 1816–1838, Which Established a Museum and Botanic Garden under Government Patronage*, Smithsonian Institution, *Bulletin 101* (Washington, D.C., 1917).

19. James K. Flack's "The Formation of the Washington Intellectual Community,

With a reputation sustained by leading Philadelphia scientists, the American Philosophical Society's membership continued to include the most outstanding men of science throughout the nation. But its regular meetings and administrative tasks reverted to local men, many of whom were social and political leaders only avocationally interested in science. Because it had an extensive exchange program for specimens and periodicals and bestowed honorary membership on notable foreign scientists, the society was probably the best-known scientific institution abroad. Then, in the 1820s, two new and dynamic organizations, the Academy of Natural Sciences and the Franklin Institute, diverted the attention of some of the most active men of science in Philadelphia. Overlapping membership restrained outright competition, but energy for research found an important new outlet in the meetings and publications of the younger groups.

The American Academy of Arts and Sciences of Boston had a less ambitious program than its Philadelphia counterpart, the American Philosophical Society. Its active membership correlated closely with the Harvard University faculty roster, and Boston's leading families were also well represented. Members met monthly, usually to hear one or more papers by members; but while it had a library, it did not develop museum holdings. By the 1830s its scientific contingent had a strong interest in the physical sciences. As a result they succeeded in having the Rumford Fund's provisions modified to allow support for research, and a substantial portion of the interest was then applied toward a new observatory for Harvard.[20] Because of this emphasis on the physical sciences, conflict with Boston's new Linnaean Society and its active successor, the Boston Society of Natural History, was minimal.[21]

In contrast to the philosophical societies' stability and prestigious membership was the proliferation of new, admittedly local scientific

---

1870–1898" (Ph.D. dissertation, Wayne State University, 1969) focuses on the Washington Philosophical Society, and includes a discussion of the early limitations of Washington as a cultural, scientific center.

20. The only book-length study of the American Academy is the non-analytical sesquicentennial history, *The American Academy of Arts and Sciences, 1780–1940* (Boston, 1941). Also see *The Rumford Fund of the American Academy of Arts and Sciences* (Boston, 1905), pp. 1–7.

21. Ralph S. Bates, *Scientific Societies in the United States*, 3rd ed. (Cambridge, Mass., 1965), p. 69, finds some decline in the academy in the 1840s and suggests the challenge of the Boston Society of Natural History as one reason. Letters of contemporaries do not seem to suggest, as they do for the Philadelphia groups, much conscious competition.

academies and societies founded in the 1820s and 1830s. During the second quarter of the nineteenth century they appeared in urban and rural areas alike as well as on college campuses. Originally an eastern phenomenon, they followed population westward toward the frontier. Having only rudimentary resources, these societies proved inefficient for advancing or even for diffusing science. The failure rate of scientific societies, like that of scientific periodicals and small colleges of the period, was high.[22] Yet even their often temporary existence reflected not only the enthusiasm for self-culture but often also an earnest desire to contribute directly to scientific progress. Inevitably such high purpose resulted in total disillusion or a retreat to general discussions and an occasional invited lecturer on science. As late as 1839 the senior Silliman concluded that most such groups existed "in name only."[23]

It was possible that outside the established scientific centers some groups might defy the pattern, not only proving themselves viable but contributing to research. But such societies were usually exceptional because of members with unusual ability.[24] John G. Anthony, a prominent

22. This phenomenon has frequently been noted, but no one has yet examined the role of these societies in relationship to the cultural or scientific development of the United States. Bates provides a summary table of them, demonstrating the increase by decade: in 1805, four societies; 1815, seven; 1825, twenty-three; 1835, twenty-six; 1845, thirty-two; 1855, thirty-five; 1865, thirty-six. This detail obscures the numerous failures indicated throughout his chapter on the period, especially in New York, where the possibility of obtaining state funds had precipitated twenty-six local societies. Nor do Bates's figures reflect the rise of local museums, often in conjunction with academies whose demise left a residue collection in some member's attic. See Bates, *Scientific Societies*, pp. 28–57, especially p. 51. Also see Daniels, *American Science*, p. 14; David D. Van Tassel and Michael G. Hall, eds., *Science and Society in the United States* (Homewood, Ill., 1966), p. 26. Harry R. Kallerup, "Bibliography of American Academies of Science," Kansas Academy of Sciences, *Transactions*, LXVI (1963), 274–281, is a helpful bibliography for the later period of state academies of science.

For a discussion of the high rate of failure in these other areas, see Donald DeB. Beaver, "Appendix I: A Statistical Study of Scientific and Technical Journals," in "The American Scientific Community, 1800–1860: A Statistical-Historical Study" (Ph.D. dissertation, Yale University, 1966), pp. 211–268; it provides an excellent discussion of journal profiles in the period and suggests that the birth and death rates of journals were roughly parallel down to 1865, both steadily increasing. For the record of failures among colleges, see Donald G. Tewksbury, *The Founding of American Colleges and Universities before the Civil War with Particular Reference to the Religious Influences Bearing upon the College Movement* (New York, 1932), p. 28; Tewksbury indicates that of 516 colleges established, only 19 percent survived. In general the 1840s felt a slackening of general reform enthusiasm with a consolidation of gains made earlier.

23. *AJS*, XXXVII (1839), 146.

24. Hendrickson, "Science and Culture," p. 33. Hendrickson suggests that the

member of the Western Academy of Natural Sciences at Cincinnati, was such a man. But, even more, he was an example of the frustrations of men of science in isolated areas, especially in the West. Unable to keep abreast of the current literature and newest discoveries but an eager student of conchology, he corresponded with several Eastern scientists, indicating his findings and asking them for information. In Cincinnati he worked to keep fellow field researchers up to date on recent developments and nomenclature.[25] Humble about his own achievements, he was the ideal of eastern researchers, who resented the sometimes exorbitant priority claims of western men of science. His complaints about isolation and inadequate communication were persistent problems for most researchers in recently settled or sparsely populated areas; not surprisingly, they were enthusiastic about a national scientific organization. As one New Yorker contributor to Silliman's *Journal* noted, it was "much easier to obtain information from Petersburg or Paris than from Cincinnati...."[26] Nonetheless these academies provided essential support for local members through their libraries and collections, also serving as sounding boards for new ideas through meetings and publications and as liaisons to larger communities. As societies, however, they followed rather than led.

Thus the most active organizations were in eastern urban areas, supported by a mixed membership of the entrepreneurial or professional classes together with men in science. When Dwight attempted a listing of scientific organizations in 1826, he found only twenty-nine of them.[27] Aside from several generated in upper New York by a legislative grant,

---

convergence of interested individuals was of critical importance and that the departure of key members often brought destructive changes.

25. Henry Shapiro in private discussion and unpublished material on the Western Academy has indicated both the contributions and the frustrations of Anthony, who with only a limited education specialized in mollusks. Anthony also maintained an extensive correspondence with such naturalists as Samuel S. Haldeman, Louis Agassiz, John Torrey, and Isaac Lea. In a letter to Haldeman, Cincinnati, 22 Feb. 1842, Haldeman MSS, ANSP, he outlined his efforts to establish uniform nomenclature among his fellow western naturalists through a detailed comparison of all specimens found in the area. A useful series of letters from Anthony is in the Benjamin Tappan MSS, Library of Congress, Washington, D.C. (hereafter LC). A general summary of the society's history is Walter B. Hendrickson, "The Western Academy of Natural Sciences of Cincinnati," *Isis*, XXXVII (July 1947), 138–145.

26. S. E. D[wight], "Notice of Scientific Societies in the United States," *AJS*, X (1826), 376.

27. *Ibid.*

all were in cities with substantial population, a reputable college, or both. Not surprisingly, Dwight singled out Philadelphia as unusually successful in its encouragement for science, suggesting the congruence of the academy's *Journal* and library, the University of Philadelphia's professorships in natural history, and the museums there. Although there was some interest in astronomy and meteorology, increasingly members of the local scientific societies were avocational geologists or naturalists. This was in part because the earth sciences were an area of recent achievement and because America remained relatively open to new discoveries.

Natural history, including geology, botany, and zoology, felt the impact of new European research in the early decades of the nineteenth century. Naturalists did not always find a place in the established societies. Even when they did, they went outside their local groups in order to corroborate findings of fellow fieldworkers and to initiate more substantive analysis in specific areas of natural history. Where there were sufficient numbers of naturalists, they banded together, founding the most scientifically significant societies of the generation following the War of 1812. Of particular importance were the Academy of Natural Sciences of Philadelphia (1812), the New York Lyceum (1817), and the Boston Society of National History (1830). Together with the Franklin Institute in Philadelphia (1824), whose expertise was in the physical sciences and technology, these societies provided an organizational model and a testing ground for the generation of men who founded the American Association for the Advancement of Science.

The three largest urban areas north of the Mason-Dixon line and along the Atlantic coast continued as scientific centers well into the nineteenth century. Here men in natural history organized societies which permitted them to develop specific, specialized collections and to discuss their research. Brief consideration of the groups in Boston, New York, and Philadelphia suggests both the positive and negative features of these societies, which provided a transition between the ideal of a philosophic society and a specialized, professional society. Other similar groups enjoyed temporary success, such as the New Haven Natural History Society, the Albany Institute, and the Literary and Philosophical Society of Charleston. Those in the South and West were important feeders, providing new descriptions and specimens to eastern closet researchers and their cabinets. Most were short-lived, however, and their records unfortunately meager. The three eastern societies established norms for the others by example and gave the founders of the Association of American Geologists valuable experience.[28] They also reflect fairly accurately how political and economic vicissitudes of the period affected the success of scientific organizations.

The three natural history societies had similar origins, each growing from a nucleus of men who in the 1810s had sought to establish research facilities in natural history. The earliest, the Academy of Natural Sciences of Philadelphia, listed as its purposes the formation of a natural history cabinet, a library of works on science, a chemical laboratory, and an experimental philosophical apparatus for the illustration and advancement of knowledge to "the common benefit of all individuals who may be admitted members of our institution."[29] Similarly the New York Lyceum developed from the efforts of Samuel L. Mitchill and a surprisingly young group of medical men who ambitiously hoped to establish a cabinet and to require regular papers from its members.[30] Although the Boston Society of Natural History was not chartered until 1830, its roots were in the vigorous Linnaean Society also formed by medical men and led by Jacob Bigelow in the 1810s; the latter disbanded in 1823 when it was financially unable to retain its collections. The Society of Natural History's constitution reiterated the goal of mutual cooperation for the collection of a cabinet and a library,[31] and within a decade had a collection rivaled only by Philadelphia.

Operationally, the three natural history societies were also similar.

28. The lack of a complete list of association members makes a specific correlation difficult, but of those who are known, approximately four out of five were regular members in one of the three societies.

29. Quoted in Max Meisel, *A Bibliography of American Natural History: The Pioneer Century, 1769–1865*, vol. II (New York, 1926). Meisel is the best and indispensable guide to early natural history in America; the three massive volumes were published in Brooklyn, 1924 to 1929. His "Index of Institutions," pp. 684–702, is exhaustive for local societies of any reasonable duration. For the Philadelphia academy, also see Samuel George Morton, "History of the Academy of Natural Sciences of Philadelphia," *American Quarterly Register*, XIII (1841), 433–438, and Ruschenberger, *Notice of the Academy*. The manuscript record of the society is available on microfilm; see Venia T. Phillips, *Guide to the Microfilm Publication of the Minutes and Correspondence of the Academy of Natural Sciences of Philadelphia, 1812–1942* (Philadelphia, 1967).

30. Herman L. Fairchild, *A History of the New York Academy of Sciences, Formerly the Lyceum of Natural History* (New York, 1887), pp. 1–15.

31. Thomas T. Bouvé, "Historical Sketch of the Boston Society of Natural History; with a Notice of the Linnaean Society, Which Preceded It," *Anniversary Memoirs of the Boston Society of Natural History, 1830–1880* (Boston, 1880), pp. 1–14. Also see Augustus A. Gould, "Notice of the Origin and Progress and Present Condition of the Boston Society of Natural History," *American Quarterly Register*, XIV (1842), 236–241.

They each required that new members be recommended by an active member and followed the practice of the philosophical societies by having membership categories for residents and nonresidents. Resident members alone paid dues and could vote, but contributions to the cabinet and library were encouraged from corresponding members, as were papers for presentation or publication. Having no limit on term of office, each society developed a small coterie of men who essentially directed local activity.[32] Meetings were held on a regular schedule, monthly or bi-monthly as recorded in published or manuscript minutes, which remain essential for estimating the pattern of success for each of the groups.

Both financial and philosophical motivations encouraged the newer societies to have a program including the total community. Popular activities also reflected contemporary optimism for education and were a response to public curiosity. Lectures, if well attended, could bring a profit for the sponsoring organization, and good public relations might establish political or philanthropic support. Museums became the focal attraction since lectures had competitive sponsors. In 1828 the academy in Philadelphia, having then the largest cabinet in the United States, with most items donated by members, opened its doors gratuitously to the public twice each week.[33] The New York Lyceum collections were also on display for the public, housed in the Common Council room.[34] Once established, the Boston society was open one day each week, free of charge.[35] None of the societies maintained a record of attendance, but the sustained programs suggest that the public did participate.

In the 1830s, buoyed by popular response and flush times economically, the societies peaked in their general activity. The Philadelphia academy, regularly receiving books and specimens from its nonresident president William Maclure, met weekly and published its *Journal* with new regularity. By 1839 it had embarked on a building program to house the growing collection and library.[36] Similarly the New York Lyceum

32. Thus the nominal head of the Philadelphia academy to 1840 was William Maclure, but in his absence John P. Wetherill and Samuel George Morton guided its activity. In the New York Lyceum Joseph Delafield retained the helm for almost the entire period from 1827 to 1865, with William Redfield dominating activity in the 1840s. The Boston Society tended to rotate its officers more frequently, but the working center was consistently George B. Emerson, Amos Binney, Benjamin D. Greene, Charles T. Jackson, A. A. Gould, and F. W. P. Greenwood.

33. Morton, "Academy of Natural Sciences," pp. 433–438.

34. Fairchild, *New York Academy of Sciences*, p. 30.

35. Gould, "Notice of the Boston Society," pp. 236–241.

36. "The New Hall of the Academy of Natural Sciences in Philadelphia," *AJS*, XXXVII (1839), 399.

began a building in 1836, and in 1837 it sponsored a series of lectures by Benjamin Silliman of Yale.[37] Its *Annals* contained, as did the Philadelphia academy's *Journal*, the transactions of the society; the proceedings of the meetings appeared irregularly in Silliman's *Journal*. Most ambitious was the young society in Boston, which auspiciously opened its first year by successfully devising and executing a plan for a zoological and botanical survey to be connected with the geological survey begun by Hitchcock that year. By sponsoring a successful series of lectures from 1832 to 1834, it demonstrated to John Lowell the popular interest in such activity. Within a decade it had an important cabinet, built from the nucleus of Amos Binney's shell collection and Charles T. Jackson's mineralogical collection, and also a library of over one thousand volumes. The Boston society even built a special hall for display and lectures, which excited both its membership and the general public.[38]

By the close of the decade, however, all three societies showed evidence of decline. Usually the deterioration was related to financial difficulties and to flagging public enthusiasm. The Boston societies were less affected by economic problems than were the others, but in that city decline in membership and discontinuation of the lectures by the Society of Natural History indicated rapidly dissipating interest.[39] The American Academy of Arts and Sciences also ceased to hold regular meetings and to publish its proceedings and only gradually rebuilt its strength on the basis of interest in the new Harvard observatory. Its new volumes of proceedings in 1846 were replete with papers presented by a rising group of physical scientists, including mathematician Benjamin Peirce and astronomer George Bond.[40]

But in New York the lyceum nearly failed completely. Unable to meet the inflated payments (resulting from the depression after the

37. John Torrey to Benjamin Silliman, New York, 7 Dec. 1836, Torrey MSS, NYBG. These lectures were directed toward patrons of the lyceum rather than the general public. Torrey's warning that there would be little profit suggests that public lectures usually were profitable.

38. F. W. P. Greenwood, "An Address Delivered before the Boston Society of Natural History," Boston Society of Natural History, *Journal*, I (1834–37), 13.

39. Bouvé, "Historical Sketch," pp. 33, 55. Bouvé indicates that the average yearly attendance at meetings fell off sharply between 1835 and 1848, possibly resuming its earlier tempo at the later date under the influence of William Barton Rogers and Louis Agassiz with his Swiss company of scientists. Also see E. C. Herrick to James Dwight Dana, New Haven, 22 June 1841, Dana MSS, BLY. Herrick noted to Dana, then on the U.S. exploring expedition, that the New Haven Natural History Society lacked both money and experience and that the Boston Society was probably the most prosperous of all similar societies in the nation.

40. American Academy of Arts and Sciences, *Proceedings*, I (May 1846), *passim*.

Panic of 1837) [41] on their new building begun in 1836, the lyceum was threatened by foreclosure in 1842, and in 1844 the structure was sold at auction. Often without a quorum, the lyceum met at Joseph Delafield's home until they accepted an offer to meet in the rooms of the New York University Medical College. The move, however, increased an already evident dissension among members, some of whom left the lyceum. [42]

In Philadelphia the American Philosophical Society and the Academy of Natural Sciences also experienced a substantial reordering. Partially to re-establish its leadership position, the American Philosophical Society had embarked on a program of expansion. In 1837 it urged the city of Philadelphia to help sponsor an astronomical observatory and in 1841 contracted for purchase of a larger, more modern building. But as the Panic of 1837 stretched into a lengthy period of depression, the society found itself overextended, especially after the city withdrew an informal offer to purchase its old building. As a result the society was censured by the city for misusing the observatory advance, and it lost the new building. Subsequent years were marked by internal dissension and basic retrenchment. The centennial celebration of 1843 could not completely mask the impending financial disaster. While maintaining that the society was never "more active, more successful," the main speaker for the occasion soberly concluded with an observation on the "darker side" of the picture and alluded to financial difficulties. He pleaded for personal sacrifice on the part of members and a new "unity of spirit" which alone could redeem the institution. [43] The society

41. Stuart Bruchey, *The Roots of American Economic Growth, 1607–1861* (New York, 1965), p. 134. Bruchey points out that after 1837 states as well as individuals suffered financial trauma, which generally curtailed funds for internal improvements; this seems to have extended to scientific projects like the state surveys as well. Also see Joseph Henry to J. S. Henslow, Princeton, 2 Dec. 1839, Misc. MSS, New York Historical Society.

42. Fairchild, *New York Academy of Sciences*, pp. 41–44; 46–47. Fairchild suggests that the split reflected the conflict between the older College of Physicians and Surgeons and the University Medical College. Earlier the lyceum had declined offers to use facilities at Columbia College and at New York University because members resented dependency on any other institution. Sharp internal dissension was evident in Melines C. Leavenworth to John Torrey, Waterbury, Conn., 6 May 1845, Torrey MSS, NYBG, and in John LeConte to Baird, New York, 1 Apr. 1844, Baird MSS, Smithsonian Institution Archives, Washington, D.C. (hereafter SIA).

43. Robert M. Patterson, *Early History of the American Philosophical Society; a Discourse Pronounced by Appointment of the Society, at the Celebration of Its Hundredth Anniversary* (Philadelphia, 1843), pp. 32–36. The storm was weathered, but the society lost income property, the museum hall, and community respect. Also see American Philosophical Society, *Proceedings at the Dinner Commemorative of the Centennial Anniversary of the Incorporation of the Society* (Phila-

weathered the financial crisis, but the internal problems were less easily resolved.[44] When President Franklin Bache in 1853 argued for reconsidering membership qualifications, suggesting that perhaps "requiring of too high a grade attainment in candidates for membership, and rules or usages too exclusive, tend to defeat the main objects of the association," he demonstrated that such internal questions of purpose were not yet resolved.[45] Events in the Academy of National Sciences were less dramatic, but recorded attendance also dropped, with several meetings in the mid-1840s canceled for lack of a quorum.[46]

In this period of contraction and redefinition, divisions within the societies were intensified. Frequently they reflected only personality differences,[47] but at least some of them reflected changing expectations of the organization of science. If some members were content with older standards, others were pleased that their reputation was changing from "busy triflers" to working scientists.

One other, somewhat exceptional, laboratory for organizational ex-

---

delphia, 1880), pp. 25–28, and *Proceedings*, III (1843), 34–35. A similar point is made by Whitfield J. Bell, Jr., "Astronomical Observatories of the American Philosophical Society, 1769–1843," APS, *Proceedings*, CVIII (1964), 12–13.

44. In the American Philosophical Society, for example, the internal squabbles stemmed from member dissatisfaction with the society's directions, the financial difficulties, and simply personality clashes. There has been no study of the "time of troubles," but clues are found in the John Fries Frazer MSS, APS, and the S. S. Haldeman MSS, ANSP; also see John C. Cressen to A. D. Bache, Philadelphia, 18 Jan. 1845, Coast and Geodetic Survey MSS, U.S. National Archives, Washington, D.C., (hereafter NA); James B. Rogers to William B. Rogers, Philadelphia, 20 Jan. 1848, Rogers MSS, Massachusetts Institute of Technology Archives, Cambridge (hereafter MITA). William D. Lingelbach, "The American Philosophical Society Library from 1842 to 1952 with a Survey of Its Historical Background," APS, *Proceedings*, XCVII (1953), 478, suggests the policies of the irascible George Ord, librarian from 1842 to 1848, as another factor.

45. APS, *Proceedings*, V (1848–53), 360. The only analysis of the society's decline is Bell's brief article, "American Philosophical Society." Bell suggests that the major reasons for decline after 1825 were the catholic tradition of including all knowledge in a time of increasing specialization, the transfer of Philadelphia's prestige to other urban areas, the complacency of the membership, and the enjoyment of a genteel status at home and abroad. Clearly coordinate with these reasons was declining active participation by men of science who found newer groups more interested in science.

46. Phillips, *Guide to the Microfilm Publication*, p. 26. Efforts were made to regain membership participation, including in 1849 a change in the bylaws to increase the number of standing committees for special reports. Also see Joseph Leidy to Samuel S. Haldeman, Philadelphia, 17 Feb., 2 Jan. 1849, Leidy MSS, ANSP.

47. Lack of analytical histories of even the major scientific societies makes generalization difficult, but manuscript materials on science for the period of the 1840s are replete with allusions to governing "cliques" and references to somewhat abrupt changes in leadership rosters.

perience was Philadelphia's Franklin Institute. Even more than the three natural history societies, it had originally directed its efforts toward the general public, especially through evening courses and lectures on mechanics. It never completely lost this function, but its research interests narrowed considerably under the direction of a new breed of physical scientists. Led by Alexander Dallas Bache, a small group did sophisticated research and put out a publication which attempted to disseminate scientific results rather than explicate general trends in science and technology. The institute was in this respect, however, more important as a model for the Smithsonian Institution and for the National Academy of Sciences than for the American Association for the Advancement of Science. Its activity, along with that of its Philadelphia counterparts, tapered off after 1837, primarily because its most active leaders left Philadelphia or assumed other responsibilities.[48]

Because few men had financial resources to subscribe to major European journals or even to the growing number of American publications, society libraries, built through exchange and donation, were essential tools for their members. Silliman's *Journal* remained the most widely circulated American publication, presenting not only original papers but also review articles on current European research and abstract proceedings of American scientific societies; it retained a strong natural history emphasis.[49] Quality of publication ranged widely in most society proceedings, from superficial observations on animal or fossil remains to detailed descriptions and closely argued theoretical analyses. Because publication costs were high, many journals appeared irregularly, and the number of subscribers was frequently inadequate to cover costs.[50] Thus with the response to a growing need for reputable and well-circulated journals came also more realistic awareness of the high costs of publication and the need for an editor capable of discriminating the best selection of articles.

Intimately related to the funding of publications was the question of

48. Joseph B. Sinclair has provided an excellent analysis of the early years of the institute in *Philadelphia's Philosopher Mechanics: A History of the Franklin Institute, 1824–1865* (Baltimore, 1974); also see his *Early Research at the Franklin Institute: The Investigation into the Causes of Steam Boiler Explosions, 1830–1837* (Philadelphia, 1966).

49. Edward S. Dana et al., *A Century of Science in America, with Special Reference to the American Journal of Science, 1818–1918* (New Haven, Conn., 1918), pp. 25–36.

50. *Ibid.*, pp. 32–36. Silliman was unable to pay contributors to his *Journal* and frequently published it at a personal loss. The irregularity of society publications has already been noted.

a cabinet. Here, as with a library, scientists were unable to retain a full private collection except perhaps within a narrow specialty. Popular museums like that of Charles W. Peale and of the local lyceums were simply not designed for scientific research. Therefore a cabinet was included as an essential requirement for a local natural history society because it implied a central location for regular meetings, a network of scientific correspondents capable of providing specimens from distant regions, and an active membership capable of identifying and caring for the collection.[51] Once the responsibility for a cabinet was assumed by a society, however, the group generally found voluntary curators inadequate and hired an individual to arrange and preserve specimens. A scientific society had become a business concern, complete with maintenance costs.

Membership dues and admission fees collected for lectures or entrance to the museum were used to offset the expenses of operation. Zeal was high, and when confronted by financial difficulties, members of the local societies were unwilling to curtail activities or modify their program in the 1830s. Frequently gifts from wealthier members covered embarrassing deficits, but the unbalanced budget was an ongoing difficulty and a component in the financial failures of the 1840s.

The most urgent necessity was therefore finance. Establishing a place of meeting, preserving a cabinet, and sponsoring a publication all required major expenditures. The establishment of substantial cabinets by local societies, while a drain on their resources, was crucial because it left naturalists free to consider founding a more comprehensive organization whose responsibilities might be less tied to practical demands. Undoubtedly this made the British model of a peripatetic society attractive, since it had the advantage of no maintenance expenses while simultaneously offering an opportunity to utilize existing cabinets in various areas. The lesson of finance was not lost on the founders of the Association of American Geologists, who would carefully wait until they had funds in hand before publishing their proceedings.

Despite the proliferation of societies to 1840, there was still no national spokesman for men of science. Even the most successful societies spoke only to a regional public, and none established the professional identity sought by men of science. The policies of the American Philosophical Society and the American Academy of Arts and Sciences were too ex-

51. The other alternative, of course, was for a local, state, or the national government to sponsor such facilities, as the New York legislature did for the results of its geological survey.

clusive and local. The newer natural history societies in the 1830s proved the potential dynamism of the natural sciences and of public interest, but they, too, depended on local membership and support. Their rivalry was a friendlier one, indicating a conscious awareness that local activity was important but that cooperation and exchange alone might sponsor major research in American zoology, botany, and geology.[52] Conscious of current advances in their disciplines and eager to be part of the movement, they realized that local activity alone had little effect on international reputation. They were also cautious in response to self-culture, working to guard the research integrity of collections. By their increasing interest in publication, they began to establish standards for research and editing.

Perhaps the first scheme to coordinate scientific groups in the nineteenth century was initiated at West Point in 1829. A circular for the American Association for the Promotion of Science, Literature, and the Arts indicated that it intended to work from the top, establishing smaller, local versions of itself as a parent organization. It thus differed from the informal exchange network existing between local natural history societies. Outlining an organization which sounded much like the older philosophical societies in its catholicity, the new association proposed to promote utilitarian schemes rather than to encourage scientific inquiry. Its membership was largely drawn from the New York area, especially West Point and Union College, although there were associated groups as far distant as Tennessee and Ohio. After a brief three-year existence, the association apparently ceased to exist.[53]

EUROPEAN MODELS

For the remainder of the decade men of science were caught in a quandary, eager to establish an identity for American science but unwilling to commit themselves to any contemporary organization pattern and uncertain about the caliber of men to be included. They sought better communication among themselves. The principles of the Deutscher

52. *AJS*, XXXVII (1839), 399. The editor, in noting the academy's new building in Philadelphia, commented on the friendly rivalry with New York: "the younger sister having been laudably emulous of the fame and worthy deeds of the elder."
53. Sidney Forman, "West Point and the American Association for the Advancement of Science," *Science*, CIV (19 July 1946), 47–48. Forman's argument that this group is a direct forerunner of the AAAS is untenable, except as a representation of the need for a coordinated national organization. Michele Aldrich brought this article to the author's attention.

Naturforscher Versammlung and particularly the British Association for the Advancement of Science (BAAS) were known by the middle of the decade. The BAAS, especially, provided a model capable of filling observable needs for American science: a meeting place on the national level, a component of specialization, a source for research funds, and an institution capable of representing science to the public and to the national government. Self-conscious concern about American readiness for organization and concern about international status remained nagging anxieties.

The idea of a federation of science was born in Switzerland and introduced into the German states by Lorenz Oken, political nationalist and former professor in natural history at Leipzig, in 1822. Oken's goal was as much to create a "spiritual symbol of the unity of the German people" as to emulate the Swiss "Wandergeseleschaft."[54] It was a popular success, increasing in attendance until at the 1828 meeting in Berlin President Alexander von Humboldt addressed more than 1,200 persons, with the King of Prussia attending as distinguished guest. The English inventor Charles Babbage participated at that meeting and returned to London to write an enthusiastic description of the Deutscher Naturforscher Versammlung, "a voluntary association of men in search of truth."[55] Shortly thereafter James F. W. Johnston wrote an editorial for the *Edinburgh Journal of Science* outlining the advantages of the congress; these he understood to be the drawing of public attention to science and to scientific men and also the spurring of governments to examine their scientific institutions.[56] Undoubtedly Johnston's hopes blurred his image of what German scientists accomplished through the annual meetings, but his interpretation inspired British scientists to move for their own broader-based, peripatetic association.

Sir David Brewster, editor of the *Edinburgh Journal of Science*, continued the campaign for a new organization in Britain and successfully

---

54. R. Hinton Thomas, *Liberalism, Nationalism, and the German Intellectuals, 1822–1847: An Analysis of the Academic and Scientific Conferences of the Period* (Cambridge, 1951), pp. 22–46. The constitution, printed in the standard history by Karl Sudhoff, *Hundert Jahre deutscher Naturforscher-Versammlunger* (Leipzig, 1822), pp. 47–48, limited membership to naturalists and doctors, but was sufficiently vague to allow others to join. Also see S. F. Mason, *Main Currents of Scientific Thought: A History of the Sciences* (New York, 1953), pp. 470–471.

55. Charles Babbage, "Account of the Great Congress of Philosophers at Berlin on the 18th of September 1828," *Edinburgh Journal of Science*, XX (1828–29), 31, 225–234. The society expressed, for Babbage, the German *Naturphilosophie*, the unity of nature as evidenced in all branches of physical science.

56. Thomas, *Liberalism*, p. 46.

persuaded Vernon Harcourt of the Yorkshire Philosophical Society to take the initiative in 1831.[57] In his first address to the British Association for the Advancement of Science, Harcourt indicated the purposes of the society, namely, "to give a stronger impulse and more systematic direction to scientific inquiry, to obtain a greater degree of national attention to the objects of science, and a removal of those disadvantages which impede its progress, and to promote the intercourse of the cultivators of science with one another and with foreign philosophers."[58] Harcourt's modernity was complex, however, for his address was basically a plea for catholic science in the midst of increasing specialization. Harcourt recognized that he was addressing a mixed audience that included men eager to advance science through full-time research but also men who were simply interested amateurs. He dealt with the latter seriously. While admitting the necessity of specialization, he argued that by itself the isolation of exclusive groups would eventually retard scientific advance. The British Association should in no way discard the older philosophical idea of synthesizing all science (although by now literature and arts were separated from science) even as it encouraged modern specialized research. To him, then, the BAAS was a unifying agent which would permit separate societies to "concentrate scattered forces" of endeavor and make better use of a "growing body of humble labourers" ready to give assistance.[59] The movement which began with Babbage's hope for a more professional body of scientists culminated in an association requiring minimal professional attainments.

The BAAS, minus the political overtones of the German congress, became firmly established by the middle of the decade. Enthusiasm among both men of science and the general public reflected contemporary dissatisfaction with the exclusive, aristocratic tendencies of the Royal Society in London.[60] The BAAS's peripatetic meetings stimulated

---

57. O. J. R. Howarth, *The British Association for the Advancement of Science: A Retrospect, 1831–1931* (London, 1931), pp. 12–15.

58. Harcourt's "Objects and Plans of the British Association for the Advancement of Science," as reprinted in Howard M. Jones and I. Bernard Cohen, eds., *Science before Darwin: A Nineteenth-Century Anthology* (London, 1963), pp. 189–190.

59. *Ibid.*, pp. 197–199. A clergyman himself, Harcourt indicated that membership should be open, since "it is not our desire in the general composition of the society to separate writers from readers, the professor of natural knowledge from the student. A public testimonial of reputable character and a zeal for science is the only passport into our camp we would require."

60. The relative importance of factors leading to the rise of the BAAS is debated. Williams, "The Royal Society," pp. 221–233, argues that the shift came largely over dissatisfaction with Royal Society leadership and principles, as well

provincial scientific societies and increased communication among leading men of science. The special section meetings acknowledged the
significant specializations occurring in science. The movement to encourage science partook of the contemporary reform spirit, with individuals taking both sides on the issue of whether science was declining
in England and debating the question of how to define science and
scientists.[61] Added to this was popular enthusiasm resulting from almost
yearly advances in geology and from the willingness of leading geologists to elucidate geological principles to the public as well as to debate
them in sectional meetings.[62]

In his report as junior secretary of the association in 1834, Charles
Forbes outlined again the significance of the three-year-old body, arguing forcibly that the diffusion of knowledge was not its most distinctive
purpose.[63] The Swiss and German societies, he maintained, sought
primarily scientific intercourse and the diffusion of knowledge in their
annual reunions. The BAAS, operating on a principle of extending
knowledge, had a more permanent character. Its larger purpose was accomplished through the summary *Reports* of the progress of science in
specific fields,[64] and also by financial support for specific research proj-

---

as the failure of a reform effort within that group in 1830. In contrast, Walter
Cannon, "History in Depth: The Early Victorian Period," *Journal of the History
of Science*, III (1964), 20–38, emphasizes the important vitality among provincial
societies and the need for coordination of their efforts. D. S. L. Cardwell, *The
Organization of Science in England: A Retrospect* (London, 1957), pp. 46–59, suggests that the association was the "logical consequence" because it spoke the
response of scientists who feared there might be some truth in the argument that
science was in decline in Britain.

61. Contemporary literature arguing whether or not science was in decline, as
posited by Charles Babbage, *Reflections on the Decline of Science in England, and
on Some of Its Causes* (London, 1830), provides fascinating reading. It also has
some place in current historiographical discussion, as in Foote, "Place of Science in
the British Reform Movement," pp. 192–208, and in Nathan Reingold, "Babbage
and Moll on the State of Science in Great Britain: A Note on a Document," *British Journal for the History of Science*, IV (1968), 58–64.

62. Meetings were conducted on two levels: sectional meetings for specific areas
of science and general or open meetings for the public. By 1838 there were seven
sections: mathematics and physical science, chemistry and mineralogy, geology
and physical geography, geology and botany, medical science, statistics, and
mechanical science. In the *AJS*, XXIII (1833), 182, the editor noted that when
William Buckland and Adam Sedgwick lectured, they attracted the almost un-
divided attention of their meeting, and other sectional communications were
postponed.

63. BAAS, *Report*, IV (1834), xi–xxiii.

64. *Ibid.*, p. xiv. Forbes emphatically differentiated the reports from the short,
general treatises of the popular press. Indeed, "their main object is so to classify
existing discoveries as to lead the individual who is to grapple with its difficulties, to

ects.[65] Forbes observed that the BAAS had stimulated activity in local scientific societies as well as international cooperation for specific projects. Like William Whewell, Forbes took pride in the independence of science from government. While science should seek to influence proposals and make recommendations, he commented that there was little reason to regret the absence of a national institute like that of France "so long as individual exertion can supply the stimulus which even the sunshine of wealth and patronage has failed to excite."[66] Joined by the "urban" London-Cambridge men of science, the BAAS could claim to be a representative parliament for science.[67]

Forbes touched on the disparity of the association's goals: while appealing to the broadest possible scientific interests, it hoped to advance specific areas of science. As a result its most notable contribution to advancing science was through its research funds and its sponsorship of observational science, which enlisted the aid of amateurs and specialists in accumulating data in geophysics, for example.[68]

The meeting at Newcastle in 1838 marked the peak of popular enthusiasm. Charles Lyell, famous as a popular lecturer as well as for his *Principles of Geology*, was chairman of the geological section. Daily attendance at the general meetings was from 1,000 to 1,500, and nearly 3,000 persons flocked to the seashore to hear the eloquent Adam Sedgwick.[69] In a letter encouraging Charles Darwin to support the association, Lyell demonstrated his own ambivalence toward the BAAS when he wrote, "I am convinced, although it is not the way I love to spend my own time, that in this country no importance is attached to any body of men who do not make occasional demonstrations of their strength

---

start with the most complete and accurate knowledge of what has already been done in any particular science ... [and] the report should point out the most important questions which remain for solution, whether by direct experiment or by mathematical investigation."

65. *Ibid.*, p. xx.

66. *Ibid.*, p. xxii. A similar opinion was held by G. B. Airy, who suggested, "In Science, as well as in almost everything else, our national genius inclines us to prefer voluntary associations of private persons to organizations of any kind dependent on the state." Quoted in Foote, "Place of Science in the British Reform Movement," p. 204.

67. Cannon, "History in Depth," pp. 24–25.

68. Howarth, *British Association*, p. 179. The British Association, through the efforts of Edward Sabine and John Herschel, helped sponsor an international network of meteorological observations.

69. Edward Bailey, *Charles Lyell* (New York, 1962), p. 127. Lyell was a continuing participant in the BAAS, and the society's success was to a large extent the result of frequent attendance of known men of science.

in public meetings."[70] Such crowds and enthusiasm gave full opportunity for caricature; Charles Dickens wielded his sarcastic pen in a "Full Report of the Second Meeting of the Mudfog Association for the Advancement of Everything" in *Bentley's Miscellany* as well as in doggerel verse in *Punch*.[71] His critique of the popular aura of the association was not unsympathetically received by many men of science.

CAUTIOUS RESPONSES

Americans caught and reflected the mixed attitude of the English men of science. They were aware of the BAAS's activities through publication of abstracts of its meetings in Silliman's *Journal* as well as in other periodicals; these, like English editorials on the German congress, seriously blurred the functions of the association as they translated its activity in terms of the American experience.[72] Silliman, for example, stressed the value of the reports on science and the research grants in his *Journal*, but an equally critical factor in the success of the BAAS was the encouragement given to amateurs and the quality of lectures presented at the general meetings.

By 1837 at least forty Americans had attended British Association meetings.[73] In general the response was enthusiastic, as was the coverage in the *Journal*. The gathering of distinguished scientists to promote scientific advance as well as public consciousness of science seemed ideal to men whose own institutions were local and financially powerless to

70. Quoted in Howarth, *British Association*, pp. 37–38; Bailey, *Charles Lyell*, pp. 127–128. Lyell added, "It is a country where, as Tom Moore justly complained, a most exaggerated importance is attached to the faculty of thinking on your legs, and where, as Dan O'Connell well knows, nothing is to be got in the way of homage or influence or even a fair share of power, without agitation. . . . I can also assure you, as the strongest commendation, that the illiberal party cannot conceal their dislike and in some degree their fear, of the growing strength of the Association. . . ."

71. *Bentley's Miscellany*, IV (1838), 223–224; "Labors of the British Association for the Advancement of Science," *Punch* (14 Aug. 1841), p. 57.

72. *AJS*, XXIII (1833), 182. In 1838 the BAAS had assets of nearly £7,000 and spent £900 for research. Silliman first published brief notices of the BAAS meetings in "Miscellanies" sections, along with other societal and foreign news briefs. By 1835 he was abstracting large portions of text from English periodicals and also republishing papers of particular interest. See XXI (1831), 373–374; XXIII (1832), 179–182; XXV (1834), 411–412; XXVIII (1835), 55–84; XXIX (1836), 347–358; XXXI (1837), 332–381; XXXIII (1838), 265–296; XXXIV (1838), 1–56; XXXV (1839), 275–321; XXXVIII (1840), 93–138.

73. BAAS, *Proceedings*, VII (1837), 44–46. These included Alonzo Clark, Samuel Dana, Robert Hare, Elias Loomis, Joseph Henry, J. J. Lowell, M. Vaux, Samuel St. John, Henry D. Rogers, and Theodore Sedgwick.

assume projects like those of the BAAS.[74] Americans who participated by presenting papers or taking part in discussions found their reputations and that of American science enhanced. But the Newcastle meeting gave pause to at least one American in attendance; he believed that the negative influence of amateurs might be exaggerated in America, where the spirit of deference was less operative than in England. Joseph Henry wrote to his friend Alexander Dallas Bache, "I think you will not be much impressed with the scientific character of the British Association there is such a mixture of display of ignorance and wisdom, of management in the compliments given and the honours received that the whole makes rather an unfavorable impression on a person admitted a little behind the sciences."[75] Such skepticism, projected into the American experience, proved an important component in the postponement of a similar national association for the United States for nearly a decade. Yet Joseph Henry embodied the ambivalence of fellow scientists, for he wanted some organization in which "the real working men of science in this country should make common cause and endeavor by every proper means unitedly to raise our scientific character, to make science more respected at home, to increase the facilities of scientific investigations and the inducements to scientific labours."[76] Optimistic about the potential of science itself, scientists were less confident about man's institutions for science.

Even so, considering American self-consciousness about scientific advance and the desire to establish a national organization similar to the BAAS, the delay in founding such a group until 1848 is somewhat surprising. As suggested earlier, the environment itself was not inhibiting. The public was receptive to science and participated in activities on a popular level. While not always generous financially, the community usually created no impediments or special demands, eagerly seeking institutions that symbolized American progress. In addition, the transportation revolution steadily lowered costs and shortened the time required for travel; this was especially true in eastern states where men of science already formed small enclaves and could benefit from in-

74. Elias Loomis to Edward Herrick, n.p., 12 Sept. 1836, Herrick-Loomis MSS, BLY. Loomis was fascinated by the men at the meetings and gave detailed impressions of the philosophic Sir David Brewster and the dynamic Professor William Whewell.

75. Quoted in Nathan Reingold, *Science in Nineteenth-Century America: A Documentary History* (New York, 1964), p. 87.

76. Henry to Bache, Princeton, 1 Nov. 1838, Rhees MSS, Henry L. Huntington Library, San Marino, Calif. (hereafter HHL).

creased communication.[77] Although a scientific network extending to more provincially located scientists was not yet feasible, a coordinating body of leading scientists in Boston, Albany, New Haven, New York City, Philadelphia, and Washington certainly was. The delay, it seems, was a deliberate one, wrought by the actions and attitudes of the men of science themselves.[78]

Nowhere is this better demonstrated than in three separate attempts to found a national scientific group at the close of the decade of the 1830s. In each instance the somewhat larger views of the founders were defeated or significantly limited because of either overt opposition or hesitation on the part of essential scientific leaders. The three efforts were distinct in conception and individual in formation, representing a spectrum of ideas about how a national organization might best function and represent science. Although two of them reflected awareness of the BAAS model, conceptions of membership, profiles of leadership, and activities varied considerably. Each of the three—the aborted American Institution for the Cultivation of Science (also referred to as the American Society for the Promotion of Science), the Washington-based National Instiute for the Promotion of Science, and the specialized Association of American Geologists and Naturalists—reflects the ambiguity in the slowly emerging consciousness among the scientists as they debated how they might best organize themselves and what goals they should jointly pursue.

Despite eminent sponsorship and quite concrete planning, the first attempt of the American Institution for the Cultivation of Science to found a national group was a decisive failure. The meteoric appearance of the American Institution, however, stimulated important discussion and started the process of self-definition necessary for the later successful effort of the geologists.

Initiated in New England's cultural center, this first effort to found a national scientific association began auspiciously. John Collins Warren, leading Boston surgeon, returned from a European tour in August 1838, having attended scientific meetings in Liverpool, Paris, and Naples.

77. George R. Taylor, *The Transportation Revolution, 1815–1860* (New York, 1957), pp. 84, 141. Taylor suggests that despite the economic depression for business, railroad construction continued throughout these decades. For the period Taylor discusses, traveling time decreased by almost 70 percent while the cost fell more than 50 percent; for example, from Philadelphia to New York by steamboats and stage cost $10 and took thirteen hours in 1816 while the same trip by railroad cost only $3 and took four hours in 1860.

78. Hofstadter and Metzger, *Development of Academic Freedom*, p. 262.

Apparently inspired by the popular success of the BAAS, to which he had presented a paper on the western aborigine in 1837, Warren persuaded Judge Joseph Story and Governor Edward Everett, fellow members of the American Academy of Arts and Sciences, to support him in creating an association for the promotion of science.[79] Buoyed by their positive response, Warren wrote to William E. Horner of Philadelphia in an effort to gain an informal response from the American Philosophical Society.[80] Simultaneously he wrote to scientific leaders Joseph Henry and Benjamin Silliman soliciting their cooperation and suggestions.[81] An initial positive response from Horner made success seem imminent.[82]

Opposition, or at least passive resistance, soon became evident, but Warren and his friends proceeded to work on plans for founding the new organization. Henry, whose skepticism about the BAAS has already been noted, replied that the plan of the BAAS "would require some modification in order to suit the limits and character of our country," implying that the United States was not ready for the extensive program of the British Association.[83] Then the American Philosophical Society's formal committee to consider "the expediency of the Society forming an American Association" reported that after two meetings on the subject the committee's conclusion was that it would be "inexpedient for the Society to engage in such an undertaking."[84] Similarly, a meeting of the American Academy in Boston expressed some skepticism, although it finally voted to support the proposal. Warren and interested fellows of the academy sponsored a general meeting for men ready to assist in founding such an association in October 1838.[85] A subcommittee appointed at that meeting notified the American Philosophical Society of

79. Edward Warren, *The Life of John Collins Warren, M.D., Compiled Chiefly from His Autobiography and Journals*, II (Boston, 1860), 1–2, Journal entries for 8 and 15 Sept. 1838.

80. *Ibid.*, I, 339–340; II, 1–2. Also see John C. Warren to Mason Warren, London, 2 Aug. 1837, Warren MSS, Massachusetts Historical Society, Boston (hereafter MHS).

81. Warren, *Life of John Collins Warren*, I, 339, Journal entry for 23 Sept. 1838; Warren to Henry, Boston, 29 Sept. 1838, Henry MSS, SIA; Warren to Silliman Sr., Boston, 29 Sept. 1838, Silliman MSS, BLY.

82. Warren, *Life of John Collins Warren*, I, 339–340.

83. Henry to Warren, Princeton, 10 Oct. 1838 (draft copy), Henry MSS, SIA.

84. MSS minutes, 16 Oct. 1838, American Philosophical Society Archives, APS. The committee included Horner, Henry Vethake, R. M. Patterson, A. D. Bache, and Phily H. Hicklin. Later the question was reopened but again defeated 15 Feb. 1839.

85. Warren, *Life of John Collins Warren*, II, 1. The academy did not publish its proceedings until 1846, nor was the author able to consult archives.

their action and invited other scientific gentlemen to join them through a circular which outlined their proposal to form an American Institution for the Cultivation of Science.[86] The BAAS, despite the difference in specific wording, was clearly the model; meetings were to be held successively in "different great cities of the Union," and the new societies' objects were "the advancement of physical science and literature, by . . . effecting an interchange of discoveries and improvements between the inhabitants of different parts of the country."[87] Yet the flurry of activity was without result, for the organization died aborning and left only an idea for further discussion.

In his autobiography Warren attributed the abandonment of the project to the "jealousy" of the American Philosophical Society.[88] The early favorable but unofficial response from Horner, before the committee's negative decision, supported this contention. Warren attempted to counter fears of competition by arguing that the American Institution would not publish papers and was not intended to compete with established societies. In addition, he met the unstated skepticism of Henry by observing, "They [the Committee of Correspondence] are aware that many imperfect and ostentatious communications would be made to the meetings but these might and would receive sufficient check from the sectional committee appointed to examine such papers. . . ."[89] Fear of charlatanism and American unreadiness for national organization appear to have been of major concern to others who had doubts about the BAAS as well. In August Henry had written, "I am convinced that a promiscuous assembly of those who call themselves men of science in this country would only end in our disgrace."[90] He wrote similarly to his close confidant John Torrey in New York, who echoed, "I rather think with the Philadelphians that we can hardly get up an assoc[n] for the promotion of science. There is indeed too much Charlatanism in the

86. *Ibid.*, I, 340, Journal entry for 30 Oct. 1838. The committee printed and distributed fifty circulars; a copy of the circular, dated 1 Nov. 1838, is in the Henry MSS, SIA. Silliman published a copy in the *AJS*, XXV (Jan. 1839), n.p., without comment.

87. "Circular."

88. Warren, *Life of John Collins Warren*, II, 2. Others felt similarly about the American Philosophical Society's exclusive attitudes. See Edward Herrick to Elias Loomis, New Haven, 23 Dec. 1839, Herrick-Loomis MSS, BLY.

89. Warren to Horner, Boston, 20 Nov. 1838, American Philosophical Society Archives, APS; Henry to Torrey, Princeton, [1838], Torrey MSS, NYBG.

90. Henry to Bache, Princeton, 9 Aug. 1838, Henry MSS, SIA, reprinted in Reingold, *Science in Nineteenth-Century America*, p. 87.

country—enough to overpower *us modest men* [italics his]."[91] By the time the suggestion reached westward and prompted James B. Rogers, then in Cincinnati, to inquire rhetorically, "Are we wise for such an enterprise? or does it not smack too much of pretention?"[92] the proposition was buried.

An anonymous, critical review in the *New York Review* provides keen insight into contemporary thoughts on the topic.[93] Written by someone familiar with the foreign associations, the review demonstrated again that the BAAS could inspire pessimism as well as the enthusiasm of Warren. Further, the time lapse since the founding of the German congress and even since the establishment of the BAAS allowed opportunity for reflection and evaluation of the foreign groups; the flush of initial success had opportunity to pale before the inevitable difficulties of real operation. Critical of the German congress by 1838 and skeptical about the BAAS, the author never denied the importance of national associations for men of science. His discussion of the history of the other two societies, however, was designed to modify and essentially limit the high aspirations of Warren. His successful enunciation of privately held skepticism undoubtedly reinforced the opposition in Philadelphia.

Unlike Johnston's and Brewster's earlier historical summaries of Oken's society, the author pointed out that the German congress had declined in the 1830s. At the outset it was successful largely because its design coincided with the normal pattern of German professors who used their vacations to visit new geological regions, important research collections, and university libraries in conjunction with their own research. They welcomed the additional opportunity to meet with scientific peers and to travel through new regions to attend the peripatetic meetings of the German congress. As its reputation grew, the society's membership experienced an influx of men who simply sought prestige by association. As a result,

91. Henry to Torrey, Princeton, [1838], Henry MSS, SIA; another undated letter from 1838 in Torrey MSS, NYBG; Torrey to Henry, New York, 9 Nov. 1838, Gray MSS, Gray Herbarium, Harvard University (hereafter GHH), also reprinted in Reingold, *Science in Nineteenth-Century America*, p. 91.

92. Rogers to William B. Rogers, Cincinnati, 1 Mar. 1839, Rogers MSS, MITA. He adds the indictment, "Is our science sufficiently matured to enable us to embody it in the systematic form of reports without drawing from foreign fountains: Are we not essentially a fortune seeking, money loving people, incorporating private gain with all our scientific pursuits?"

93. "Article VI: Scientific Associations," *New York Review*, VIII (Apr. 1839), 401–417.

The men whose whole life had been consecrated to minute observation of nature, to patient researches, to profound and laborious studies, felt themselves lost in a class of persons placed between the amateur and the real *savant*, in that numerous class designated by the name of *demi-savants*, who are the scourge of scientific societies. Knowing a little of every thing, in which there is no harm, and believing that they understand every thing, in which there is a vast deal, these persons drive away all the men of real science, who could pardon them for their ignorance, but cannot tolerate their pretensions. At the present time, the *scientific congress* is known among the aristocracy of German *savants*, under the name of the *congress of apothecaries*.[94]

The decline of the society, in effect, was isolated. There was no similar decline in research in science itself, for the society as such had never had a direct influence on the research or study of the German scientists.

The British Association, contended the author, by its deviation from the German model, had gone in more substantive directions. Its publication of reports and of desiderata had a notable influence in suggesting fruitful areas for research and in stimulating such projects as an international survey in terrestrial magnetism.[95] The assignment of reports to respected men of science provided accurate compilations of English, Continental, and American publications, thus establishing priorities and eliminating overlapping research. Such efforts suggested to the author that the only possible object of the BAAS was "to extend and purify the cultivation of science."[96] Yet it was open, as was the German society, to the difficulties of unsolicited participation; the author's single critical comment on the BAAS was a note that its meetings seemed "calculated to encourage and to bring into notice forward and shallow declaimers, to the exclusion of modest merit; and to lower the character and true dignity of scientific men, by giving them inducements and opportunities to make public displays, for which their habits are not at all suited."[97] He added, parenthetically, that "this remark is not without its application here [in the United States]."[98]

Having elaborated his estimation of the two foreign groups, the author turned to the proposal circulated by Warren's committee in Boston. Without questioning the legitimacy of an American national association, the author addressed himself to the problem of the purpose and methods indicated in the circular; his specific suggestions were

94. *Ibid.*, p. 405. Italics in the original.
95. Howarth, *British Association*, pp. 179–183.
96. "Scientific Associations," *New York Review*, p. 409.
97. *Ibid.*, p. 417.
98. *Ibid.*

designed to limit and clarify the plan proposed by the Bostonians. Gently but firmly he distinguished between philologists and philosophers and suggested that the word "literature" had probably been "inadvertently" inserted in the circular, for the organization was clearly directed toward men of science.[99] Thus he quickly modified the broad "philosophical" implications, which would have created another general organization, and focused instead on a more modern conception of science.

An American society, he suggested, must be broader than the German congress, which met principally to further acquaintances of men of science, and also avoid conflict with the British Association, whose reports already met the need for suggesting fruitful areas of research. Rather, it should endeavor to be

> American, first of all things. It should at once form a kind of national tribunal, in which the labors of all the savants of the union should be annually represented, and their success be made known in the face of the whole country; it should be a focal point, in which the men, who resisting the whirlwind which sweeps along in the pursuit of gain the great mass of the community ... would come every year to give new temper to their courage, by the approbation of their fellow-laborers—a source, too, from whence the young man who brings with him from college a vague desire to learn the mysteries which regulate the universe, would repair to draw forth that holy ambition which prompts man to forget himself, in his strivings for the cause of humanity.[100]

As if aware of the Americans' sensitivity to the European estimation of the young nation's scientific achievement, he added that they must find "their way within the walls of the European academies, before they perceived the rewards of their labors."[101]

Without being harshly critical, the author alluded to the limits of American science and posited an even narrower scope for the new organization. It was essential that the association "acquire the rank of an authority, which the old world will consult, even as we are accustomed to have recourse to it, in a multitude of scientific questions"; moreover, "the philosopher must consider that it is an honor to belong to this national institution."[102] To achieve such ends, the society had to be carefully planned and move with deliberation. His suggestion was that the organization comprehend initially only those areas in which

99. *Ibid.*, p. 412.
100. *Ibid.*, p. 413.
101. *Ibid.*
102. *Ibid.*, p. 412.

American men of science had made important contributions, pointing specifically to geology, zoology, and meteorology, the "observational" sciences. With competent work in these areas and steady reports made through the organization, credibility and a reputation for expertise could be established at home and abroad.[103]

Both as a statement directed toward the Bostonians and as an expression of attitudes of men of science, this review is revealing. The author, even while assuming the inevitability and utility of an American scientific association, was clearly without confidence in the proposed enterprise. His reasons for caution were legitimate and his pleas for a judicious plan of systematic development understandable in a nation which had no precedent for Warren's proposed society. The effect of his negative attitude was to forestall, if only indirectly, a comprehensive society.

Two separate and quite different proposals for a national scientific organization also had their origins in private discussion beginning in 1838. One was sponsored by a Washington group whose vision was less a community of scholars than a viable sponsor and patron for a national scientific museum. The other was a New England product, created in the mind of Edward Hitchcock of Amherst College. Each group took two years of desultory discussion before its actual formation, but the fact of their existence in 1840 indicates the ample interest in national-level science.

Joel R. Poinsett formulated the National Institute for the Promotion of Science in Washington because of his concern for the results of the U.S. exploring expedition which was then underway in the Pacific region. When the Secretary of the Navy had been dilatory in implementing a congressional appropriation for an expedition, Poinsett, then the Secretary of War, had willingly assumed the responsibility.[104] Avocationally interested in science himself and a member of an early,

103. *Ibid.,* p. 415. There was a distinctive denigration of "observational sciences," as distinguished in this instance by the author from "experimental sciences." The author selected the areas of geology and zoology because the two British Association reports relating to the United States were on these areas.

104. The following discussion on the National Institute is largely derived from this author's "A Step toward Scientific Self-Identity in the United States: The Failure of the National Institute, 1844," *Isis,* LXII, part 3 (Fall 1971), 339–362. Also see Madge D. Pickard, "Government and Science in the United States: Historical Backgrounds," *Journal of the History of Medicine and Allied Sciences,* I (July 1946), 265–289. For the narrative of the expedition itself, see David B. Tyler, *The Wilkes Expedition: The First United States Exploring Expedition, 1838–1842* (Philadelphia, 1968), and Daniel Haskell, *The United States Exploring Expedition, 1838–1842, and Its Publications, 1844–1874* (New York, 1942).

unsuccessful philosophical society in Washington, the Columbian Institute for the Promotion of Science, Poinsett was eager to make the nation's capital a true cultural center. His aspirations were enhanced by an unusual bequest from an Englishman, James Smithson, who left a half-million dollars to the young United States to found "an establishment for the increase and diffusion of knowledge among men."[105] To Poinsett's fertile mind the two purposes could and should be joined, with the funds being used to build a natural history museum whose nucleus would be the specimens and other materials brought back by the exploring expedition under Charles Wilkes.[106]

To accomplish his plan, Poinsett recognized that group sponsorship was essential both to provide substantive evidence of support for the idea itself and to establish management for the enterprise. The National Institute accepted its constitution in May 1840 and promptly elected Poinsett as its first head. The project of a museum to display scientific specimens from domestic and foreign surveys was not without competition in its bid for the Smithsonian bequest, however. Among other proposals, those for an astronomical observatory, the "Lighthouse of the skies" long sought by John Q. Adams, and a renewed effort for a national university had significant congressional backing.[107] As a result, the Washington-based National Institute expanded its initial concentration on political membership to solicit support from men of science as well as from the broader community.

Initially scientific men did, in turn, respond favorably to Poinsett's proposal for a museum. From the Academy of Natural Sciences of

105. For detailed accounts of the Smithsonian Institution's early history, see George Brown Goode, ed., *The Smithsonian Institution, 1846–1896: The History of Its First Half Century* (Washington, D.C., 1897), and William J. Rhees, ed., *The Smithsonian Institution: Documents Relative to Its Origin and History, 1835–1899*, 2 vols. (Washington, D.C., 1901).

106. For an account, with limited analysis, of Poinsett's importance, see George Brown Goode, "The Genesis of the United States National Museum," in *A Memorial of George Brown Goode*, Smithsonian Institution, *Annual Report...for 1897*, part II (Washington, D.C., 1901), pp. 98–103. Attempts to make his purpose a constitutional reality were unsuccessful; for legislative analysis, see Pickard, "Government and Science," pp. 265–289.

107. The most recent analysis of the debate over the Smithsonian bequest is by A. Hunter Dupree, *Science in the Federal Government: A History of Policies and Activities to 1940* (Cambridge, Mass., 1957), pp. 66–90. Also see Wilcomb E. Washburn, ed., *The Great Design: Two Lectures on the Smithsonian Bequest by John Quincy Adams* (Washington, D.C., 1965), pp. 13–41. David Madsen's *The National University: Enduring Dream of the U.S.A.* (Detroit, Mich., 1966), pp. 57–63, adds nothing new to the earlier cited accounts of the Smithsonian Institution.

Philadelphia as well as from individual scientists from Virginia to Massachusetts, men with an interest in natural science concurred in the desirability of a national cabinet to preserve survey specimens.[108] At the same time, in Washington, responsibility increasingly fell to Francis Markoe, a State Department clerk; his inability to establish a pattern of responsibility among the scientists led the institute into difficulty. The institute was successful, however, in seeing that results relayed back by other naval ships to Washington from the exploring expedition were safely stored.

Given the institute's primary reason for being, concern for the disposition of these materials was appropriate and necessary. Yet in this initial action it laid the ground for discontent. Daniel Webster, then Secretary of State, assigned the institute space in the basement of the Patent Office. When this arrangement proved unsatisfactory because of the dampness, a move to the upper Exhibition Hall aroused the anger of Henry Ellsworth, commissioner of the Patent Office. In addition, when Charles Wilkes returned to Washington and found that the National Institute had assumed responsibility for his collections and had even acquired the initial funds appropriated for the preservation of specimens to hire its own curator, he took deliberate steps to supervise all subsequent funding. Frequently impolitic, Wilkes allowed his displeasure to be made known by his corps of scientists on the expedition. Without clear evidence or any discussion with its leadership, Wilkes somewhat arbitrarily assumed that the institute was attempting to usurp responsibility for caring for and publishing the results of the expedition, a task he felt belonged to his own carefully selected and experienced corps. When he successfully secured support for his position from the powerful head of the Library Committee of the U.S. Senate, Benjamin Tappan, it became clear that some of the scientific men had grown skeptical of the institute's power.[109]

Critical opposition, however, came only later and for somewhat different reasons. The National Institute, which in 1842 secured a twenty-year charter from the federal government, was born in the mind of a

108. See, for example, Samuel G. Morton to John K. Kane, Philadelphia, 23 July 1839, American Philosophical Society Archives, APS; Edward Hitchcock to Francis Markoe, Amherst, 13 Nov. 1840, Galloway-Maxcy-Markoe MSS, LC; William B. Rogers to H. C. Williams, University of Virginia, 8 Sept. 1840, in Emma Rogers, ed., *Life and Letters of William Barton Rogers*, I (Cambridge, Mass., 1896), 171.

109. For evidence of the difficulties, see the author's "Step toward Scientific Self-Identity."

politician, and his projected *modus operandi* for establishing a direct link between government and science was not readily accepted by all men of science. When, in addition, the institute suggested a more ambitious scheme than the museum enterprise and presumed to speak for men of science on the national level, several factors combined to defeat its entire program.

The most organized opposition to the institute came from the third group, which had grown from the idea planted in 1838 by Edward Hitchcock, president of Amherst College and former head of the Massachusetts Geological Survey. Nurtured by what was probably the most visible and self-conscious subgroup of American scientists of the period, the Association of American Geologists and Naturalists successfully emerged after two years of discussion, although in a considerably more limited form than that projected by Hitchcock. Because it moved deliberately and with due speed, with the probable goal of founding a national, comprehensive organization, no other such projects were initiated until it emerged as the American Association for the Advancement of Science in 1848.

Caution and self-doubt figured predominantly among reasons for the delay throughout the 1830s and into the 1840s. As suggested, external factors were not sufficient, because the general public was responsive to, if not actually eager for, such organization. Nationalism and popular enthusiasm did, ironically, serve as curbs to the effort in that men of science feared that unrestrained and ill-educated amateurs would only perpetuate the low reputation of men of science in America. It is clear that the men of science were dragging their feet as the entire community moved inevitably toward a national organization (and spokesman) for science.

Sustaining the doubts of these men were several other factors. The self-culture movement, whose most dreaded results of quackery and mediocre research were probably never realized, reinforced the doubts of many scientists about the dangers of popularization.[110] The difficulties experienced by the established local groups aroused skepticism about establishing yet another society requiring personal and financial support, particularly as economic depression deepened late in the 1830s. Moreover, a certain regional loyalty and rivalry between groups

110. It must be kept in mind that the response to self-culture was not uniform throughout the nation and that divisions of opinion were not clearly along lines of geography, generation, or even areas of scientific interest.

mitigated against cooperation. An inevitable component in resistance was simply inertia, as well as the lack of a leader capable of or interested in initiating or sustaining such a project.

Perhaps the most important, if unconscious, reason for delay was the introspective tendency of the scientists, taken as a community. Nationalistic bluster only partially masked the diffidence of American visitors to Europe who were frequently reminded of the weakness of American science. Not surprisingly such encounters perpetuated self-doubt and skepticism among peers; a national organization could come only when successful research had brought mutual confidence and professional stature to individual men and thus to the nation. This uncertainty about peer achievement, a problem compounded by limited acquaintanceship not only between men in different fields but also between workers in the same area separated geographically, maintained the timidity toward engagement in joint projects.

The effect of delay is difficult to assess.[111] For the 1830s it was probably less significant than for the following decade. During the 1840s, even as public enthusiasm and local societies ebbed, the positive gains in science were consolidated into other institutions, like the universities, government agencies, and the new Smithsonian Institution established under the authoritative direction of Joseph Henry. Some of the institutions that developed during the period undoubtedly assumed functions and a support which would otherwise have accrued to a national voluntary organization. Yet it took those years and the successful example of the Association of American Geologists and Naturalists to establish community self-definition and thus a foundation for a truly national organization.

111. I. Bernard Cohen, "Science in America: The Nineteenth Century," in Arthur M. Schlesinger, Jr., and Morton White, eds., *Paths of American Thought* (Boston, 1963), p. 176. Cohen, who accepts Tocqueville's indictment concerning basic science in America, suggests that the "impotence of American science" in the pre–Civil War years was related to the lack of viable institutions.

III

The Association of American Geologists and Naturalists

Through the 1830s geology reigned ascendant among the sciences.[1] Considered part of the broadly defined area of natural history, it enjoyed support in most local scientific societies. Sciences like zoology, botany, and mineralogy required extensive data gathering and were accessible to amateur participation. The data were familiar, and most theory could be understood in the vernacular. The natural sciences appealed to the practically minded as well as to the romantics. Papers and research in the natural sciences and chemistry were common in both local agricultural and scientific societies, whether rural or urban.[2] Farmers looked to agricultural chemists for advice, and mining entrepreneurs in Pennsylvania and the Great Lakes region soon recognized the chemical and mineralogical talents of trained geologists for evaluating commercial potential in unexplored regions. At the same time natural studies had an aesthetic or artistic quality. A sense of awe for the "wonders of creation" inspired many of the early field researchers. By the mid-nineteenth century geology emerged as a distinct scientific field as capably studied in the United States as abroad.[3] Its practitioners

1. Material in this chapter, in rather different form and with more attention to the geological nature of the society, is found in "The Geologists' Model for National Science, 1840–1847," American Philosophical Society, *Proceedings*, CXVIII (Apr. 1974), 179–195.

2. Bates, *Scientific Societies*, pp. 38–64.

3. Two men associated with the Smithsonian Institution at the turn of the century laid important groundwork for the study of early nineteenth-century science which may partially explain why natural history and geology can use the additional attention. George Brown Goode was primarily interested in institutional history, but his own background in botany led him to emphasize natural

were therefore a logical group to found the first viable national association for scientists.

The Anglo-American cultural and intellectual connection remained a strong one, despite political and diplomatic differences. During the first decades of the nineteenth century the English were in the vanguard of geological research. William Smith, a surveyor and canal engineer, demonstrated that fossils could be used to identify stratigraphical succession.[4] Although economic setbacks prevented him from pursuing the research that might have established him as a leader in geological science, Smith's observations led to extensive fieldwork and influenced the establishment of the first major specialized English scientific society, the Geological Society of London.[5] Advances created by the intensive interest in new data and conflicting theories were synthesized in the early 1830s. Charles Lyell argued in his three-volume *Principles of Geology* that the processes currently in operation were responsible for changes in the earth throughout time.[6]

Enthusiasm for the developing science was contagious, and four young Americans were infected while studying and traveling in Europe. During his year of independent study in Europe after graduation from Yale, Benjamin Silliman learned geology in a cursory way. On his return he continued his interest by collecting local New England minerals. William Maclure, who more systematically studied the work of Smith, returned after one of several trips abroad and produced the first major stratigraphic map of the United States. Although limited and based on the controversial concepts of Abram G. Werner, it provided the basis

history. His major relevant work is reprinted in Goode, *Memorial*, previously cited, including "The Genesis of the United States National Museum," pp. 85–191; "The Origin of the National Scientific and Educational Institutions of the United States," pp. 263–354; "The Beginnings of Natural History in America," pp. 357–406; and "The Beginnings of American Science: The Third Century," pp. 409–466. The other man was geologist George P. Merrill, who edited and compiled *Contributions to a History of American State Geological and Natural History Surveys, Smithsonian Institution, Bulletin 109* (Washington, D.C., 1920), and *First One Hundred Years of American Geology*, previously cited.

4. Joan M. Eyles, "William Smith: Some Aspects of His Life and Work," in Cecil J. Schneer, ed., *Toward a History of Geology* (Cambridge, Mass., 1969), pp. 142–158.

5. M. J. S. Rudwick, "The Foundation of the Geological Society of London: Its Scheme for Co-operative Research and Its Struggle for Independence," *British Journal for the History of Science*, I (Dec. 1963), 325–355.

6. Leonard G. Wilson, "The Intellectual Background to Charles Lyell's Principles of Geology, 1830–1833," in Schneer, ed., *Toward a History of Geology*, pp. 426–443.

for later structural work.[7] Elected president of the Academy of Natural Sciences of Philadelphia, Maclure was the most well known American geologist of the early period. Probably the first systematic training in geology was that acquired by Lardner Vanuxem at the École des Mines. He returned to the United States in 1819 to become professor of chemistry and mineralogy at South Carolina College and participated in the geological survey of that state. A somewhat pithy paper in 1829, which specified errors promulgated by American geologists, brought him into some prominence among fellow scientists. After work as a mining consultant in Mexico he retired to a farm in Pennsylvania and remained there except for the years in which he participated in the New York Natural History Survey. Henry Darwin Rogers had accompanied Robert Owen to England in 1832–33, where Rogers's initial enthusiasm for philanthropic reform was rechanneled toward scientific investigation. He attended meetings of the Geological Society of London, used their extensive collections, and met several leading geologists, including Lyell. The experience of these four students suggests that the sophistication and intensity of interest in geology in the young United States paralleled developments abroad.

Geology sustained its central position through the 1830s in part because it correlated well with the progressive outlook and the utilitarian expectations of science by society in general. Often legislative bills authorizing research specifically indicated that the surveys were to pursue economically significant results for the agricultural, industrial, and transportation concerns of the state as well as to have educational, scientific value.[8] The first state-sponsored geological surveys were those of Denison Olmsted in North Carolina (1821) and Vanuxem in South

7. Schneer, "Introduction," *Toward a History of Geology*, p. 14.
8. Hendrickson, "Nineteenth-Century State Geological Surveys," pp. 357–371. Hendrickson argues quite persuasively that although "liberal" legislators were interested in promoting the educational, scientific aspects of the surveys, the final determination of support was economic; the number of surveys at a given time could be correlated loosely to the general state of the economy. Source materials demonstrating that surveys were often justified in economic or other practical terms is presented in Merrill, *Contributions*. Also see William Back, "The Emergence of Geology as a Public Function, 1800–1879," Washington Academy of Sciences, *Journal*, XLIX (July 1959), 205–209. The geologists involved in the surveys for the most part were willing to justify themselves by pointing out economic advantages, with a few exceptions, and only with the establishment of "professional" attitudes did conflict of the magnitude outlined in White arise. See Gerald Nash, "The Conflict between Pure and Applied Science in Nineteenth-Century Public Policy: The California State Geological Survey, 1860–1874," *Isis*, LIV (Sept. 1963), 217–228.

Carolina (1823), and the directors had simultaneous appointments and responsibilities in their respective state colleges.[9] Olmsted's summer labor and reports in particular demonstrated the relevance of such surveys for internal improvements. Edward Hitchcock, geologist of Massachusetts, astutely subtitled his first report "The Economic Geology of the State." In describing regional rock strata he pointed out minerals "useful in the arts."[10] Two later surveys (conducted by specialists and with substantially more funding) produced the most notable results. It was active local societies memorializing their state legislatures that created both the New York and Pennsylvania surveys in 1836. The function and very existence of such groups suggest a new consciousness of the scientific lobby.

The New York survey was the more extensive, with the legislature appropriating $104,000 to be spent over a four-year period. After some administrative shifts the state was divided into four districts, with William W. Mather of West Point, Ebenezer Emmons of Williams College, young James Hall of Albany, and Lardner Vanuxem of Bristol, Pennsylvania, each responsible for a district. Topical assignments went to Lewis C. Beck of Rutgers College, John Torrey of New York City, James E. DeKay of Long Island, and Timothy Conrad of Philadelphia. In 1840 Hall replaced Conrad as state paleontologist, and his volumes on New York paleontology, not fully published until 1884, are the greatest monument of the state survey. John A. Dix, secretary of state for New York, in 1836 suggested that the men of the survey meet annually as a board of geologists to discuss classifications and to match descriptions among the several districts. Vanuxem found that such sessions provided "more concert of action in our official reports, and for better understanding the general geology of the state."[11]

The initiative for forming a geological society to fulfill the same coordinating function for all geologists came from Edward Hitchcock. His situation as professor and later president of Amherst College cur-

9. Charles S. Sydnor, "State Geological Surveys in the Old South," in David Kelly Jackson, ed., *American Studies in Honor of William Kenneth Boyd* (Durham, N.C., 1940), pp. 86–109.

10. "Reports on the Geology of Massachusetts," *AJS*, XXII (Apr. 1832), 1–7.

11. Michele Aldrich pointed out that Dix made the initial recommendation. Vanuxem to Hall, Bristol, Pa., 16 Mar. 1838, Hall MSS, New York State Library, Albany (hereafter NYSL). For the New York survey, see Merrill, *Contributions*, pp. 327–329, and his *First One Hundred Years*, pp. 187–188, 223–236, as well as John M. Clarke, *James Hall of Albany, Geologist and Paleontologist, 1811–1898*, 2nd ed. (Albany. N.Y., 1923), pp. 48–54.

tailed his time for geology, but Hitchcock believed that an equal handicap was his limited opportunity to interact with men in his scientific field.[12] In 1819 he had eagerly joined the American Geological Society, which met until 1826 in New Haven, and he had been a corresponding member of a local geological society in Pennsylvania. These two relatively short-lived groups had suggested to Hitchcock the importance and the difficulties of a specialized organization for geology. Their existence indicated that his frustration and desire for interaction was not an isolated one.

In 1819 the first truly national, specialized society had been founded at Yale through the initiative of George Gibbs and Benjamin Silliman.[13] Following the example of the older societies for science in America as well as the relatively young Geological Society of London, formed in 1807, the American Geological Society's founders intended to promote the mutual education of members and to build a collection of books and specimens for membership reference use. Silliman's *Journal* provided a publication outlet for papers presented to the society, as it did later for similar nonpublishing groups.[14] William Maclure was elected president at the first meeting in September 1819, but Gibbs took the chair because Maclure was absent in England. The members hoped to have an annual meeting in September, as well as quarterly meetings, with the society "provisionally" located in New Haven.[15]

The deposit of Gibbs's mineral collection at Yale and the building of rooms to house it fostered enthusiasm for the enterprise.[16] Maclure re-

12. Hitchcock's own autobiography provides the best survey of his personal history: *Reminiscences of Amherst College, Historical, Scientific, Biographical and Autobiographical; Also of Other and Wider Life Experiences* (Northampton, Mass. 1863).

13. John Fulton and Elizabeth Thomson, *Benjamin Silliman, 1779–1864: Pathfinder in American Science* (New York, 1968), pp. 70–76. Bates, *Scientific Societies*, pp. 55–56, suggests another attempt, the American Mineralogy Society in New York in 1798, but finds no evidence of any real activity.

14. Gibbs had been influential in the founding of Silliman's *Journal* in 1818, which rapidly became the leading scientific periodical in the United States. The subtitle of the first volume, later dropped, added "especially of mineralogy, geology and other branches of natural history...."

15. *AJS*, II (Apr. 1820), 139–144. Silliman believed that Maclure was most appropriate because "no man (with the exception of Col. Gibbs) has so good a right to give advice on American geology." *AJS*, III (May 1821), 363.

16. Gibbs deposited the collection at Yale in 1812, and in 1819 it was moved to a new building for better protection and display. In 1825 Gibbs sold the magnificent collection to Yale for $20,000. See Fulton and Thomson, *Benjamin Silliman*, pp. 70–72.

sponded to his election by sending books and specimens from Europe to add to the society's collection. Although brief notes in the *Journal* mention annual meetings every year but 1823, there were few quarterly meetings and virtually the same officers were elected each year.[17] The record in Silliman's *Journal* indicates that although the membership nominally included men from every state, participation was limited, and after 1826 the meetings ceased.[18] A local contingent of interested men was too small to justify formal meetings, and the inability of other members to travel easily to the meetings prevented the society from becoming truly national.[19] Within twenty years the transportation revolution allowed scientists to travel more rapidly and in relative comfort and permitted specimens to be exchanged with greater ease and assurance of safe arrival. Coincidentally, the construction of railroads and canals created cross-sections of stratigraphy useful for research.

A second geological society, founded initially for local residents in Philadelphia in 1834, had a precise goal. The Geological Society of Pennsylvania was organized specifically to promote a mineralogical and geological survey of the state. Its political overtones are evident in the selection of leading politicians as major officers, and in the society's proposal to the legislature that it supervise the survey.[20] A single volume of *Transactions* was published, listing nearly one hundred resident and one hundred nonresident members.[21] But when the survey was finally established in 1836 under Henry Darwin Rogers, the lobby organization dropped from sight.

Hitchcock was a member of both of these societies, as were most of the older state geologists. Hitchcock, relatively isolated from scientific centers, sought some regular opportunity to meet with fellow scientists. During the winter of 1837–38 he visited friends in Boston and

---

17. Meisel, *Bibliography*, II, 392–393.

18. Merrill, *First One Hundred Years*, p. 61, gives 1828 as the date of demise but without documentation.

19. Edward Hitchcock to Silliman Sr., Conway, N.H., 6 Aug. 1821, and Silliman to Hitchcock, New Haven, 11 Aug. 1821, both in Hitchcock MSS, Amherst College Archives, Amherst, Mass. (hereafter ACA).

20. Meisel, *Bibliography*, II, 537–539. George Featherstonehaugh's first and only volume of the *Monthly American Journal of Geology and Natural Science*, I (1832–33), 425–428, announced the organization of the society and elaborated basic questions on geology which might be answered by residents throughout the state.

21. Geological Society of Pennsylvania, *Transactions*, I (1834–35), vii–ix. The editor hinted at the purpose of a state survey by later noting that they existed in Maryland, Massachusetts, North Carolina, and South Carolina, and more recently in Virginia, New Jersey, Connecticut, and Maine, pp. 171–174, 411.

discussed with them the possibility of founding a national organization. Their approval encouraged him to write to Benjamin Silliman, Sr., and Henry Rogers, who were especially well known among the geologists.[22] Silliman concurred in the idea but declined to assume the initiative, perhaps because of his prior experience with the New Haven American Geological Society. As a result of his European contacts and his American reputation, Rogers's support for Hitchcock's idea was essential. After Rogers returned from England, he participated in the Geological Society of Pennsylvania, became a lecturer at the Franklin Institute, and taught mineralogy and geology at the University of Pennsylvania. His appointment to be head of the New Jersey survey established his reputation as a competent geologist and helped stimulate the Pennsylvania legislature into an even more extensive survey in 1836, also with Rogers at its head. In the meantime Henry's brother William Rogers became head of the Virginia survey and professor at the University of Virginia.[23]

Rogers's experience in England suggested to him the importance of a national scientific association, and in December 1834 he wrote about his interests to Silliman. In reply to Rogers's inquiry about forming a "scientific meeting" in the United States, Silliman expressed caution about attempting too much too soon. Issues and prospective participants were still diverse: "a large number of mathematicians and engineers—some natural philosophers and chemists and a considerable corps of painters, provided colors were thought desirable to set out the meeting."[24] Silliman felt that two additional obstacles to a national meeting

22. Hitchcock to Henry Rogers, Amherst, 4 Apr. 1888, Rogers, ed., *Life and Letters*, I, 154–155; Hitchcock to Silliman, Amherst, 9 Apr. 1838, and Silliman to Hitchcock, New Haven, 14 Apr. 1838, both in Hitchcock MSS, ACA; Hitchcock to William W. Mather, n.p., 12 Oct. 1848, cited in Hitchcock, *Reminiscences*, p. 370.

23. Rogers, ed., *Life and Letters*, I, 91–102. There were four active Roger brothers, James Blythe (1802–52), William Barton (1804–82), Henry Darwin (1808–66), and Robert Empie (1813–84), born to Patrick Kerr Rogers, an Irish immigrant who brought with him to the United States an interest in science and taught his sons the rudimentary skills necessary to practicing men of science. Henry and William were the most prestigious and were interested in both research and the administration of science. The volumes by Emma Rogers are indispensable for understanding the Rogers family, because some of the letters produced there are no longer in the Rogers MSS, MITA, or in the collection in the Virginia State Library at Richmond.

24. Rogers's letter was not found, but its recommendations are suggested by Silliman's systematic reply to suggestions. See Silliman to Rogers, 22 Dec. 1834, Rogers MSS, MITA. Subsequent quotations are also from this letter.

in the United States were the distances scientists would have to travel and the necessity for adequate leisure time and money to attend. Silliman dropped the idea and the busy Rogers apparently took no action.

Four years later, in the spring of 1838, Hitchcock wrote to Rogers urging him to call together the New York and Pennsylvania geologists before the summer fieldwork began.[25] Again Rogers procrastinated. When Hitchcock and Rogers met in September, they discussed the proposed society in more detail. Hitchcock shrewdly argued that the society would provide a forum for Rogers's new theoretical ideas, a place for them to be articulated and then tested by fellow researchers.[26] Many of the state surveys had been authorized in the mid-1830s, and the geologists were formulating final reports at the close of the decade. It was a critical time to establish uniform nomenclature and classification systems.

Impressed by the British Association's popular success and scientific benefits, Hitchcock evidently hoped to form a similar organization. Henry Rogers, however, had absorbed Silliman's caution and considered aloud to his brother "whether it were better to delay the movement until a General Association for all the sciences can be brought about or to make it now for geology merely."[27] Other agencies were already taking action for a less comprehensive society. At their spring meeting in 1839 the New York geologists recommended calling a meeting for 1840, to which only men involved in state surveys were to be invited.[28] Positive response to Vanuxem's letters of invitation indicated the readiness for such an organization. Although frustrated that he and William did not have enough time to formulate a general memoir in time for the meeting, Henry Rogers nonetheless helped plan the meetings in Philadelphia. Table 1 lists meeting dates, places, and presidents of the AAGN through 1847.

Edward Hitchcock presided over the three-day meeting, attended by geologists representing seven state surveys.[29] Informal presentations

25. Hitchcock to Rogers, Amherst, 4 Apr. 1838, Rogers, ed., *Life and Letters*, I, 154. Hitchcock said that he had discussed the plan with other New England men and mentioned the individuals who might be interested, suggesting that any such meeting be held in New York City or Philadelphia.

26. Henry Rogers to William Rogers, Philadelphia, 26 Sept. 1838, Rogers, ed., *Life and Letters*, I, 155–156.

27. *Ibid.*

28. *Ibid.*, pp. 163–164.

29. There are no comprehensive files for the Association of American Geologists and Naturalists, but the ANSP has some manuscript minutes and correspondence, chiefly for the years from 1845 to 1847, when Silliman Jr. was secretary. The as-

The AAGN

TABLE 1

Meeting Dates, Places, and Presidents of the
Association of American Geologists and Naturalists

The following table is derived from published accounts of the AAGN meetings.
The date is the one on which the general session opened.

Date	Location	President
2 Apr. 1840	Philadelphia	Edward Hitchcock, Sr.
25 Apr. 1841	Philadelphia	Benjamin Silliman, Sr.
25 Apr. 1842	Boston	Samuel G. Morton
25 Apr. 1843	Albany	Henry D. Rogers
8 May 1844	Washington	John Locke
30 Apr. 1845	New Haven	Chester M. Dewey[a]
2 Sept. 1846	New York	Charles T. Jackson
20 Sept. 1847	Boston	William B. Rogers[b]

[a] When the elected president, William B. Rogers, could not attend, Dewey took the chair for the general sessions.

[b] Rogers presided because the elected president, naturalist Amos Binney, died before the Boston meeting opened.

and discussions demonstrated that the new society could provide the forum needed for the emerging geological profession. Several state surveys—New Jersey, Michigan, and Pennsylvania—presented brief reports on their progress. The attendance of Douglass Houghton and Bela Hubbard of the Michigan survey demonstrated that a truly national society was possible.[30] Although the 1840 session was an organizational meeting, the men limited discussion to mutual problems and specific questions to be considered in preparation for another session. Membership remained limited to geologists and men active in closely related areas of science. The intention to be small and well defined was explicit:

> Resolved, that no person shall be considered as qualified to become a member of this Association who is not devoted to Geological research with scientific views and objects.
> Resolved, that not more than 10 gentlemen, in addition to those already invited be requested to attend the next annual meeting as members.[31]

sociation itself published the addresses of retiring presidents Hitchcock (1842) and Rogers (1844) with a brief abstract of those respective meetings as well as a single volume of *Reports* in 1843 and an abstract of the meeting in 1845. See Association of American Geologists and Naturalists (hereafter AAGN), *Reports of the Transactions of the Association of American Geologists and Naturalists, at Philadelphia in 1840 anl 1841, and in Boston, 1842. Embracing Its Proceedings and Transactions* (Boston, 1843).

30. Abraham Sager to James Hall, Detroit, 18 Mar. 1840, Hall MSS, NYSL.

31. Written by Vanuxem, Henry Rogers, and Mather, two resolutions in 1840

The additional appointment of a nominating committee suggests that voluntarism was not the original mode. When a constitution was formally adopted in 1842, it more closely resembled the constitution of the BAAS than that of the Geological Society of London.[32] The migratory character of the former was most appealing to persons who anticipated benefits from visiting new regions and using the unfamiliar collections of their colleagues. The small session, unnoted by the local public press, had established enough mutual confidence to continue and to assume the name Association of American Geologists, a title that was expanded in 1842 to include "naturalists."

The gradualism suggested by Silliman, Rogers, and the *New York Review* writer became the cautious theme of the new association's expansion toward a more inclusive organization. The delay in forming even the initial small group of established, employed geologists indicates the diffidence of men of science. The fear of failure resulted in an underestimation of their own abilities, and it dwarfed initiative in institution-building. Considering the experiences of the local societies, the hesitation was also realistic.

At the second meeting in 1841, again held in Philadelphia, Edward Hitchcock gave an address as retiring president. His presentation surveyed geological progress in the United States and brought American scientists up to date on European glacial studies and theories, information of vital importance to the geological surveyors present. He unselfconsciously admitted the difficulty of a self-trained geologist who now confronted specialization and competition from full-time researchers. Evidently pleased with the new organization, he urged members to pursue science diligently and assist fellow scientists in resolving mutual geological questions. Hitchcock's address was published in Silliman's *Journal* and also separately, but the individual papers presented to the meeting appeared only in the *Journal*, if at all.[33] Silliman, pleased with the activity of the state geological surveys and the Philadelphia meeting, reported to Gideon Mantell, "Everything in science in the country is

specifically limited membership; these were not published in 1842, the intentions of the society having evidently changed. This information is in handwritten minutes at the ANSP.

32. AAGN, *Reports*, pp. 9–10. Because there were no formal proceedings, the secretary simply prepared an abstract of the meeting for the *AJS*, XXXIX (June 1840), 189–191.

33. "Anniversary Address," *AJS*, XLI (July 1841), 232–275. This arrangement of publication was less costly, for the association had additional copies printed from the *AJS* typeset at cost.

rather looking up. More interest is taken in the labors of scientific men and more laborers are in the field."[34]

The Boston meeting of 1842 constituted a milestone in the history of the association. Attendance was high, with about forty members present. Benjamin Silliman's presidential address, given in an evening session, attracted an audience of nearly 500 persons.[35] The star of the meeting, however, was the lion of contemporary geology, Charles Lyell, who was touring the United States. Lyell presented no formal paper but established his presence through extensive commentary on other reports.[36] Also at that meeting Henry and William Rogers presented probably the single most important paper read before the Association of American Geologists and Naturalists. While working on their respective surveys of Pennsylvania and Virginia, they had discussed and analyzed the changes along the Appalachian Mountain chain. In 1842 they revealed their conclusions "On the Physical Structure of the Appalachian Chain, as Exemplified by the Laws Which Have Regulated the Elevation of Great Mountains Generally."[37] Like William Maclure and their contemporary James Hall, the Rogers brothers were attempting to establish some systematic overview for American geology, and their efforts were widely discussed in the following decades.

Despite their desire to publish their proceedings, the society was cautious. Publication was a means of establishing research priority, an essential medium for transmitting new ideas and classifications, as well as evidence of the competency of the publishing organization. These concerns were characteristic of a rising professionalism. Unfortunately, reported sales of Hitchcock's address in 1841 were not encouraging, and additional financing beyond annual dues was essential before commitment to a full journal could be made.[38] Then, in 1842, Nathan Appleton, a wealthy Boston textile manufacturer, pledged his personal support and together with friends raised $500 toward publication of

34. Silliman to Gideon Mantell, New Haven, 14 Apr. 1841, Silliman-Mantell MSS, BLY.
35. Silliman Jr. to Gideon Mantell, New Haven, 14 May 1842, Silliman-Mantell MSS, BLY.
36. These were not submitted in writing to the secretary and were therefore not recorded; see AAGN, *Report*, I (1843), 146. Lyell remained in the United States from July 1841 to Aug. 1842 and summarized his experience at the meeting in his *Travels in North America: with Geological Observations on the United States, Canada, and Nova Scotia*, II (London, 1845), 261–262.
37. AAGN, *Report*, I (1843), 70–71, 474–513; *AJS*, XLIII (July 1842), 177–178.
38. Benjamin Silliman, 26 Apr. 1842, "Report on Prof. Hitchcock's Address &c.," AAGN MSS minutes, ANSP. All members had received a free copy, but outside sales netted only $9.

association proceedings.[39] The *Report*, like the two preceding presidential addresses, did not sell well. Amos Binney, who supervised the Boston publication, accurately warned, "If we can promptly sell this edition we shall have the means of printing the Albany papers at once. If not I fear they will not be printed unless in the public journals."[40] No subsequent full report of association meetings was published, although Appleton later, in 1845, pledged additional support.[41]

The Boston meeting also adopted, after some discussion and amendment, the constitution and bylaws prepared by a committee consisting of Lewis C. Beck, Henry Rogers, Edward Hitchcock, John Locke, and Charles T. Jackson.[42] The success of the third meeting seemed to ensure the new association a permanent status that required a formal constitution. It was not modeled directly on that of either the Geological Society of London or the British Association, although, like the latter, it did provide for applications as well as nominations to membership and also created both a standing committee to conduct the meetings and a local committee to take care of physical arrangements.[43] The requirement that all communications to the association be written attempted to ensure that papers presented be of publishable quality, and discussion was to be reported only if submitted by the critic. The deep commitment to advancing geological science was evident as discussions became more heated and division along lines of theoretical difference became discernible. At one point of intense debate the association resolved that the remainder of the session "be strictly confined to the reading of papers" and that "no discussion be allowed thereon."[44] Contention might be avoided by such temporary measures, but generally debate was open and substantive. The regular membership represented

39. Silliman Jr. to Nathan Appleton, New Haven, 8 June 1842, Appleton MSS, Massachusetts Historical Society, Boston (hereafter MHS); Silliman Jr. to Gideon Mantell, New Haven, 14 May 1842, Silliman-Mantell MSS, BLY. Two free copies were given to each person contributing to the publication fund.

40. AAGN MSS, ANSP; Amos Binney to James Hall, Boston, 17 June 1843, Merrill Collection, LC.

41. Silliman Jr. to Nathan Appleton, Yale College Cabinet, 5 May 1845, Appleton MSS, MHS. Thanking Appleton for his liberal offer, Silliman observed, "The Association are fully aware of the vital importance of a regular publication of transactions by them in order to fully secure the benefits to be derived from such an organization."

42. AAGN, *Report*, I (1843), 39, 77–78.

43. The constitution is printed in the annual reports of the BAAS. Also see Horace B. Woodward, *The History of the Geological Society of London* (London, 1907).

44. AAGN, *Report*, I (1843), 173.

most leading American geologists (seventy-seven persons were on the 1842 roster) and seemed capable of broadening to include men interested in other earth sciences as well. Local as well as international prestige was enhanced by the approbation of so eminent a geologist as Lyell and by the thick volume of proceedings. If discussions were occasionally heated, the sheer necessity of resolving strong disagreements among American geologists was reason enough to justify them. Honorary leadership accrued to established men like Silliman and Hitchcock, whose private correspondence suggests their almost paternal interest in the young society, while the somewhat younger men, including the Rogers brothers and James Hall, found ample reason to use its podium and to perform its more routine labors. So long as the major function of the society was discussion on problems of generally mutual interest, no insurmountable breaches developed. The need for communication among geologists, a catalyst in bringing about the association, was well fulfilled. William Rogers keenly felt that requirements of teaching limited his scientific research, and he asserted: "For us such reunions of the scientific brethren as our Association of Geologists are of precious value and form the best compensation we can enjoy for the prolonged restraints of our vocation. What new impulses to exertion, what encouragement and guidance do they not give? and then in our hours of lonely meditation to how many cheering and delightful *social* recollections. . . ."[45]

Eager to promote further local support and to show off their results, James Hall and the other state surveyors offered to be hosts at Albany for the 1843 meeting. The successes of the Boston meeting encouraged membership and local attendance. Hall was exuberant, having recently accepted the position of New York state paleontologist and being eager to exhibit the survey collections recently arranged in the Old State Hall.[46] James Dwight Dana, a member of the Wilkes exploring expedition, presented three papers whose theories on continental subsistence and ocean temperatures aroused lively debate at the meeting. In meaningful contrast, a paper by Edward Hitchcock on wine production received little attention. He had been prime mover for an organization which moved rapidly beyond him; his continued interest in "curiosities," while acceptable in small, local societies, was not relevant to the new practicing geologists. Hitchcock's stature and eminence

---

45. William B. Rogers to J. W. Bailey, University of Virginia, 22 Oct. 1843, Rogers MSS, MITA.
46. Clarke, *James Hall*, pp. 134–135.

precluded any attempt to silence him, and his less relevant presentations were simply bypassed; his work on geology continued to be respected. Assessing the meeting for his absent brother, Henry Rogers wrote that there was an excellent spirit and more "solid work" than the previous year.[47]

The slowly developing organization was not rigorous enough to satisfy all naturalists, however. Asa Gray and his close friend and mentor, John Torrey, rarely if ever attended, but their absence did not seriously undermine the group.[48] Henry Rogers presided over the Albany meeting, but the apparently disinterested Samuel George Morton failed to appear with the expected retiring address.[49] New efforts made to solicit local sponsorship for the publication of proceedings were ineffectual.[50]

In the years following 1843 the active members tried harder to promote the organization. This effort was in part a response to the threat of competition from another organization, also founded in 1840, the National Institute in Washington, D.C. The geologists had responded positively to early information about the institute and even made plans to hold future meetings in Washington, in conjunction with the institute. When some members received a notice in 1842 suggesting that the institute was trying to become more than a museum and in fact a national scientific organization, resistance developed.[51] A circular printed in 1843, however, established direct opposition. The institute, abandoning the concept of a joint meeting, simply invited the Association of American Geologists and Naturalists along with other major societies to its meeting in April 1844. Quietly but deliberately the geologists voted at the 1843 meeting not to change their plans to go to Washington the following year but simply to establish a different date for their

47. Henry Rogers to William Rogers, Albany, 30 Apr. 1843, Rogers, ed., *Life and Letters*, I, 222–223. Rogers felt that attendance was small because Albany itself had almost "no taste for such matters." Although initially disheartened by the absence of Morton, Silliman Sr., A. A. Gould, Locke, and Ducatel, the attending members enjoyed seeing the New York collections and assigned numerous reports to ensure "first rate" papers for the next meeting in Washington in 1844.

48. Gray to Torrey, Cambridge, 7 May 1843, Torrey MSS, NYBG.

49. His disinterest in the new organization is suggested in Patsy Ann Gerstner, "The 'Philadelphia School' of Paleontology: 1820–1845" (Ph.D. dissertation, Case Western Reserve University, 1967), pp. 275–276.

50. Henry Rogers to James Hall, Philadelphia, 11 May 1843, Merrill MSS, LC: Rogers felt that "the future prosperity of the Association will in a good measure depend on our continuing to publish our transactions."

51. The following discussion relating to the National Institute is condensed from the author's "Step toward Scientific Self-Identity."

own meeting. The association met for a full week's meeting in May, a month after the somewhat pompous and ineffectual convention of the institute, whose few scientific papers were described as "old and superficial" by Dana.[52] The association was clearly more representative, if only of a subgroup, of science. Silliman Jr. concluded that their meeting "compared very well in scientific character with the other annual sessions of the same body."[53]

Henry Rogers's retiring address as president of the association was, following tradition, a thorough review of geological research in the United States. To evident advances made during the decade he linked the association's meetings, where personal intercourse had strengthened or discarded hypotheses and through which zeal and cooperation made possible a summary of "the geology of three fourths of the vast region between the Atlantic and Mississippi."[54] It was their singularity of purpose which established the group dynamic. Taken in this context, the fiercest priority struggles and scientific debates denote a certain group strength. The nucleus of leading geologists brought out of the welter of new research not conformity but a dialogue and clarification of issues which were not resolvable on the basis of existing research alone. Experience proved that acquaintance might, but also might not, bring close friendship. Simply meeting the need for better communication, however, did establish a firm base for the association, if not the expansion hoped for by its founders. The almost inevitable tension in such a group was not readily accepted by the founders like Silliman and Hitchcock, but they were rapidly being displaced as scientific leaders by younger and better-trained persons.

While the first three years of the association concentrated on establishing the organization, the two following the pivotal Boston meeting raised the serious question of how best to expand. Despite the inclusion of naturalists as part of the association's title, geologists continued to dominate the proceedings, indicating that the gradual inclusion technique was not completely satisfactory. In addition, the clear, if only informal, regional distinctions among members had not been resolved

52. Dana to A. A. Gould, Washington, 11 Apr. 1844, Gould MSS, Houghton Library, Harvard University (hereafter HLH).

53. Silliman Jr. to Gideon Mantell, New Haven, 28 May 1844, Silliman-Mantell MSS, BLY.

54. Henry D. Rogers, *Address Delivered at the Meeting of the Association of American Geologists and Naturalists, Held in Washington, May, 1844, with an Abstract of the Proceedings of Their Meeting* (New York, 1844). Rogers was not happy with the New Haven publication and had another copy prepared. See Rogers to Nathan Appleton, Philadelphia, 14 Oct. 1844, Appleton MSS, MHS.

satisfactorily. The South and West remained uninvolved except as a few individuals were personally interested in participation. Public involvement was casual, although meetings were open to the public. Some members recognized that more general sessions directed at the public or a systematic policy of expanding membership to include amateurs might bring more participants, more financial support, and an enhanced public reputation. Such expansion also raised the serious questions of ineptitude and showmanship made plain in the self-culture movement. Yet unless they did so privately at the annual meetings, the leadership of the association made no positive commitment to any process of expansion. Rather, it developed pragmatically, based on the actions of the local committees or the various officers. Thus the younger Silliman, although he held rather exclusive ideas about the purposes of the association, was quite willing to expand local committee planning functions and sometimes membership to involve leading social and political figures. Before the meeting of 1846 William Rogers confided to Silliman his hope that the association would "enlarge its ranks while expanding its usefulness and reputation."[55]

Obviously some men were eager to fulfill the larger expectation that the association would become a national organization. But, simultaneously, the geologists felt unready to expand into areas of science where the response of men of science was unknown or the value of association less high. As a result a motion by Henry D. Rogers to expand into "an association for the promotion of science" was tabled.[56] The fact that Silliman retained the motion in his abstract of the meetings seems evidence that eventually the association did hope to open its doors widely "for all cultivators of science and the arts," and to this end the constitution reduced the terms of membership to a "mere formality of signing that instrument."[57]

Other issues also surfaced as the association struggled to define the limits of its interest and the scope of its activity. Although Hitchcock and others occasionally made references to the power and wisdom of

55. Rogers to Silliman, University of Virginia, 18 Jan. 1845, AAGN MSS, ANSP.

56. *New York Tribune*, 8 May 1845. Rogers proposed that it be named American Society for the Promotion of Science and be incorporated and that a fund be raised for it.

57. *AJS*, XLIX (Oct. 1845), 219. Silliman concluded that if the meeting was small compared to its European counterparts, its numbers were encouraging when contrasted to the "small band of votaries who follow subjects of science in this country."

God parenthetically while discussing a topic of science, the question of religion's relationship to science achieved the floor for the first time at the New Haven meeting, 1845, when a paper by Dr. Webber quoted from the controversial *Vestiges of Creation* on the doctrine of generation.[58] In the ensuing debate the elder Silliman condemned the book for its "dangerous and irreligious tendency," while Charles T. Jackson added his skepticism about the experiments cited. On the other side, Rogers and Samuel S. Haldeman stated that while they did not accept all the premises of the author, they felt that the problem of generation was a profound one and scientists must remain open to all possibilities.[59] In general, the discussions concentrated on more orthodox research topics.

An ongoing issue was the Hall-Emmons debate on the Taconic system. Emmons had first announced his Taconic system at the 1841 meeting and elaborated on it in his report on the second district of New York in 1842.[60] Supported only by Vanuxem, against the skepticism of Henry Rogers, Hitchcock, Mather, and Hall, Emmons became an adamant proponent of his system, and the question reappeared at several subsequent AAGN meetings.[61] When Emmons reiterated his views with new evidence in 1846, the subsequent debate resulted in another committee investigation, headed by Haldeman; in 1848 the committee report noted that evidence was scanty but vindicated Emmons.[62] This question, in fact, like several others raised by the new research, could not be immediately resolved. The fact was a frustrating development for men who had hoped that open discussion would provide more rapid and satisfactory resolutions to difficult problems.

Nonetheless, the annual meetings had been important for helping American geologists to attain European recognition and to establish their own standards. Heated debates waged at meetings and reflected

58. *New York Herald*, 9 May 1845.

59. *Ibid.*

60. The most thorough discussion of the continuing debate on the "Taconic Question" is in Merrill, *First One Hundred Years*, pp. 594–614. Also see Cecil J. Schneer, "Ebenezer Emmons and the Foundations of American Geology," *Isis*, LX (Winter 1969), 439–450.

61. Hall to Joachim Barrande, Albany, 15 Oct. 1861, and to Louis Agassiz, 16 Dec. 1861, both cited in Clarke, *James Hall*, pp. 372–373, 377.

62. Debate was especially volatile at the 1846 meeting, and Silliman Jr., as secretary, attempted to keep controversy out of the press by suggesting that the conversation might be attributed to the "heat of the weather." See *New York Evening Express*, 7 Sept. 1846. Also see S. S. Haldeman to Baird, near Columbia, Pa., 1 Oct. 1846, Baird Personal MSS, SIA.

in the *Journal* were indicators that scientific advance was not a happily progressive procedure but fraught with blind alleys and misleading data—yet out of discussion and research, advance occurred.

In 1847, proud of their accomplishments and ready to broaden their scope even further, the association members welcomed the internationally recognized zoologist and geologist, Louis Agassiz. For a second time Boston became host for the association; the members evidently hoped to recapture the spirit of the first meeting there. Fortuitously, Agassiz arrived from Switzerland during the previous winter to present a series of lectures at the Lowell Institute.[63] Prepared for his trip by American enthusiast Charles Lyell, Agassiz planned a year of study and travel in North America. Evidently pleased by what he saw and by his reception, Agassiz accepted an appointment at the new Lawrence Scientific School at Harvard, a position largely designed for him.[64] Agassiz became, by sheer power of personality and reputation, a central figure to the politics of American science. During the 1847 meeting he presented three scientific papers and also expounded at some length about European national associations.[65] His implications were not lost on the more ambitious members of the association. In 1847 the geologists and naturalists voted to become the "American Association for the Promotion of Science."

The Association of American Geologists and Naturalists had not been an unqualified success, but the determined efforts of such leaders as Edward Hitchcock, the Rogers brothers, and the Sillimans had preserved the national (at least to the Mississippi) character of the membership. Participation at meetings remained steady and there was a growing roster of members. Most significant, scientists from other areas found the example sufficient and were ready for a larger organization; in 1848 physicist Joseph Henry noted, "It is time that we should have among us a scientific *esprit de corps*."[66] Gradual expansion had met with limited success—a survey of the papers published in Silliman's *Journal* suggests that more chemistry and meteorology were included than natural history, despite the duality symbolized by the association's title. Botanists and zoologists were simply superimposed on the geologists' organizational structure. As late as 1854 Dana commented to Hall, "Geology is

63. See Edward Lurie's excellent biography, *Louis Agassiz: A Life in Science* (Chicago, 1960).
64. Margaret W. Rossiter, "Louis Agassiz and the Lawrence Scientific School" (A.B. thesis, Radcliffe College, 1966).
65. *Boston Journal*, 25 Sept. 1847.
66. Henry to Elias Loomis, Washington, 28 Dec. 1848, Loomis MSS, BLY.

almost the only science in which progress has been made in the country except some departments of zoology."[67] The geological society contributed to that primacy and to the institutionalization of science by bringing important scientific issues into open discussion and thereby giving birth to an interacting scientific community on the national level. The dramatic decision to reformulate the association to include all sciences was based on a positive assessment of the geologists' efforts.

67. Quoted in Clarke, *James Hall*, p. 318.

IV

Formation of a Viable
Organization for National Science

The American Association for the Advancement of Science was a culmination of the aspirations of men in the Association of American Geologists and Naturalists, the continuing interest in a national scientific organization, and a rising hope for *esprit de corps* among American men of science. From the geologists the AAAS inherited a stable membership and a substantial reputation; AAGN sponsorship, even more than that of the Yorkshire Philosophical Society for the BAAS, ensured a firm foundation for the new organization.[1] From its formation in 1840 the AAGN had developed an identity similar to that of the BAAS and had cautiously expanded its scope of activity in expectation of becoming essentially a joint meeting for coordinating scientific men in various fields. When the decision to become the broadly national AAAS was made in 1847, however, it reflected the readiness for a national organization and an absence of any marked opposition compa-

1. There is abundant evidence that the AAGN was in fact not only the precursor but the actual founder of the AAAS. William Redfield, whose efforts helped precipitate the change, was elected first president. Within the powerhouse of the new organization, the standing committee, ten of fourteen members were geologists and naturalists in 1848. The Rogers brothers, whose participation in the AAGN had helped establish its early reputation, were in several ways responsible for its nucleus of organization. Some participants believed that the legitimate founding date was 1840 because the AAGN was the planned foundation of the AAAS. See Hitchcock's *Reminiscences*, pp. 371–373, and Denison Olmsted, *Address on the Scientific Life and Labors of William C. Redfield, A.M., First President of the American Association for the Advancement of Science* (New Haven, Conn., 1857). Olmsted credits Redfield as "first to suggest the idea of the American Association on its present comprehensive plan." The role of the AAGN was not formally acknowledged until 1874, when the AAAS *Proceedings* listed along with former AAAS officers those of the AAGN.

rable to that in 1838, rather than the triumph of the geologists' experimental organization. As a result the spirit of its formative years was one of determination rather than jubilation. The founders believed that, assisted by outstanding men with distinguished scientific reputations, a viable organization could be created with the potential to coordinate scientific inquiry and to establish science as a true and visible profession in the United States. The critical step was to formulate a constitution and practical procedures to that end.

When the AAGN resolved itself into the AAAS in 1847, it committed present and future membership to an expanded, national, open organization for persons interested in science. At least one reason for the success of the AAGN was its specialization in geology; the AAAS was obligated to recognize discipline needs through sectional meetings while coordinating scientists in general sessions. Skepticism and fears were temporarily checked by a general concurrence in the need for an association, and the constitution was deliberately loose in its construction. Neither the goals nor their practical implications were necessarily synonymous in Washington, New York, and Charleston, among naturalists and physicists, or even within disciplinary areas. The need for working harmony during the early years of self-definition masked serious differences. Thus the first four years, 1848 through 1851, were distinguished by a basic agreement among the members, a self-image reflected in the approving eyes of the American press and casual visitors to association meetings. The transition from AAGN to AAAS was smooth, and progress seemed self-evident in the expanding membership, an increasing level of participation at association meetings, and public approbation. Everyone's focus was on the future and on the goal of a working, reputable national organization. Disillusionment and dissension occurred only gradually, as significant differences in both scientific and social philosophy became evident. The formative years did allow, however, enough time to establish a stable, if loosely coordinated, national organization for scientific discussion. All participants tentatively agreed that the function of the AAAS was both to diffuse and advance knowledge. That this dual purpose might foster conflicting efforts was not publicly discussed.

The decision to transform the AAGN into the AAAS was made at the 1847 Boston meeting, a strategically sound choice, because in 1842 the AAGN had held its most strikingly successful session in the city whose claim to primacy as a scientific center was no longer so seriously rivaled by Philadelphia. In addition, Boston surpassed the latter in enthusiasm for the new enterprise.[2] Joseph Henry came to the 1847 meeting, apparently overcoming earlier scruples about popular organization, and he used the AAGN as a podium from which to elaborate the goals of the Smithsonian Instituion and to chart its proposed course. Although Henry was aware of a current burst of criticism against the "Congress of Lilliputian Savans" in England,[3] the opportunity to meet with established scientists outside his own field reinforced his tentative approval of AAGN expansion.[4] He joined fellow physical scientist Benjamin Peirce, a mathematician at Harvard, in encouraging the geologists' resolution to change the name and the constitution of the AAGN.

With an "almost unanimous vote" to extend the activities in 1847, the AAGN selected three Bostonians to revise the constitution and rules. The three—chemist-geologist Henry D. Rogers, mathematician-astronomer Benjamin Peirce, and naturalist Louis Agassiz—represented among themselves all major areas of scientific research in America. Apparently the task of actually writing a constitution fell to Rogers, who by 10 May 1848 distributed a circular over the names of the three committee members containing the proposed "Rules and Objects of the Association" as well as a statement of purpose.

Henry D. Rogers, on whom the responsibility devolved for altering the AAGN constitution, had been intimately involved with the initiation and progress of the older organization. His national and international prestige was important in sustaining the reputation of the geological group, and his theoretical paper on the Appalachian chain made the association's single volume of *Reports* an important publication. He moved to Boston in 1845 and subsequently was an unsuccessful candidate for the Rumford professorship at Harvard.[5] He was also active in

2. Thus the American Academy of Arts and Sciences in Boston sponsored Warren's efforts in 1838, only to have the project rejected by the American Philosophical Society. When the association did meet in Philadelphia in 1848, an unsigned letter in the local *Public Ledger*, 23 Sept. 1848, suggested that although the meeting was attended by "considerable numbers of the most distinguished Professors of Science from distant colleges," many local men treated it with "coldness, not to say contempt."

3. Henry to [Joseph B. Varnum], Princeton, 22 June 1847, Henry MSS, SIA.

4. Henry's skepticism of such a project in 1838 has already been noted. By 1848, however, he was enthusiastic. See Henry to Nathan Appleton and Jeffries Wyman, Princeton, 8 July 1847, and Henry to (?), Princeton, 7 Aug. 1847, all draft copies in Henry MSS, SIA; also Henry to his wife Harriet, Cambridge, [24 Sept. 1847], and Boston, 25 Sept. 1847, both in Henry MSS, SIA. Henry later claimed that he opposed the organization from the outset. See Henry to Chester Dewey, Washington, D.C., 7 Nov. 1859, Dewey MSS, University of Rochester Library, New York.

5. Rogers, ed., *Life and Letters*, I, 256–267.

trying to persuade John A. Lowell and other Boston philanthropists to establish a "polytechnic school of the useful arts."[6] Not unfavorable to popular "scientific" lectures if presented as an educational series, Rogers lectured at the Lowell Institute in 1846 and also in New Hampshire. The constitution he devised demonstrated his hopes for unifying science. He specifically included engineers, and his prefatory comments suggested an interest in relating the efforts of "experimenters" and "observers," by which he seems to have meant something roughly parallel to what today might be termed "theoreticians" and "experimentalists." Because of his attendance at BAAS meetings, he was perhaps the most conscious of the three men of the need to unite American men of science and to present science in understandable form to the public.

Benjamin Peirce differed from Henry Rogers in personality as well as in his attitudes toward popular science. Peirce took delight in the abstruseness of higher mathematics and in-group repartee and was not particularly interested in the designing of a constitution. Although he had participated in popular lectures in 1843 when the Great Comet excited general attention, his purpose then was specifically to gain support for an observatory at Harvard.[7] Peirce was skeptical of group participation in matters of science, for mathematics required largely individual research and astronomy depended on accurate observations by well-trained astronomers. Later his experience at AAAS meetings and his intimate relationship with Bache and the scientific Lazzaroni intensified his antipathy toward amateurs. In 1847, however, his skepticism was not evident, and Peirce left no notable imprint on the new constitution.

The enthusiasm of the popular Louis Agassiz was an important component in the initial success of the AAAS.[8] Proud that the Swiss had founded the first "federation" of science and himself a participant at British Association meetings, Agassiz hoped to raise what he believed to be a necessary and justifiable self-confidence among American men of science.[9] He had come to the United States to deliver a course of lectures at the Lowell Institute and seemed to delight in popular response. Yet he, like Peirce, would later become restive when the public attempted to

6. *Ibid.*, pp. 256–257.

7. Victor Lenzen, *Benjamin Peirce and the Coast Survey* (San Francisco, 1969), p. 5.

8. A model scientific biography is Lurie's *Louis Agassiz*; also see Agassiz, ed., *Louis Agassiz.*

9. It is not obvious that the Swiss federation influenced Oken, but the paternal chain of national associations is loosely linked. More important was Agassiz's forceful enthusiasm and his verbal impressions of both the Swiss and the English successes.

participate in scientific investigations without proper training. As a naturalist Agassiz depended on a network of fieldworkers to supply specimens, but he recognized that even in his field untrained observers and collectors might be more bother than worth.[10] In 1847, however, Agassiz was excited by the project and felt that the new society might unify American men of science and redress, as he conceived it, an unbalanced emphasis on geology and utilitarian science.[11]

Later in the 1850s the AAAS would suffer from an absence of articulate spokesmen to elaborate its goals enthusiastically, but in 1848 its purposes were eloquently outlined in a two-page statement of the constitution committee, apparently also written by Rogers. Persuaded that the "time has now fully come" for broadening the AAGN into a truly national organization of science, the committee invited the attention of all men of science to their proposed organization.[12]

The declaration of intention was balanced and enthusiastic. Rogers suggested that science in the United States had enjoyed vigorous development and now lacked "nothing but a systematic organization." Echoed were complaints, heard since the time of Franklin, over the lack of a national group of scientists representing in their activity all the creative and sustaining aspects of scientific endeavor. Henry Rogers was sophisticated enough to realize that in America such an organization must be suited to both the "high wants and the controlling practical tendencies of the country and the age." Having alluded to the earthbound limitations of contemporary vision, however, he quickly pointed out the intellectual progress of scientific culture. In the improved quality of scientific books imported, the more original and experimental character of American treatises, and the increased activity of local societies and institutions, he found evidence to support his contention that recent advances made a national organization imperative.

Beginning with the assumption that the AAGN was the logical ini-

10. Agassiz, ed., *Louis Agassiz*, II, 437. In a letter to Milne Edwards, 31 Dec. 1846, Agassiz indicated the success of the AAGN by commenting, "The geologists and mineralogists form the most numerous class among the savans of the country.... [This fact tends] to the detriment of other branches [and] ... the utilitarian tendency thus impressed on the work of American geologists will retard their progress."

11. Bache to Peirce, Washington, 22 Apr. 1855, Peirce MSS, HLH. In frustration Bache criticized (or blamed) Agassiz for his leading role: "Why did he (Agassiz) get up the association!!!"

12. A copy of the printed "Circular," dated 10 May 1848, is in the AAAS Library, Washington, D.C. All subsequent comments and quotations relating to the proposed constitution and the statement of purpose are from the "Circular."

PROCEEDINGS

OF

THE AMERICAN ASSOCIATION

FOR THE

ADVANCEMENT OF SCIENCE.

FIRST MEETING,

HELD AT PHILADELPHIA, SEPTEMBER, 1848.

PHILADELPHIA :

PRINTED BY JOHN C. CLARK, 60 DOCK STREET.

1849.

REPRINTED AT SALEM PRESS, 1874.

Title page from the *AAAS Proceedings*, vol. I (Philadelphia, 1849).

PROCEEDINGS,

&c.

———•———

First Day, September 20, 1848.

In conformity with a resolution of the "Association of American Geologists and Naturalists," adopted during its session at Boston, in September, 1847, that body agreed to resolve itself into the American Association for the Advancement of Science, and that the first meeting, under the new organization, should be held in the City of Philadelphia, on the third Wednesday (20th day) of September, 1848; and, agreeably to the arrangements and invitation of the Local Committee then appointed, the new Association held its first regular meeting this day, September 20, 1848, at the hour of 10, A. M., in the library room of the Academy of Natural Sciences of Philadelphia.

At 12, M., the meeting was called to order by Prof. WM. B. ROGERS, of Virginia, Chairman of the last Annual Meeting, who, after some preliminary remarks, read the draft of a Constitution and Rules of Order, which had been prepared by a Committee, appointed for the purpose, at the meeting in Boston, in 1847.

The Chairman of the Local Committee then submitted letters, which had been received in reply to the letters of invitation issued by that Committee, *accepting* the invitation, from Lieut. J. M. Gillies, of Washington; Samuel Henry Dickson, of New York; Z. Allen, Esq., of Providence, R. I.; Prof. J. C. Booth, of Philadelphia; Prof. J. S. Hubbard, of Washington, D. C., and J. H. C. Coffin, of Washington, D. C.

Letters, accompanied by promises to make communications to the

Introductory proceedings from the first meeting, *AAAS Proceedings*, vol. I (Philadelphia, 1849).

Henry Darwin Rogers, 1808–66 (courtesy of Massachusetts Institute of Technology Collections).

William Barton Rogers, 1804–82 (courtesy of Massachusetts Institute of Technology Collections).

Joseph Henry, 1797–1878 (courtesy of the Smithsonian Institution).

Alexander Dallas Bache, 1806–67.

Benjamin Apthorp Gould, 1824–96 (courtesy of Benjamin A. G. Thorndike).

Louis Agassiz (*left*), 1807–73, and Benjamin Peirce, 1809–80 (courtesy of the Museum of Comparative Zoology, Harvard University).

James Hall, 1811–98 (courtesy of the
National Gallery of Art).

James Dwight Dana, 1813–95 (courtesy
of Yale University Art Gallery).

William Redfield, 1789–1857.

Dedication of the Dudley Observatory, held while the AAAS was in session at Albany in 1857 (courtesy of the Albany Institute of History and Art).

Spencer F. Baird, 1823–88, first permanent secretary, 1851–54.

Joseph Lovering, 1813–92, second permanent secretary, 1854–73.

tiator for the AAAS, Rogers explained its migratory character, its republican organization, and its steady growth. The eight years of annual meetings in the major Atlantic cities and scientific centers had familiarized geologists and naturalists with one another as well as expanded their geological experience. Simultaneously the meetings had aroused local interest in science by their sponsorship of known scientists to lecture on special topics. The open-membership policy, as operative in the later years, attracted men who could and did contribute as fieldworkers in geological and meteorological investigations. Authority had gravitated naturally to men of established reputation and enthusiasm for organization, but, again, the opportunity to attend and participate was open and committee reports were solicited from every interested scientist. Although the growth suggested by Rogers had leveled off by 1843, the AAGN was nonetheless a reputable and established vehicle for presenting new ideas and challenging questionable data or hypotheses. The capstone of his argument was that more than any other learned society, the AAGN "has had its hand on the throbbing pulse of the young and ardent scientific intellect of the land, and has thus experimentally been able to estimate the energy of the vital power which is herein." Rogers knew firsthand the intense enthusiasm which brought young men of the geological surveys often hundreds of miles to attend annual meetings; it was unmatched by even the local activity in other major societies. Geology, with its chemical and natural history adjuncts, was the dominant science of the decade, and the association had served as a forum for continuing discussion and a clearinghouse for new ideas. Its expansion offered a similar opportunity for other areas of scientific interest.

In writing the constitution, Rogers kept in mind the original characteristics of the founding organization while also using the BAAS constitution as a model. The British Association had been formulated in response to the needs of provincial societies and because many British scientists felt that the alternative national organization, the Royal Society, was inadequate.[13] The American Association was less interested in linking societies than in creating a meeting place for individual scientists. Unlike London, the young American capital in Washington did not dominate science and offered little leadership at all. It was national pride and a desire to improve communication that propelled the

---

13. A. D. Orange, "The British Association for the Advancement of Science: The Provincial Background," *Science Studies*, I (1917), 315–329, stresses the young organizations's efforts to reach and to coordinate the activities of the middle-class provincial societies and to awaken curiosity about philosophical objects.

Americans toward organization. Reinforced by English attitudes toward them which had "softened," lost much of their "former asperity," and become even "respectful,"[14] the Americans wanted a stage on which to present their research. In the format of their institutions and in the general organizational structure the two associations were similar, although the American's structure was less detailed than the amended British Association constitution which had been in force for fifteen years.[15]

Some differences, however, are suggestive. There was no disclaimer arguing noninterference with existing societies, for the AAAS believed itself to be sufficiently different from other established organizations in America. It stated that membership was offered not only to members of scientific societies but also to "Collegiate Professors of Natural History, Physics, Chemistry, Mathematics, and Political Economy, and of the Theoretical and Applied Sciences generally; also Civil Engineers and Architects. . . ." The association was to be headed by three major officers, elected annually, a president, a secretary, and a treasurer. Composed of the three current and three retiring officers, the chairmen of sections, and six members elected at large, a standing committee made all major decisions and arrangements at and between the annual meetings. The standing committee roughly corresponded to the council of the BAAS. Sections, designed to divide the active membership into special interest groups, were not specifically named and were to be somewhat more autonomous in electing officers and in making their own arrangements than their English counterparts. By giving the sections distinct power over their own activities, Rogers implied that they were federal branches of the organization whose interests might be uniquely expressed in their meetings and represented on the standing committee. There was no suggestion that the AAAS was a federation of local societies. Probably in response to the inaccurate reporting and popularized accounts of AAGN affairs in the daily press,[16] the new constitution

14. Joseph Reed Ingersoll to A. D. Bache, Philadelphia, 21 Nov. 1853, Rhees MSS, HHL.

15. The following comparison is based on constitutions found in AAGN, *Reports* (1840–42), pp. 77–78; AAAS, *Proceedings*, I (1848), 8–12; and BAAS, *Report*, I (1831), ix–x, and XVIII (1848), v–vii.

16. Silliman Jr. to Redfield, New Haven, 18 Aug. 1846, and Redfield to Emmons, New York, 6 Nov. 1845, and to the editors of the *AJS*, New York, 1 Dec. 1845, all in Redfield MSS, BLY. H. D. Rogers to B. Silliman, Jr., Philadelphia, 14 May 1845, AAGN MSS, ANSP, suggested that the "burlesque imparted to our proceedings overlook the special value of the communications made."

specifically recommended that the proceedings be recorded by professional stenographers.

Little in the statement of purpose was new, and two of the three goals were similar to the stated objects of both the AAGN and the BAAS. First, "by migratory and periodical meetings" the association intended to "promote intercourse between those who are cultivating science in different parts of the country." This national union offered men of science an opportunity to become acquainted, exchange information, and debate questions of interpretation with peers. In addition, the association hoped "to give a stronger and more general impulse, and a more systematic direction to scientific research in our country." Linked with these and somewhat arbitrarily joined together were the goals of stimulating a broad interest in science while at the same time providing guidance and direction for the impulse. Rogers could not resist registering the major complaint of men of science against their lack of support; he added that the association would seek to procure for the labors of scientific men "increased facilities," which he suggested would bring "wider usefulness" as well. In sum, the statement of purpose conveyed the pervasive belief that through association men of science would enhance and systematize scientific research in the United States.

While laboring with the constitution, Henry Rogers commented in a letter to his brother William that his document would be, among other things, democratic.[17] He seems to have meant the term loosely, implying an opportunity for participation in decision-making; the constitution embodied this hope in two deviations from that of the BAAS. Administratively, the standing committee, the section officers, and the committee members were all elected rather than appointed. Moreover, the general membership was, in effect, open to anyone who indicated a strong interest in science through membership in another scientific society or a science-related occupation, or involvement with another established scientist willing to recommend him; and any member could vote. By 1848 the BAAS had established several distinct categories of membership and limited voting privileges to a general committee whose membership included only officers, publishing members, officers of other scientific societies, specifically designated delegates from such

17. Henry Rogers to William Rogers, Boston, 16 May 1848, Rogers, ed., *Life and Letters*, I, 287–288. His assessment was that the constitution was "democratic, federal, flexible and expansive, progressive, with all the true conservatism these features imply."

societies, and foreign visitors specially recommended each year.[18] This group, in turn, appointed the various special committees. Thus in the proposed AAAS attending members did have an equal voice in decision-making which was not available in the BAAS. Membership was individual rather than contingent on any other affiliation.

While the sectional meetings were intended to maintain the caliber and style of papers presented before the AAGN, the general meetings were to serve as an arena for discussion of general matters of scientific procedure and for the popularization of science within the larger community. The AAAS tried to handle simultaneously two issues—that of elitism and that of focusing interests within a diffuse organization. First, its membership was more open than had been the tradition of other prestigious American societies, which seemed almost a move away from self-selecting professionalism. In addition, the emphasis on sectional meetings was perhaps the necessary transitional step toward specialization and the development of a professional spirit within discrete groups. Some national group was essential as a scientific forum to engender not only a disciplined loyalty but also a commitment to the nascent scientific community and its interdisciplinary needs.

The Rogers family was central in establishing the new association. While Henry Rogers worked to prepare a constitution, his brother James made local arrangements for the association in Philadelphia.[19] Although the circulars went out over the signature of Samuel George Morton, as chairman of the local committee, and Walter R. Johnson, as secretary of the AAAS, they were written by Henry Rogers and distributed through William Rogers.[20] The latter sent bundles of circulars to scientific friends in the East, Midwest, and South to encourage the widest possible participation. In New England Rogers placed Benjamin Peirce in charge of the distribution and similarly requested the help of William Redfield in New York and Joseph Henry in Washington.[21]

18. BAAS, *Report*, XVIII (1848), vi. There were four basic classes of membership, each of whose privileges and limitations were carefully delineated: life members, annual members, associates for the year, and corresponding members.

19. James Rogers to William Rogers, Philadelphia, 9 Sept. 1848, Rogers MSS, MITA. The local committee had no precedent on which to make plans, and responses to the circular of 10 May were scanty, although the committee was persuaded that the "wide scope of the new organization would surely result in large attendance."

20. Henry Rogers to William Rogers, Boston, 16, 30 May 1848, Rogers, ed., *Life and Letters*, I, 288.

21. William Rogers to William Redfield, Boston, 24 Aug. 1848, Redfield MSS, BLY; William Rogers to Joseph Henry, Boston, 24 Aug. 1848, Henry MSS, SIA.

The latter was asked to send circulars to interested Smithsonian correspondents in the Ohio and "Northwest" regions. Henry reassigned the
task to Charles Page, who left it uncompleted when Walter R. Johnson
protested that such wholesale distribution was contrary to the rules
of the association.[22] Rogers's open policy prevailed, however, for after
the 1848 meeting a list of 462 men was sent to each proposed member,
which indicated that they need only accept the rules of the association to
retain membership status.[23]

Because Henry Rogers left for Scotland shortly before the meeting,
his brother William introduced the proposed constitution at the Philadelphia meeting in 1848 and explained its provisions. After limited discussion it was unanimously accepted.[24] Attending scientists were familiar
with the working pattern of the AAGN and the BAAS and readily
concurred in the general purpose as well as the administrative format
of the proposed association. The transformation in 1848 was therefore
relatively easy, not only because of the available models but also because
its new members confidently believed that the association could act as
an annual convention with little interim planning or highly centralized
control. When this assumption proved untrue, the association experienced organizational trauma over the method of providing responsible
and continuing leadership. Unity and harmony were hallmarks of the
first years of the association because the members worked to make it
so, intent on creating a friendly, working relationship among American
men of science.

When news of the proposed change to an inclusive American association reached Washington, certain National Institute leaders once again
made overtures concerning a joint organization.[25] In 1847 the institute,

22. Henry to Page, Princeton, 11 Sept. 1848, and Page to Henry, Washington,
14 Sept. 1848, both in Henry MSS, SIA. Page apparently wanted to interpret
strictly the membership clause granting membership only to persons with an
expressed interest in science.
23. AAAS, *Proceedings*, I (1848), 144–156, lists only 461, which is the number
usually cited by subsequent accounts of the meeting. Not all of these participated
in the formation, as is evident in the puzzled inquiry to Edward C. Herrick, secretary of the AAAS, by Samuel J. Parker: "What is the value of this new society
into which the American Association of Geologists and Naturalists is merged? I
read a printed circular with my name attached; but not knowing whether it was
a mere sham...or a valuable Society, I have as yet made no reply." Ithaca, N.Y.,
11 Jan. 1849, Herrick MSS, BLY.
24. AAAS, *Proceedings*, I (1848), 8.
25. This is evident from the manuscript minutes of the National Institute and
from Charles F. Stansbury, *Report of the Recording Secretary of the National
Institute for the Year 1850* (Washington, D.C., 1850).

under somewhat changed leadership, had attempted a revival. A committee to "ameliorate the conditions of the National Institute," ostensibly aided by Joseph Henry, proposed affiliation with the Smithsonian Institution.[26] Charles Wilkes, finding himself without legislative backing, became an active member and reversed his earlier position in order to support the institute's hopes for creating a national museum on his exploring expedition collections.[27] Wilkes suggested coordinating the American Association and the National Institute to Congressman Joseph Ingersoll in 1848. Without any obvious encouragement, he pointed out the prestigious leadership of the new organization and suggested, optimistically, that it might be persuaded to assume the name of National Institute and to take charge of the present institute and the government collections. He concluded, "I think I may answer for the Association of Geologists, that they would readily accede to the wishes of Congress on this subject. . . ."[28] Although he took his scheme for reorganization of the institute to the Philadelphia AAAS meeting and it was apparently discussed in private, once again the men of science were unimpressed by the institute, and Wilkes's suggestion was not even brought up for formal debate.[29] The previously defeated institute offered nothing to the association, for its purposes remained local and quasipolitical.

Unencumbered by any specific obligation except the self-imposed determination to publish regular proceedings, the young organization concentrated on the creation of a working administration for its larger

26. Henry became actively involved in the institute during this period in an apparent effort to control its destiny and to prevent the Smithsonian Institution from assuming financial responsibility for the collections. By 1848 this was evident even to the institute, for J. J. Abert wrote angrily to Francis Markoe [Washington], 2 Sept. 1848, that a letter by Henry showed "the error of his views, if not his hostility to our Institute." A copy of the letter is in Henry MSS, SIA, and there is a series of letters through 1847 and 1848 indicating Henry's participation.

27. Edward L. Towle, "Science, Commerce and the Navy on the Seafaring Frontier (1842–1861)—The Role of Lieutenant M. F. Maury and the U.S. Navy Hydrographic Office in Naval Exploration, Commercial Expansion and Oceanography before the Civil War" (Ph.D. dissertation, University of Rochester, 1966), pp. 97–98, records Wilkes's gradual fall from legislative favor.

28. Wilkes to Ingersoll, Washington, 2 June 1848, Rhees MSS, HHL. For a description of this confusingly arranged collection, see Nathan Reingold, "The Anatomy of a Collection: The Rhees Papers," *American Archivist* (Apr. 1964), pp. 251–259. Wilkes was also drawing up a plan for reorganizing the institute which would make it the recipient of public lands. See William Preston to Joseph Henry, Columbia, S.C., 30 [Apr. 1848], Henry MSS, SIA.

29. Louis Agassiz to A. D. Bache, Cambridge, 8 Oct. 1848, Rhees MSS, HHL. Agassiz simply asked Bache's opinion of Wilkes's plan "of combining our association with the National Institute, in order to obtain the benefit of the support from the government."

and more complex structure. Simply molding a cohesive spirit was difficult in a mixed, diffuse group which had no common center, no ongoing principle with specific design, and no basis for consensus. The difficulty of retaining a sustained commitment by the membership until such cohesion took place was enlarged by the peripatetic and discontinuous nature of the annual meetings. Most immediately, however, finances and publication questions raised difficulties which forced creation of a more permanent structure of secondary officers. Wisely the AAGN had avoided financial difficulty by collecting dues at its meetings sufficient to cover expenses and by resisting the temptation to publish proceedings. The AAAS, however, felt it needed a regular publication and that its larger membership would guarantee the funding. From the outset publication was a difficult administrative problem. The geologists had wanted to report their proceedings in order to promote better exchange of ideas among practitioners; the new officers sought a publication that would represent American science.[30] One result of this intention was to formulate and impose more uniform standards for reports.

The committee appointed by the association to oversee publication in 1848 made Robert W. Gibbes chairman of the geographically dispersed group. The South Carolina physician proved a persistent and faithful worker throughout the winter, but his location delayed mail delivery and many members simply did not respond to his request that they send copies of papers presented at the meeting. By December he had received only three papers and no minute books from the sectional secretaries, Benjamin Silliman, Jr., and Walter R. Johnson.[31] By March 1849 he had not, despite numerous letters applying direct and indirect pressure on delinquent members, received papers from such crucial scientists as Agassiz, William Rogers, or Thomas S. Hunt. William Redfield, as outgoing president, assisted the frustrated Gibbes, who finally in May sent off a draft copy to Alfred E. Elwyn for publication.[32] When

---

30. When publication appeared questionable, the Cambridge publication committee, for example, requested additional funds in a circular which declared that the publication "will confer distinction upon the Association, not only here but abroad. It would, on the other hand, be seriously injurious to the prosperity, and in the highest degree derogatory to the honor of the Association, (the faith of which has been formally pledged) if the promised volume of the proceedings of the late meeting should not appear." 5 Oct. 1849, Henry MSS, SIA.

31. Robert Gibbes to William Redfield, Columbia, S.C., 10 Dec. 1848, Redfield MSS, BLY.

32. Robert Gibbes to William Redfield, Columbia, S.C., 21 Mar., 21 May, 5 June 1849, all in Redfield MSS, BLY; Gibbes to Joseph Henry, Columbia, S.C., 3 Feb. 1849, Henry MSS, SIA; Gibbes to James Hall, Columbia, S.C., 11 Apr. 1849,

Elwyn reported back that the treasury had only $200, representing the dues of less than half of the listed numbers, and that 500 copies would cost $325, Gibbes personally advanced $100 toward the publication.[33] That the proceedings appeared at all in semicomplete form was credit to the exhaustive efforts of Gibbes. Fearful of a current cholera threat, Gibbes did not come north to the Cambridge meeting in 1849, but he urged Redfield to point out to members the difficulties incurred and to recommend that a stenographic reporter be assigned to each section to take verbatim notes. Having the proceedings published and circulated quickly was important for sustaining interest and encouraging sales, and newspaper reports were "imperfect and often erroneous."[34] Moreover, reporters relished the sensational and "failed to distinguish between a discussion and a disputation."[35] When the *New York Herald* offered to publish their account in pamphlet form, the AAAS leaders were undoubtedly appalled.[36]

In response to the difficulties, a six-member committee from Cambridge was appointed the following year to oversee the publication, including men from both the natural history and physical science sections.[37] Publication was completed in Cambridge under their direct supervision, with Eben Horsford, newly appointed Rumford Professor at Harvard, assuming major responsibility for editing. To facilitate his task, he sent pre-printed circulars to all who had presented papers and also wrote personal letters to key individuals urging them to stir their friends into submitting manuscripts early.[38] The format of the report, giving not only the papers but also extended criticisms and comments,

Hall MSS, NYSL; Gibbes to William B. Rogers, Columbia, S.C., 3 Nov. 1848, Rogers MSS, MITA.

33. Robert Gibbes to William Redfield, Columbia, S.C., 21 May, 5 June 1849, Redfield MSS, BLY.

34. Robert Gibbes to William Redfield, Columbia, S.C., 10 July 1849, Redfield MSS, BLY; Gibbes to Joseph Henry, Columbia, S.C., 20 July 1849, Henry MSS, SIA. Although the *Daily Evening Traveller* (Boston), 25 Aug. 1850, felt the threat of cholera was "subsiding," the weekly mortality tabulation indicated that 75 of 215 adult deaths in Boston were attributable to the dreaded disease.

35. William Redfield to Charles H. Davis, New York, 30 Oct. 1848, Redfield MSS, BLY.

36. [J. Kempston] to Joseph Henry, New York, [2 Sept. 1850], Henry MSS, SIA.

37. AAAS, *Proceedings*, II (1849), v. All were Cambridge residents: Jeffries Wyman, Louis Agassiz, Benjamin Peirce, Charles H. Davis, Asa Gray, and Eben Horsford.

38. Horsford to William Redfield, Cambridge, Mass., 12 Oct. 1849, Redfield MSS, BLY. Copies of the forms used are chronologically filed for 22 Aug. 1849, Bache MSS, SIA, and Oct. 1849, Henry MSS, SIA.

suggests that reporters were also employed. The proceedings were again late, but this time the delay was due to a printers' strike, and once again financing was a problem.[39]

The Charleston City Council bestowed a crucial blessing when it offered to pay $500 toward the publication of the proceedings of the semiannual meeting held there in March 1850.[40] Lewis R. Gibbes, a Charleston naturalist who had served as secretary of the meeting, took charge of the publication.[41] Both attendance and the number of papers presented at the southern session were smaller than at the first two meetings, so that all papers were presented to a single general assembly.[42] Alexander Dallas Bache, who had presided in the absence of Joseph Henry, persistently prodded Gibbes to insure that the volume would be available for distribution five months later at the annual meeting in August, which it was.[43] The Charleston local committee apparently retained the dues which it collected from local members to offset costs of the meeting; lack of understanding over the matter resulted in some friction, particularly when the new secretary called for the money, stating that it was needed to fund the New Haven proceedings.[44]

Confusion, delays, and misunderstanding plagued secretaries as they attempted to send out initial circulars, arrange local organizational matters, and supervise publication of the proceedings. Lacking such administrative aids as an address file, ongoing records, and familiarity with publication procedures, the officers were dissatisfied, and com-

---

39. Horsford to William Redfield, Cambridge, 14 Nov. 1849, Redfield MSS, BLY; Elwyn to John P. Norton, Philadelphia, 29 Oct. 1849, Norton MSS, YUA. Elwyn grumbled, "You express a hope that by the promptness and liberality of the members the Proceedings will soon be published, & if their publication depends on those two circumstances there is very little hope, of their ever being given to the world. If men of science were actuated by the same spirit which you seem to be, there would be no delay, but poverty, indifference, business, and a variety of other et ceteras, seem to express the larger picture, as not one fourth have ever paid anything."

40. The early spring meeting was officially designated the "third meeting," since Horsford observed, "To call it a semi-annual meeting would intimate that we had regular half-yearly meetings—which is not established. To call it an Extra meeting would apparently take from its rank...." Horsford to Lewis R. Gibbes, Cambridge, 13 May 1850, and A. Bache to Lewis R. Gibbes, Washington, 30 May 1850, both in Gibbes MSS, LC.

41. AAAS, *Proceedings*, III (1850), iv–vi.

42. A. Bache to Lewis Gibbes, Washington, 30 May 1850, Gibbes MSS, LC.

43. A. Bache to Lewis Gibbes, Washington, 1 June, 8 Aug. 1850, Gibbes MSS, LC; Bache to E. C. Herrick, Baltimore, 2 July 1850, Herrick MSS, BLY.

44. Eben Horsford to S. F. Baird, Cambridge, 18 June 1851, Baird Personal MSS, SIA; A. E. Elwyn to Lewis Gibbes, Philadelphia, 21 June 1852, and Baird to Gibbes, Smithsonian Institution, 3 Apr. 1851, 10 Dec. 1850, all in Gibbes MSS, LC.

plaints by the general membership implied that organization was inefficient. Undoubtedly it was the direct result of such frustrations that the positions of permanent secretary and permanent treasurer were established at the fourth meeting. In March 1851 at Charleston the standing committee discussed how to develop a "more permanent and *transportable* nucleus of organization."[45] A scheme to establish a complex, geographically more representative core group which would be consistently responsible seemed unlikely, however.[46] Permanent administrative officers appeared to be the solution, and in August the posts of permanent secretary and treasurer were created by resolution.[47]

In 1849 Alfred E. Elwyn, a University of Pennsylvania M.D. who never practiced medicine but was active in philanthropy in Philadelphia, accepted the position of treasurer. He was reelected at New Haven in 1850, and his nomination by the standing committee was accompanied by the observation that the position required some permanence; Elwyn proved an effective officer and he remained treasurer until 1870.[48] Interested in science only as an observer, he was content to listen and avoided embroilment in controversial matters of science or administration. Because he disliked such unpleasant responsibilities as sending out notices to delinquent members, the task fell to the secretary. Elwyn was meticulous in detail, and his orderly financial records were always approved by auditing committees.[49]

Similarly the association selected Spencer F. Baird, an active zoologist, as first permanent secretary because of his evident administrative ability.

45. A. Bache to E. C. Herrick, Baltimore, 2 July 1850, Herrick MSS, BLY.

46. *Ibid.* Bache suggested that the nucleus consist of men whose terms of office would expire in rotation; the major difficulty was, he felt, "to find those who would pledge themselves to attendance, during the term of service."

47. In the Bache portion of the Rhees MSS, HHL, is a "Memorandum" written in a neat, unfamiliar handwriting, possibly from Henry and copied by a Smithsonian assistant, calling for an early meeting of the standing committee; an annotation by Bache marks it as "rec'd 11 August 1850." The "Memorandum" simply poses questions about organization and wonders about "the means to be adopted to secure a large sum of money for the use of the publishing com^e? Shall we have a permanent Secretary & how shall he be paid? Without such an officer the affairs of the association undergo an annual death & resuscitation." Also see Henry to Bache, Smithsonian Institution, 14 Aug. 1850, copy in Henry MSS, SIA, in which Henry suggests that such a change would create a "final organization" for the association.

48. Elwyn apparently did not attend any meeting after the Civil War, that is, from 1866 to 1870.

49. Elwyn papers have not been located. At the Historical Society of Pennsylvania an intriguing yellowed file card suggested a "bundle of treasury accounts" under his name, but a subsequent search failed to uncover the material.

Located in Washington as assistant secretary of the Smithsonian Institution, he was familiar with publication procedures and maintained a regular correspondence with other leading naturalists. Baird was young, ambitious, and eager to supplement his salary from the Smithsonian. As token recognition of the large responsibility of the secretarial office, the association resolved to pay $300 annually to that officer.[50] Baird spoke sarcastically of whether the salary would be paid wtih the association in financial difficulty. Yet he seems to have collected regularly.[51] The task was onerous: the secretary's job only began at the meeting, where he was expected to maintain records of those present and of papers given, and was only half-completed when the volumes representing the annual meeting were finally published. In addition, he was responsible for the distribution of proceedings through booksellers or the mails, with the list of recipients dependent on the ever-changing record of paid membership with the treasurer in Philadelphia. He also inherited from Elwyn the task of warning delinquent members and assumed responsibility for informing new members of their election.[52] Although Baird extended Eben Horsford's practice of printed notices for certain routine matters, his correspondence files clearly indicate that he was compelled to write numerous individual letters to members to answer particular questions or to urge action on specific projects.[53]

50. AAAS, *Proceedings*, IV (1850), 390. In 1856 the salary was raised to $500.
51. Baird to O. C. Marsh, Washington, 9 Feb. 1851, Baird Personal Papers, SIA. Baird wrote to Marsh that "the Association perpetuated an excellent joke in voting me $300.00 per annum for my services. The reality of the thing is that there are not funds enough to pay for half of the volume, much less the 300 dollars additional." Also see William H. Dall, *Spencer Fullerton Baird, a Biography, Including Selections from His Correspondence with Audubon, Agassiz, Dana, and Others* (Philadelphia, 1915), p. 217. Financial records of Baird, filed under AAAS Incoming, suggest that Baird was paid in full, plus his expenses for travel and a double room at meetings. Because there is essentially no archive for the AAAS during these formative years and neither Alfred Elwyn nor Baird's successor left any records thus far discovered, the Baird MSS at the Smithsonian Institution are invaluable for the three years of Baird's tenure. The Baird MSS are divided into Baird's official and personal correspondence (hereafter designated OFF and PP respectively), and letters relative to the AAAS are found scattered through both without any apparent system. In addition, the Smithsonian Institution Library has Baird's copies of the printed proceedings, which often have penciled corrections and marginal notes.
52. Elwyn could not tolerate unpleasantness, and he replied to an evidently hostile letter of Lewis R. Gibbes that Prof. Baird, permanent secretary, made "an *arbitrary* [italics added] rule at New Haven that he would not send copies of the proceedings except to those who had paid three dollars in that year." Elwyn to Gibbes, Philadelphia, 21 June 1852, Gibbes MSS, LC.
53. Letterpress volumes suggest that when working with the publications he frequently wrote several letters a day for the association.

Although Edward C. Herrick was secretary for 1850, he did not have time to supervise the publication of proceedings, and so Baird assumed that responsibility immediately after his election in 1850. Herrick completed the task of notifying new members of their election and forwarded all other matters to Baird.[54] Throughout the winter Baird worked to organize the proceedings, feeling hampered by Herrick's slowness in forwarding information as well as by the evident indifference of members to association business. By diligent effort, however, he was able to see the New Haven volume through the press in time for distribution at the Cincinnati meeting. In a letter of relief mixed with frustration, he admitted to Lewis R. Gibbes, who had experienced the difficulties firsthand, "Of all unpleasant jobs, this is the worst I ever knew."[55] Producing volumes from manuscripts in various handwritings and a frequently incomplete record of activities was difficult, and the printer echoed Baird's sentiment: "Of all work I have ever had in hand this volume has been the most troublesome, it is really an uphill task to wade through it."[56] Baird was exceptionally diligent in his efforts, and numerous letters to his printer indicated his concern for detail. When he received the first press copy, for example, he corrected it and concluded that although this would involve extra expense, it should be reprinted.[57]

Thus it was undoubtedly a relief when Cincinnati offered to pay for the proceedings and stipulated that the work be contracted to a local printer. James W. Ward supervised the publication, although Baird arranged for the distribution of copies to members.[58] Baird found, however, that not only did local supervision appear to slow down the publication process but the Cincinnati publication committee's failure to return proofs to him for correction resulted in several errors.[59] So at the

54. E. C. Herrick to S. F. Baird, New Haven, 31 Aug., 7 Oct. 1850, Baird PP, SIA.
55. Baird to Lewis Gibbes, Washington, 26 Apr. 1851, Gibbes MSS, LC.
56. E. O. Jenkins to S. F. Baird, New York, 18 Apr. 1851, Baird PP, SIA.
57. Baird to E. O. Jenkins, Washington, 9 Apr. 1851 (?); also see 5 and 11 Feb., 6, 17, and 19 Mar., all in Baird OFF, SIA.
58. S. F. Baird to James W. Ward, Smithsonian Institution, 30 Apr. 1852, Baird OFF, SIA.
59. Baird to W. H. Emory, Smithsonian Institution, 14 May, 11 June 1852, Baird PP, SIA. Baird replied to Emory's complaint that his article as published in the *Proceedings* contained errors and that the men in "Cincinnati had paid the most profound neglect to my most urgent solicitations" for information and proof sheets. Even more serious was the decision not to accept a locally supervised publication of the Cleveland *Proceedings*. A discussion is found in John D. Holmfeld, "From Amateurs to Professionals in American Science: The Controversy over the Proceedings of an 1853 Scientific Meeting," American Philosophical Society,

annual Albany meeting Baird once again assumed the responsibility for publication. Because the city governments of Charleston and Cincinnati as well as the members had paid for the proceedings, the financial strain eased after 1851, and for the rest of the decade the budget was in the black.[60]

In 1853 the AAAS, encouraged by an unrequested report from the standing committee, voted to limit financial responsibilities. Benjamin A. Gould, Jr., apparently initiated a five-page statement which argued that only enough money should accumulate to pay annual expenses and publishing costs. Unhappy that the funds from Charleston and Cincinnati had not benefited members, the committee observed that the cities had not intended "a distinction between the Association as a corporate body and its members as individuals."[61] In an *obiter dictum* to the resolution against the accumulation of a permanent fund, it added:

> In the opinion of the Committee the American Association was founded for the purpose of bringing together scientific men from different parts of the country, for the discussion upon scientific subjects and further cultivation of a friendly and social feeling between those devoting their lives to the pursuit of science and those who feel an earnest interest in its progress. It is believed that the organization of the Association had seriously these objects in view, that it was never intended that the Association should accumulate money.... There are no objects to be attained by the Association according to its present organization which required a fund. The Association has not adopted the practice prevalent in the British of appropriating sums of money to individuals or to Committees to aid them in the prosecution of particular investigations. The committee is not aware that any proposal to do so has ever even been brought forward.

Arguing that most members had difficulty simply finding funds to attend AAAS meetings, the committee recommended that copies of proceedings be distributed at cost or free if a host city paid for publication and that free copies were to be sent to learned societies and public libraries abroad. The association then unanimously resolved not to create a permanent fund.[62]

Proceedings, CXIV (16 Feb. 1970), 22–36. Unfortunately Holmfeld did not have access to sources which suggest Baird's anxiety over the volume and his initiative.

60. Zadock Thompson to Baird, Burlington, Vt., 19 Sept. 1851, Baird OFF, SIA.

61. A transcript of the unpublished resolution is in the AAAS Library, Washington, D.C. It was not printed in the *Proceedings* for 1853 because the secretary indicated in a note that "after all possible effort [he] could not obtain the resolutions upon which this report was grounded." Subsequent quotations are from the document.

62. *New York Times*, 5 Aug. 1853.

The tendency toward a more efficient, centralized structure suggested by the election of permanent officers was encouraged by certain leaders who recognized the need for a more closely knit organization. No man more consciously scrutinized the developing association than Alexander D. Bache. Bache himself embodied new professional attitudes. After graduation from West Point he had returned to his native Philadelphia, rapidly imprinting his views of science on the American Philosophical Society, the Franklin Institute, and Girard College.[63] Although anxious to establish close working associations with fellow scientists, as in his "club" of the mid-1830s, Bache was on the committee of the American Philosophical Society which rejected Warren's proposal in 1838 and was skeptical about broad, inclusive organizations. Yet he recognized the importance of scientific institutions and therefore toyed with the idea of reshaping the National Institute in 1844. He subsequently took the initiative for the National Academy of Sciences in 1863.[64] Once Bache became a participating member of the AAAS, he quickly moved to a central position in that organization. In fact, he presided over three of the first six meetings, as president-elect in Henry's absence in Charleston and then at the annual New Haven session in 1850, as well as the intermediate Cincinnati meeting in the spring of 1851. The articulate Bache used his position as head of the association and of the standing committee to direct the attention of the members toward his concepts of the organization's goals.

Impatient with evident inefficiency and with the undisciplined nature of the association, Bache concentrated first on centralizing and standardizing its procedures. In part, this step was necessary simply to enforce the rules already in existence regarding the nomination of new members, the handling of sections, and the election of officers. That the actual changes proposed through the standing committee, with the clear recommendation of the permanent officers, were more far reaching than arguments for them might suggest was only gradually evident. The spirit of professionalism of the leaders did not become part of the AAAS until they began the imposition of new standards. But the series of resolutions passed in 1850 and 1851 served initially to tighten the loosely constructed constitution and to rationalize organizational procedures.

63. Nathan Reingold is planning a biography of Bache; see his article in the *Dictionary of Scientific Biography*, ed. Charles C. Gillispie, I (New York, 1970), 363–365, and "Alexander Dallas Bache: Science and Technology in the American Idiom," *Technology and Culture*, XI (Apr. 1970), 163–177.
64. Dupree, *Science in the Federal Government*, pp. 135–146.

In 1848 Joseph Henry noted it was "the active contributors on whom the reputation of the country really depends"[65] and that the papers presented at the AAAS were, in this broad sense, representing American science. Initially anyone could present a paper before the association, the only requirement being that its title be listed so that a program for the week's meeting might be arranged. Certain members were increasingly unhappy with the association's open policy and determined to eliminate "all matters not strictly new, or belonging to the progress of science," since the primary purpose of the organization was to advance science, not to diffuse it.[66] Dissatisfied with the casual presentations of some authors the standing committee assumed more direct supervision in the session after Cincinnati, passing a resolution "that no paper be read before the future meetings of this Association unless an abstract of it has been previously presented to the secretary."[67] Ostensibly the resolution simply assured that a topic might be placed before the appropriate section meeting, but such submission also allowed re-editing and the subtle exclusion of papers whose value or validity was questioned. Moreover, the secretary was authorized to question errors or repetition and to resubmit papers to the author or the standing committee before allowing publication.[68] A resolution at the Albany meeting was even more specific: members were to give complete titles, an estimation of the time required to read their papers, and abstracts of the contents. To be accepted, papers had to be so registered before the first day of regular section meetings.[69] One member felt that only supervision could improve meetings "too slow in beginning, & in getting to work."[70]

In a move for efficiency the association passed two resolutions relative to sectional organization. Unlike the BAAS, whose sectional chairmen held office during the entire annual meeting, the first sectional chairmen of the AAAS were typically, although not always, elected for only a daily session. This practice may have evolved from the suggestion of Agassiz that no leader would want to be committed to one section for an entire annual meeting. In any case, the changing of officers allowed more men to hold office. Because the sections were lax in planning and

65. [Henry to Francis Markoe], Washington, 16 Aug. 1848, copy in Henry MSS, SIA.
66. AAAS, *Proceedings*, V (1851), 249.
67. *Ibid.*, II (1849), 272, and V (1851), 249.
68. *Ibid.*, IV (1850), 391.
69. *Ibid.*, VI (1851), 402.
70. Chester Dewey to John Torrey, Rochester, 9 June 1854, Torrey MSS, NYBG.

reporting their proceedings, however, new resolutions required that the chairman of the standing committee appoint a chairman each day and a secretary to report all proceedings and discussions.[71]

In such indirect ways the standing committee accumulated power throughout the formative years of the association in its attempt to create a continuous and responsible governing agency. The committee provided for continuity from meeting to meeting because its membership consisted of officers from both the preceding and the forthcoming years. From the constitution it derived three broad responsibilities: to manage the general business of the association during the meeting, to conduct any necessary business between meetings, and to nominate officers for the coming year and persons for admission to membership. Initially it had simply arranged the program and organized the sections "to vary in conformity to the wishes and the scientific business of the Association."[72] When the position of permanent secretary was created, the standing committee extended its reach by the authorization to establish the duties of the post. In 1851 the standing committee, despite its implied power, explicitly sought "the full power to complete and finish any outstanding business of the Association, in their name."[73] By nominating officers as well as the at-large members of the standing committee, it became self-perpetuating. When these powers, extended by resolution and by precedents established by 1851, appeared to be exercised arbitrarily, a dissident voice rose to challenge its authority. At first the responses were individual, but gradually a self-conscious opposition challenged the apparent power bloc controlling the association.

Before 1851, however, the association members worked mutually to formulate the somewhat loosely constructed group into an efficient and functioning organization. Concentration on growth and general reputation worked to the positive good. David Wells stated in the first volume of his *Annual of Scientific Discovery* that "the Association had now become truly national in its character, and had taken deep hold of the feelings of men of science and investigators in all departments of knowledge."[74] The changes in structure were necessary and logical, for this considerably larger organization could not operate with the limited forethought given to AAGN meetings. The operative patterns established by 1851 carried through the remainder of the century even as the

71. AAAS, *Proceedings*, VI (1851), 405.
72. *Ibid.*, I (1848), 9.
73. *Ibid.*, IV (1850), 341.
74. (Boston, 1850), p. 361.

formal constitution and constituency of the association changed. With the more centralized, bureaucratic structure came a concentration of power, and the perhaps inevitable result was a seedbed for dissatisfactions when the power seemed arbitrarily exercised. If the early years concentrated on the procedural matters of organization, the later years found the more significant matters of function far less easy to resolve.

Once the more formal matters of internal organization were settled, the AAAS became more introspective about its role in relationship to the general society, to those whose interests in science were real but avocational or intermittent, and to researchers of science. In each case the question related back to the still uncertain definition of a "true man of science" or professional. Often the discussion centered on the intermediate group whose status was least clear. Experience taught the AAAS that addressing and educating the public was a responsibility which might be limited but never wholly denied.

V

The Congress of Savants

Edward Everett's comments at a dinner meeting of the AAAS at Cambridge in 1849 reflected an interested layman's view of the new group. As observer of the local AAAS, he concluded,

> Even if it were true that Scientific Associations had no tendency to promote discovery, in either sense of the word, it might still be a matter of great importance, that they furnish occasions and facilities for illustrating and diffusing more widely the great laws of nature.... This work is to lift the mass of the scientific community, and no one can reasonably deny that an association like ours is an approved and effective part of that system of concerted action, by which men advantageously unite themselves to accomplish desirable ends.[1]

Like the newspaper accounts which frequently characterized the association meetings as a Congress of Savants (or Savans), Everett recognized that a major purpose of the meeting was to promote science through the interaction of men of science. He confidently assumed that open meetings diffused knowledge and concerted action prompted science. The AAAS was committed to the diffusion as well as to the advancement of knowledge, although the two goals were not necessarily complementary. In the minds of most of the scientific members, especially the geologists and a new contingent of physicists, the long-standing need for promoting science through research and discussion had top priority. Simultaneously they recognized that bringing science to the level of public understanding was intimately related to financial support and

1. Edward Everett, "Remarks of Mr. Edward Everett, at the Dinner Table at Cambridge, 21st of August 1849, being the last day of the Session of 'American Association for the Advancement of Science' for that year," MSS copy in Everett MSS, MHS.

respect for the profession. From the beginning the AAAS's open membership and annual conventions ensured its visibility as the national scientific society. As such, it might enhance local societies, but it also might limit their effectiveness.

The dilemma of balancing professional goals with public needs and expectations is a persistent one. In the late 1840s there were vague public anticipations concerning the "useful arts," and men of science had not yet agreed on their response to popular culture. On questions regarding discrete scientific phenomena such as in meteorology, men of science could clearly synthesize and relate new findings for the general public. While the Patent Office was competent to consider the merits of a new invention and the Smithsonian Institution might recommend a scientific contingent to accompany any Army or Navy survey, the problem of simplifying scientific theory for general public consumption took considerable forethought. Just as scientists differed on the desirability of popular lectures, so, too, they varied in their sense of responsibility or even their competence to handle specific problems and technological applications. Their ambivalence was reflected in the AAAS. Some involvement was implicit: local committee planning offered potential for community involvement, the committee structure invited promotional activities, and questions involving the relation of science to race and to religion were a concern of many individual members. Still, many members worried that these concerns distracted researchers from the advancement of science.

PARTICIPATION PATTERN

Active participation in the AAAS by the best-known American men of science assured recognition of the organization by the general public.[2] The naturalists Silliman, Dana, and Hall, as well as Joseph Henry, Benjamin Peirce, and others reputable in the physical sciences, attended the first meeting in Philadelphia; only Henry Rogers, on a trip to Europe, and Alexander Dallas Bache were notably absent. Evening sessions,

2. With the exception of a few men like John W. Draper, the roster of AAAS members in the 1850s included all leading scientists, although their attendance at meetings might not have been regular. Draper's nonparticipation is not accounted for in Donald H. Fleming, *John William Draper and the Religion of Science* (Philadelphia, 1950). Public opinion of science in the early nineteenth century has never been systematically studied. Historical commentary on the supposed indifference to basic research in the period usually refers ambiguously to scientists and laymen alike.

specifically open to the public, offered lectures by men who would normally command lecture fees. Thus the AAAS commenced with an enthusiastic scientific contingent and became popularly known as the meeting of savants. As the failure of the National Institute had proven in 1844, prominent scientists alone could maintain association claims as a scientific body, whatever the general character of the membership might be.

In fact, the AAAS membership was essentially open. The founders based initial requirements on the British model, which required that members be affiliated with a local society having similar interests. They added the category of scientific faculty and engineers, recognizing that colleges somewhat isolated geographically from scientific centers might have practicing scientists.[3] From a study of the membership roster, however, it appears that any person nominated by an AAAS member and seconded by the standing committee might join. Interested in affiliating all persons concerned with furthering science, the leadership exercised some discrimination in selection but rarely denied anyone membership. Membership grew without regular pattern, the yearly increase suggesting more about the sponsoring community and the initiative of members at a particular meeting than about the association itself.[4] Only when it became clear that the membership roster had a substantial number of delinquents did the association vote to delete members for nonpayment of dues, and then a two-year period of grace allowed ample time for up to four notices. Only through this means were members excluded.[5] An apparent effort to differentiate levels of membership was never functional. Henry Rogers initially intended a "democratic" constitution, but he returned from Europe in the early 1850s prepared to modify the charter and allow honorary memberships.[6] The association discussed his proposal for several years, and in 1857 members created a category of associate member, although they never utilized it.[7]

3. This recognition that the college had made a home for men of science is curiously juxtaposed to the fact that colleges were not themselves initiators of scientific activity. Even the establishment of scientific schools resulted from enthusiasm of outside benefactors and efforts of individual scientists interested in research rather than from action on the part of a united faculty or board of trustees. Stanley Guralnick's helpful study of the old-time college outlines the problem, but his focus on the progressive inclusion of science does not adequately explain the lack of innovation there and the dependence on European examples.

4. See the Appendix for a list of general members. Table 6 will show regional tendencies of persons joining the AAAS.

5. AAAS, *Proceedings*, IV (1850), 341.

6. *Ibid.*, II (1849), 179.

7. *Ibid.*, XI (1857), xxii.

Women were admitted as members, following the precedent established by the American Academy of Arts and Sciences in accepting the
astronomer Maria Mitchell. Mitchell worked with her father as he observed from their home on Nantucket, and in 1847 she independently
discovered a new comet.[8] Recognized by her peers as precise in mathematical computation and thorough in her observations, she maintained a
somewhat aloof, analytical posture toward the new association. Only
one woman, a nonmember, presented a paper.[9] Nonmembers seriously
interested in science, whether men or women, might attend, but most
women accompanied male members and were regarded as "ornaments"
in a drab crowd of suited men.[10] Although three women members were
admitted before 1860, discussion about the general admission of women
to sessions demonstrated a divided opinion as late as 1858. In the last year
Henry Rogers proposed to follow the BAAS policy of admitting women
at "half price" to encourage their attendance, and someone else suggested
that women be categorized as "associate members." In reply, Baltimore's
Lewis Stainer observed wryly that women already accompanied family
members and that was "annoyance enough."[11] No vote was taken on the
matter and the contemporary feminist issue was skirted. Thereafter
women continued to attend, both as members and as nonmembers.[12]

Because the founders did not reject the concept of a larger, inclusive
organization, the recruitment of new members was continuous in the
peripatetic organization. Not only did the AAAS hold regular annual
meetings in major urban scientific centers such as Philadelphia and Albany, but there were also semiannual or intermediate meetings at
Charleston in 1850 and at Cincinnati the following year. These trips to
the South and West were belated acknowledgment that even cities geo-

8. *Notable American Women*, III (Cambridge, Mass., 1971), 554–556.

9. Mrs. Eunice Foote of New York presented a paper at the AAAS meeting in
1857; she was introduced by Henry, who suggested it was proper for women to
be active in the association.

10. Women received an open invitation to attend in 1848, but the resolution was
not included in the *Proceedings*. See *Philadelphia Public Ledger and Daily Transcript*, 21 Sept. 1848. The women on the membership roster were Maria Mitchell,
Margaretta Morris, and Almira Lincoln Phelps. When the latter was admitted, the
president noted in an *obiter dictum* that women were not prohibited by the AAAS
constitution from joining but suggested that "it is probable that no others will
consent to be named, lest it should be deemed a challenge to the public to admire their scientific acquirements." *New York Times*, 10 Aug. 1859. There may
have been other women among the unidentified members listed only by initial.

11. *Baltimore Sun*, 4 May 1858.

12. A resolution in 1859 finally declared, "No action is necessary in regard to
the motion to admit ladies as members, in as much as two ladies have already been
admitted." AAAS, *Proceedings*, XIII, 364.

graphically isolated from scientific research centers had individuals capable of contributing to national science.[13] In each instance public response proved, on the amateur level, particularly notable; substantial numbers of men from the region were invited to membership, and this in turn secured a more truly national composition for the AAAS.[14] The decision to make these more distant locations hosts for meetings intermediate to regular annual conventions presupposed the inability of stabilizing the association without regular meetings accessible to the eastern men who led in science. But two semiannual meetings proved a financial strain on the members, few of whom had time for more than one such convention. The only subsequent meeting held outside the middle and northeastern states before 1868 was the annual meeting at Cleveland in 1853.

The results of efforts to allow public participation at general lectures, to provide membership to legitimately interested amateurs, and to vary meeting places was that the AAAS became the recognized spokesman for national science. Coverage in newspapers increased, and local communities were genuinely proud of a visit by the company of savants. The only competing contenders for national status were the scientifically oriented federal agencies and the new Smithsonian Institution in Washington, D.C. With leading scientists from Washington—Bache of the Coast Survey, Henry and Baird of the Smithsonian, and Maury of the Naval Observatory—as active AAAS participants, the threat of competition was avoided.

Yet the very existence of these groups did, in fact, modify the operational aspects of the AAAS. The Smithsonian Institution, especially, became the coordinating unit for foreign exchange and for national observation systems. When Spencer F. Baird arrived as assistant secretary, he brought with him not only a collector's instinct, later to manifest itself in a natural history museum, but also a scientist's conscious need for exchange of specimens and periodicals, for which he expanded the Smithsonian's system of exchange. By 1851 the Smithsonian was sending

13. A handwritten note on the AAAS circular sent by Thomas Rainey from Cincinnati observed, "This is the first meeting that the Association has held in the West and may be the last during many years. It is therefore earnestly desired that Westerners attend this meeting numerous [sic]." Copy in Breckenridge Family MSS, LC.

14. There is a high correlation between the place of meeting and the new membership for that year. See Table 5 which also indicates that people joining at a local meeting in the flush of enthusiasm for the AAAS more frequently dropped membership within two or three years than did those traveling from a distance to attend.

American journals and other articles to 210 European institutions.[15] The result was enhanced recognition for American efforts. Shortly after assuming the post as secretary at the Smithsonian in 1846, Henry initiated a system for coordinating meterological observations throughout the United States, centered in his institution. Most local societies and scientific schools wisely did not presume to compete with the well-funded activities in Washington.[16]

Perhaps the most evident indicator of the AAAS's national stature was the positive local response to individual meetings. By 1850 more than one city bid for the opportunity to host an annual meeting. Often an invitation emanated from a local group of scientists, but sometimes the offer came from the city itself, acting through a mayor or council. Following the precedent of the AAGN in 1846, the composition of the local committee came more and more to include leading citizens who were not members. The socially and politically prominent hosts were responsive and at least nominally involved in the meeting. After a somewhat cool reception at Philadelphia in 1848, the AAAS was usually feted; each city attempted to surpass the hospitality of the previous host. Typically, a local scientific society would invite the association membership to its regular meetings and to visit any cabinet or library it possessed. Wealthy citizens became hosts at evening receptions, and the local committee might even sponsor a tour in the vicinity. Frequently, too, community leaders would provide housing for the most eminent scientific leaders, making the latter truly distinguished guests for the one triumphant week. At the conclusion of such meetings an elaborate vote of thanks, recorded in the *Proceedings*, outlined the favors shown to visiting AAAS members.

The sixth meeting, held in Albany in 1851, exemplified the intense, almost competitive nature of the sponsorship. Local arrangements were so elaborate and time-consuming that a few members wondered about their original purposes. Between 1820 and 1860 Albany, together with Troy and Schenectady, constituted a cultural center of growing importance.[17] As the state capital, it was headquarters for the state agricultural society, the state geological survey, and the state library, and served as regional center for such educational facilities as the Albany Academy, the Albany Institute, the Albany Medical and Law Colleges, as well as

15. Smithsonian Institution, *Collections*, XVII (1879), 731. The number steadily accelerated and by 1860 totaled 525 institutions, with 5,617 items sent abroad.
16. Dupree, *Science in the Federal Government*, p. 109.
17. Samuel Rezneck, "The Emergence of a Scientific Community in New York State a Century Ago," *New York History*, LXIII (July 1962), 211–238.

nearby Rensselaer Polytechnic Institute and Union College. Albany had successfully entertained the AAGN in 1843, and once again in 1851 James Hall became an active promoter for the annual meeting of scientists. This time he and several leading citizens had a specific project which they hoped would gain support from a meeting of the scientists, a national university built in Albany. Civic leaders, as well as a number of prominent scientists, were eager to see a "true" university established in the United States, and the AAAS meeting seemed an excellent opportunity to prod an interested but reticent legislature into action.[18] John Pruyn worked diligently at planning during the six months before the meeting, and the results were nearly overwhelming.[19] The Albany Academy offered its rooms for the general meetings of the association, while especially large sessions were held in the State House. Nearly all the important educational institutions in the city proffered similar invitations to use facilities.[20] As arranged by the local committee, the association spent a full day in Troy and held one session at Rensselaer Polytechnic Institute.[21] Forty carriages were furnished for a drive through the Albany Cemetery, which was an elaborately designed garden, as well as for the drive to the U.S. Arsenal and to Troy; members returned by railroad to Albany.[22] Every evening, except the one on which Bache delivered his presidential address, visiting members were entertained by parties in the most prestigious homes in Albany.[23] As in 1843, local railroads provided reduced fare to association members.

Although the AAAS meeting was a popular success, the movement for a national university did not reach full bloom. Agassiz, whose early enthusiasm sparked the enterprise, refused to commit himself without financial guarantees, and leading New Yorkers grew cautious as they recognized the level of independence demanded by the men of science. The AAAS was not capable of bolstering the project, which needed funding and local support, and took no formal position on the proposal.[24]

18. Silverman and Beach, "A National University," pp. 705–707.
19. Pruyn to James Hall, Albany, 18 Nov. 1850, and George R. Parkins to James Hall, Albany, 10 June, 30 July 1851, all in Hall MSS, NYSL.
20. AAAS, *Proceedings*, VI (1851), 392.
21. Under the leadership of Benjamin F. Green, the institute was reasserting itself as a science center. See Samuel Rezneck, *Education for a Technological Society: A Sesquicentennial History of Rensselaer Polytechnic Institute* (Troy, N.Y., 1968), pp. 78–131.
22. An excellent running account of the informal events of the meeting is John P. Norton's diary, 20–26 Aug. 1851, Norton MSS, YUA.
23. AAAS, *Proceedings*, VI (1851), 389–390; Thomas Hun to Joseph Henry, 4 Aug. 1851, Baird Family MSS, SIA.
24. Silverman and Beach, "A National University," pp. 706–707.

The community's response to the association, however, was generous. Several citizens raised a subscription for the publication of the proceedings following the precedent set by the city governments of Charleston and Cincinnati. In addition, separate daily programs of the papers to be presented were printed for the use of members.[25] Albany, having a purpose in its efforts, was perhaps unique in its response, but only in degree, as local committees became increasingly eager to earn a good reputation as host to the scientists. At Providence, for example, private citizens offered their homes when local hotel accommodations proved inadequate. Every evening brought a reception, first by the mayor and then by such local luminaries as Zachariah Allen, John C. Brown, and Francis Wayland.[26]

It is difficult to assess the effect of the AAAS on local scientific societies. In most cities a local group helped sponsor the meeting, and the immediate response was enthusiastic. On a few occasions, as in New Haven, local societies responded with indifference. More typical was Charleston, where the AAAS helped spark a new local museum and society. But the visit from the AAAS also provided an impossible model, and the local groups were inching past their prosperous, scientifically active periods. The national meetings provided a false optimism. Once the savants had gone, local citizens and amateur researchers were more aware than ever about the limits of their efforts. In Charleston the new society did not survive the Civil War period, and the Cincinnati society faded shortly after the AAAS meeting there. Perhaps it is ironic that in the hope of stimulating local interest in science, the association in fact undermined the viability of the local scientific societies.[27]

ISSUES OF SCIENCE AND SOCIETY

For an interested public as well as for the scientific community, the AAAS, as the only voluntary and widely representative scientific group,

25. AAAS, *Proceedings*, VI (1851), 404. Some of the daily programs are in the Hotchkiss MSS, LC. The list of papers to be presented does not always correlate with the printed *Proceedings*, indicating that the papers either were not read at all or were eliminated by the editor.

26. *Providence Journal*, 15, 17, 20, 21 Aug. 1855. Also see J. D. Whitney to William Whitney, Providence, 22 Aug. 1855, Whitney MSS, YUA.

27. This valuable observation came from discussion with Henry Shapiro after reading his unpublished paper, "The Western Academy of Natural Sciences of Cincinnati and the Structure of Science in the Ohio Valley, 1810–1850," presented for the Conference on the Early History of Societies for the Promotion of Knowledge in the United States, June 1973.

became spokesman and leading arbiter for science. Both in association meetings and in assigned committee activity, it assumed a status of authority essentially unchallenged by scientists and laymen alike. Almost every major issue appeared, if not in the printed proceedings, in private conversation among men who could thus clarify their own thinking through discussion with peers.

Like the question of membership and status, the more difficult problem of fundamental purpose was evident from the early years of the association, and the solutions were essentially pragmatic. In responding to the need for national organization, the founders had not been clear on how the goal of advancing science was to be effected through the sectional meetings. At the first meeting in 1848, after a particularly elaborate paper, chemist Robert Hare questioned whether such papers should not be presented with only a major outline and the most interesting examples, to avoid the reading of detail. In response, Henry Rogers agreed with the implied recommendation to condense papers, but he maintained that the facts in the particular instance were specifically of interest to several persons at the session. A local Philadelphian then observed that the association "by necessity" must popularize subjects under discussion, and he suggested that such papers should be handled in a special meeting on geology.[28] An evening meeting was scheduled, with "ladies" specifically invited to attend. The question of type or caliber of papers to be presented was temporarily resolved along the lines of the BAAS.[29] Reading of specialized papers continued in the section meetings, and a Philadelphia newspaper reported, "Though the proceedings of this body have been too scientific to interest the generality of persons, there is no doubt that much good will result from the labors of the Association."[30] Disparity of experience between men of science and the general public could only be overcome with generalized papers lacking the detail essential for dialogue and critical assessment of particular research. The working decision by the association to continue detailed papers acknowledged and even established the increasingly evident distinction between scientist and interested layman.

As suggested in Chapter I, the self-culture movement raised expecta-

---

28. *Philadelphia Public Ledger and Daily Transcript*, 21 Sept. 1848. The newspaper states that the discussion followed the reading of a paper of Rogers, but the date and general proceedings suggest it occurred after the reading of a particularly long and detailed letter from John W. Van Cleve of Dayton, Ohio. See AAAS, *Proceedings*, I (1848), 19–24.

29. AAAS, *Proceedings*, I (1848), 25.

30. *Philadelphia Public Ledger and Daily Transcript*, 26 Sept. 1848.

tions about science that were elaborated in the press and in popular textbooks. When a national, peripatetic scientific organization, which included the leading men of science, was formed, public response was positive and enthusiastic. Not surprisingly, questions of race and religion relevant to popular conceptions of the nature of science and of scientists were asked. Public recognition was encouraging, but gradually the group which had earlier resisted popular lecturing reasserted itself. "Young America" scientists were restive about the implied responsibility and eager to make science independent of outside demands.[31]

In some ways the still amorphous AAAS was less a leader than a reflection of where science stood on issues. Press coverage suggested that the AAAS was expected to be the spokesman for science on particular issues and, further, to demonstrate applications of science. Such anticipations grew rapidly as the AAAS established itself in the early 1850s and tapered as failure to speak on issues, evident elitism, and internal dissension brought disillusionment with science and with the AAAS. The professional tendency which sought to define and limit science itself also blurred the public image. Because individual scientists continued to respond, inside and outside the AAAS framework, however, outright rejection was avoided. For the 1850s, however, the AAAS exemplified the contemporary tension which was creating two cultures.

In 1850 Joseph Henry presented the customary retiring presidential address. He opened by observing that the address was important, "a duty to be faithfully discharged, and not a form which might be dispensed with at the option of the officer to whom it is entrusted."[32] Outlining topics to cover, he suggested that presidents should trace the past history of the association, elaborate its purposes, "vindicate the claims of science to public respect and encouragement, and set forth the nature and dignity of the pursuit. Give instruction from past experience in the methods of making new acquisitions over nature; and point out new objects and new paths of research. They should expose the wiles of the pretender and suggest the means of diminishing the impediments to more rapid scientific progress." Henry was suggesting

31. The phrase "Young America" was used frequently and not with consistency; it usually connoted an anti-establishment posture and was undoubtedly borrowed from the "Young Europe" associations of republican agitators in various nations after 1830. See, for example, James Hall to James D. Dana, Albany, 15 Mar. 1854, Dana MSS, BLY. For the young scientists it implied camaraderie, singularity of purpose, and a kind of chauvinism for science.

32. Draft copy of AAAS address, undated, Henry MSS, SIA. Subsequent quotations are from this source.

topics for public appeal and asserted that scientific leaders should "call attention to the relation of science to the moral part of our nature . . . and point out all things which may have a bearing through science on the material and spiritual improvement of man." Although not every president did so, those who prepared formal addresses recognized the occasion as an opportunity to consider science in a broader context.[33] Bache followed Henry in 1851 and welcomed the opportunity to review and evaluate the progress of American science in perhaps the most important address of the decade. In 1854 Dana summarized recent advances in geology, and at Montreal in 1857 Hall used the podium to present a new and controversial geological theory.[34] Such men took seriously their position because the AAAS presidency was and remained "the highest compliment which can be paid to an American man of science."[35] Audience and speaker alike paid tribute to the office by preparation for the annual address, which was included in the proceedings and usually published separately as well.

Despite some skepticism there was no overt opposition to the popularization of science, at least in a context where it might be controlled and directed. The evening receptions and excursions sponsored by local committees facilitated interaction with the community. Drawing on the BAAS experience, the AAAS offered evening sessions open to the public; popular scientific lecturers, including Agassiz, Silliman, and Bache, spoke without charge. In 1851 Bache suggested that there should be even more attention given to the local activity and urged that open evening sessions be a regular feature.[36] He distinguished, as did others, between the kind of "entertainment" sponsored by local lyceums and professional summaries of scientific advances in a particular field which updated general familiarity with correct scientific outlooks. Thus, even when the heavy social schedule of excursions and evening receptions came under attack, the continuation of popular lectures was not questioned.

Scientific leaders quickly recognized that open sessions offered an opportunity to publicize their institutions and researches. The astronomer

33. This was also true for the BAAS, as shown in the reprinted addresses. See George Basalla et al., eds., *Victorian Science: A Self Portrait from the Presidential Addresses of the British Association for the Advancement of Science* (New York, 1970).

34. Hall's address, although listed in the index for the 1857 meeting, was not printed at that time because he failed to deliver the promised manuscript to the editor before the deadline of publication.

35. J. W. Bailey to J. Torrey, West Point, 6 Sept. 1855, Torrey, MSS, NYBG.

36. AAAS, *Proceedings*, IV (1850), 153.

Maria Mitchell noted with quiet humor, "The leaders make it pay pretty well. My friend, Professor Bache, makes the occasions the opportunity for working sundry little wheels, pulleys and levers; the result of all which is that he gets his enormous appropriations of $400,000 out of the Congress every winter for the maintenance of the United States Coast Survey."[37] Similarly, when token support seemed important, the association appointed committees to recommend the Coast Survey, substantiating the claim of competence and achievement by the approbation of learned men.[38]

Certainly public involvement also brought personal satisfaction. The straightforward Maria Mitchell observed after one meeting, "For a few days Science reigns supreme—we are feted and complimented to the top of our bent, and although complimenters and complimented must feel that it is only a sort of theatrical performance for a few days and over, one does enjoy acting the part of greatness for awhile! I was tired after three days of it, and glad to take the cars and run away."[39] The experience could be overwhelming and disconcerting for members who felt their purpose in coming was to discuss serious matters of science; as one such member told a newspaper reporter, "Think of it! Men who have been sleeping on rocks, eating out cold victuals off of logs, digging in the earth, and breaking stone nine months of the year—bringing us to such brilliant *soirees* as these and making us (unintentionally of course) laughing stocks!"[40] Such men endured, however, and most participants relished the public enthusiasm.

The association found itself, usually less by design than by personal inclination of its membership, involved in pervasive but loosely defined issues relating science to society. Public expectations often impressed this involvement on the AAAS because of philosophical or practical concerns for which there was no final arbiter. Matters of race, religion, and speculative medicine found their way into scientific societies and publications of the period and, inevitably, into the open meetings of the AAAS as well. Because scientists of established reputation became personally involved in discussing such matters, the distinction between legitimate topics of science and peripheral social applications was not always clear, especially to the persons concerned.

37. Helen Wright, *Sweeper in the Sky: The Life of Maria Mitchell, First Woman Astronomer in America* (New York, 1949), p. 69.

38. Bache or one of his assistants presented a paper relating results of the Coast Survey every year from 1848 to 1860.

39. Wright, *Sweeper in the Sky*, p. 69.

40. *New York Times*, 3 May 1858.

Abolitionists' efforts and international moral indignation throughout the second quarter of the nineteenth century created an atmosphere so tense that "the slave question" was for a time a forbidden topic in the U.S. Congress.[41] By the 1840s a tide of southern journalism was rising in defense of the peculiar institution, using arguments that ranged from biblical to scientific.[42] In an admixture of the two polar arguments, defenders of the southern way of life found their most plausible defense in the argument of polygenism. This theory of the multiplicity of race seemed substantiated by certain interpretations of the Bible as well as by the American school of anthropology headed by Samuel George Morton. Louis Agassiz, who as a recent immigrant seemed particularly without prejudice on the race issue, gave the problem of the unity of races enormous publicity;[43] not surprisingly, the coverage was considerably greater in the South than in the North. The North, certain that slavery was immoral but less certain about the place of blacks in society, debated how to limit slavery and what to do with the freed black people. The South, desperately defending their total social and economic structure, continued to garner all evidence which appeared to justify continuing the order of things.

The Charleston semiannual meeting in 1850 provided an almost foreseeable forum for the question. From the beginning the argument was basically one-sided, primarily because personal restraint and geographical setting inhibited the most competent supporter for the unity doctrine. John Bachman, naturalist and Lutheran minister in Charleston, headed that side and was joined almost exclusively by his fellow theologians in the limited verbal encounter. As his pamphlet *The Doctrine of the Unity of the Human Race Examined on the Principles of Science*, published in early 1850, demonstrated, Bachman stood in opposition to Morton, Agassiz, and the journalist Josiah C. Nott—ironically, the southerner Bachman defended the humanity of black men against northerners. Morton's perspective was represented by Samuel S. Kneeland, whose paper on "Characteristics of the Hindoo Skull" confirmed Mor-

41. Samuel Flagg Bemis, *John Q. Adams and the Union* (New York, 1956), traces the battle waged by "old man eloquent" against the so-called "gag rule" from 1840 to 1844.

42. William R. Stanton, *The Leopard's Spots: Scientific Attitudes toward Race in America, 1815–1859* (Chicago, 1960). Stanton's volume is an excellent analysis of both northern and southern perspectives.

43. Edward Lurie, "Louis Agassiz and the Races of Man," *Isis*, XLV (May 1954), 227–242, offers insight into the motivations and arguments of Agassiz and suggests that the Charleston meeting occurred precisely when the controversy between the schools of "unity" and "plurality" were reaching an explosive stage.

ton's measurements of Hindu crania and suggestively noted size parallels within caste groupings.[44] On Friday of the sessions, Nott gave a paper which he argued proved the unity and distinctiveness of the Jewish race biblically. In response, Agassiz stated that all races possessed the "attributes of humanity," but he remained ambiguous on the question of slavery even as he reaffirmed that zoologically the races were "well marked and distinct."[45] Bachman was chairman for the day and heard not only this paper and subsequent comments, but also a paper by Peter Browne, who had microscopically examined hair to prove that the Negro albino, although white, was really a black man. Bachman, however, offered little comment.[46] The lack of debate did not indicate agreement; Bachman made plain his opposition to the proposition of distinct origins of the races but simply suggested that discussion was more appropriate in written form. He recognized that the association was not a proper place to debate questions whose answers were obviously complex and about which bitter arguments might ensue.[47] Significantly, Agassiz, who had not submitted a paper relating to race, although commenting at length and for publication on the topic, after the AAAS meeting confined his statements on race to the popular lecture platform and to theological journals. The AAAS had seen the first and last of the race question in scientific guise. Most rejoinders came from theologians in attendance who generally drifted away from the scientific arguments originally made. Later, as Gibbes prepared the proceedings for publication, Bache suggested, probably in reference to the race discussion, that he publish only papers indicating research and not "comments of a more popular nature."[48] In following years the question of race was not raised in the association until after publication of Darwin's *Origin of the Species* (1859); the scientists carefully avoided the polemical topic.[49] Bache, who obviously disapproved of the general comments,

---

44. AAAS, *Proceedings*, III (1850), 34–36.

45. *Ibid.*, pp. 98–107.

46. *Ibid.*, pp. 108–114.

47. *Charleston Courier*, 18 Mar. 1850. Agassiz's comments created "no little sensation," but Bachman stated that the question was one "less suitable for open debate ... before a promiscuous audience, than for deliberate investigation by the advocates of opposing theories, through the press," noting his own recently published pamphlet. His remarks seemed almost to chide Agassiz.

48. A. D. Bache to Gibbes, Washington, 30 May 1850, Gibbes MSS, LC.

49. For studies of social Darwinism in America, see Richard Hofstadter, *Social Darwinism in American Thought* (New York, 1955), and John S. Haller, *Outcasts from Evolution: Scientific Attitudes of Racial Inferiority, 1859–1900* (Urbana, Ill., 1971).

probably pointed out to Agassiz the need to refrain from engaging in such topics at association meetings.

The relationship between science and religion provided another frequent topic, universal because of its personal and philosophical implications and persistent because of the number of theologians attending the AAAS meetings. Once again it would be Darwin's theory that led to warfare fought in full battle dress. Yet even earlier the anxiety raised by new geological discoveries (which suggested far earlier dates for the earth's formation than that put forth by biblical scholars), by Charles Lyell's theory of uniformitarianism, and by the *Vestiges of Creation* debate at the Association of Geologists' meeting demonstrated growing tension about the relationship between science and religion.[50] But discussions, somewhat belatedly following debates in England, were conducted without rancor and often with a quiet desperation by concerned theologians. To men of science, many of whom were active church members, fears of conflict seemed groundless because a study of the physical world better revealed the existence and nature of God. Nevertheless such religious scientists formulated responses designed to ease the doubts of others and to rationalize their own dual loyalty.

Inevitably the AAAS reflected the ambiguity experienced by member scientists who affirmed personal religious orthodoxy while insisting on the right of science to seek its own truth independent of revelation. American society in the 1850s held a belief in divine providence in conjunction with an acceptance of science as a historically progressive force. Although the Protestant sector in America was uneasy about what science might be saying in religious or philosophical terms, the evident orthodoxy and assurances of leading men of science undoubtedly maintained the stable equilibrium. Later, when consensus on natural theology was no longer operative, debate even among scientists would be vehement. The AAAS, as visible spokesman for science in the 1850s, served as a blending force, compromising both sides by its unspoken policy of permitting personal religious commentary without allowing presentations at meetings to cloud the research methods of science.

50. Most historical analysis has concentrated on evolutionary theories in the geological or natural sciences, perhaps because contemporary commentary was most extensive in them. See Charles Gillispie's *Genesis and Geology: A Study in the Relations of Scientific Thought, Natural Theology, and Social Opinion in Great Britain, 1790–1850* (Cambridge, Mass., 1951); John A. DeJong, "American Attitudes toward Evolution before Darwin" (Ph.D. dissertation, State University of Iowa, 1962). A sound beginning in relating religious and social thinking to evolutionary astronomy is Ronald L. Numbers, "The Nebular Hypothesis in American Thought" (Ph.D. dissertation, University of California at Berkeley, 1969).

Evolution as a principle raised the specter of controversy throughout the first half of the nineteenth century in both astronomy and in the geological and biological sciences. Edward Hitchcock, Congregational minister and also a professor of science and moral philosophy at Amherst College, took seriously the responsibility for reconciling religion and the science of geology. His synthetic concept of nature was probably most explicitly expressed in his *Religion of Geology and Its Connected Sciences* (1851), but was also clear in his geological textbooks. Hitchcock ultimately subordinated science to theology and for a time resisted Charles Lyell's work as atheistic because it failed to relate science to religious truth.[51] Yet his geological observations were usually scientifically sound and regarded as contributions to the field. The number of printings of his work may be attributed to contemporary interest in science itself but was certainly also related to his reconciliation of science and religion for an uneasy generation.

James Dwight Dana, although a generation younger than Hitchcock, was committed both to his study of mineralogy and chemistry and to his personal religious tenets. Dana emerged as an eloquent defender of science and geology in response to Taylor Lewis's *The Six Days of Creation* (1855). Lewis's book was a frontal attack on the science of geology, the efforts of geologists, and the piety of scientists. The two disagreed more in method than in substance; according to one analyst, Dana's system was the application of scientific conclusions and scientific language to Genesis, while Lewis's approach was a philological interpretation of the same chapter.[52] Several of Dana's scientific friends, including Benjamin Peirce, George P. Bond, and Louis Agassiz, applauded his exposition in reviews of the book.[53] Dana's defense, an attempted reconciliation of science and religion, was an exercise more reminiscent of earlier decades than suggestive of future tactics. His younger colleagues were content simply to state that they found no conflict personally while admitting the difficulty of correlating science and religion.

Within the AAAS there were no elaborate papers on the argument from design, but a certain religiosity was increasingly evident. The issue

51. See Hitchcock's own introspective *Reminiscences* and Guralnick, "Edward Hitchcock."

52. Morgan B. Sherwood, "Genesis, Evolution, and Geology in America before Darwin: The Dana-Lewis Controversy, 1856–1857," in Schneer, ed., *Toward a History of Geology*, pp. 304–316, esp. p. 315.

53. Agassiz to Dana, Cambridge, 30 Jan. 1856, Gilman MSS, Johns Hopkins University Archives, Baltimore.

surfaced during the Charleston meeting in 1850 when Lewis R. Gibbes felt it necessary to vindicate geology from all suspicion that it led to infidelity. Bache, in reviewing the events of the meeting, observed that Gibbes's remarks were a painful and singularly appropriate necessity:

> I believed the time had long gone by, when the study of God's work could be supposed to lead away from the revelations of his work; that the language which I heard last Sunday, from a pulpit in this city, was of common concent and acceptance—"science is no ally to skepticism." It is, nevertheless, true that a lesser wave of the same class... which rose to overwhelm geology some twenty years since, sweeps with considerable force over the Southern portion of our Union, and requires to be stayed with judgement to subsidence. . . .[54]

Significantly, Bache was to hear several subsequent meetings even outside the South opened with prayer.[55]

Such leading scientists as Agassiz, Dana, Hitchcock, and Peirce were personally religious and attended local churches while at AAAS conventions, a fact which led many local ministers to present sermons designed for the visiting men of science. Men of science believed that, kept in proper perspective, religion was a philosophical and spiritual necessity. Agassiz saw no disparity and he asked, "When the spirit that moves is not self-glorification but an humble desire to learn the truth to be sought by nature, to read the deeds and the will of God in his works, what do minor discrepancies in the reading of both Bible and Nature impart?"[56] His objection was to "conceited theology" which sought to find minor discrepancies between the Bible and nature, instead of simply reading the will of God in his works. The great majority could agree with Joseph Henry that "religion and science were coordinate branches of human inquiry. Science relates to laws of nature within us and without us—laws in which the controlling power of God is perpetually manifest."[57]

Although Louis Agassiz was persuaded of a strong antiscientific bias among clerics, little evident opposition to science existed aside from a

---

54. AAAS, *Proceedings*, IV (1850), 164. Concern for natural theology lingered perhaps longer in the United States than in Great Britain, where it was something "of an anachronism" by the later 1830s, according to John J. Dahm, "Science and Religion in Eighteenth-Century England: The Early Boyle Lectures and the Bridgewater Treatises" (Ph.D. dissertation, Case Western Reserve University, 1969).

55. AAAS, *Proceedings*, XI (1857), 162; XII (1858), 290–291; XIII (1859), 358; XIV (1860), 229.

56. Agassiz to Dana, Cambridge, 30 Jan. 1856, Gilman MSS, Johns Hopkins Archives.

57. *Cleveland Plain Dealer*, 29 July 1853.

few men like Lewis.[58] The *cause célèbre* of the decade, the failure of Wolcott Gibbs to be appointed to the chair of chemistry at Columbia University, seems to have been more related to the power tactics used by leading scientists than to Gibbs's Unitarian affiliation.[59] Nonetheless, the case was construed as a defiance of what today would be considered "academic freedom,"[60] and many scientists agreed with Agassiz's assessment of a pervasive hostility toward science. In response to the doubts expressed by Lewis and this conscious belief that the clergy found contemporary science antireligious, religious allusions became more common in the AAAS.

At the Albany meeting in 1856 the local committee again planned a comprehensive program for the AAAS. Not only did they arrange for a group portrait of the leading scientists by R. Van Dein, but they also prepared to dedicate the State Geological Hall and the Dudley Observatory while the distinguished savants were all present. Somewhat belatedly but with evident good intention, they also invited three leading theologians to speak at local churches on Sunday, 24 August.[61] Because the addresses were invited in conjunction with the AAAS, some members subsequently moved that they be printed by the association. The matter went to the standing committee, which quietly voted not to publish them, "leaving the matter to local citizens."[62]

The three discourses, delivered at various churches, were all directed toward the relationship between science and religion. Bolstered by the

58. Hofstadter and Metzger, *Development of Academic Freedom*, pp. 198–220.

59. Milton H. Thomas, "The Gibbs Affair at Columbia in 1854" (M.A. thesis, Columbia University, 1942), pp. 59–96.

60. Samuel Ruggles to G. Peirce, New York, 12 Feb. 1856, Peirce MSS, HLH. Ruggles observed, "The war will be long and bitter but *a University* will spring out of it, and surely, as religious toleration sprang from the Thirty Years War of Germany." Also see Bancroft to Bache, New York, 17 July 1858, Bache MSS, LC.

61. The men and their sermons were Edward Hitchcock, *Religious Bearings of Man's Creation, a Discourse Delivered at the Second Presbyterian Church, Albany, on Sabbath Morning* (Albany, N.Y., 1856); J. H. Hopkins, *Relations of Science and Religion, a Discourse Delivered in Albany during the Session of the American Association for the Advancement of Science* (Albany, N.Y. 1856); and Mark Hopkins, *Science and Religion, a Sermon Delivered in the Second Presbyterian Church, Albany, on Sabbath Afternoon, August 24, 1856* (Albany, N.Y., 1856).

62. J. D. Whitney to E. Desor, Northampton, 30 Dec. 1856, in Edwin T. Brewster, *Life and Letters of Josiah Dwight Whitney* (Boston, 1909), p. 168. Also see Hitchcock to S. B. Woolworth, Albany, 13 Sept. 1856, and a note dated 4 Oct. 1858, Hitchcock MSS, ACA. The note states, "A subject bearing upon the relations of science to religion, was chosen, at the suggestion of Rev. Dr. Sprague, pastor of the Second Presbyterian Church. It is published by the request of the Local Committee of the Association."

statement in their invitation that "Science worships at the Shrine of religion," the ministers all recommended humility and deference as the proper stance for science, with John Hopkins, Episcopal Bishop of Vermont, and Mark Hopkins, Congregational president of Williams College, both choosing their text from 1 Timothy. There St. Paul exhorted the early Christian church to "avoid profane and vain babblings and oppositions of science falsely so called; which some professing, have erred concerning the faith."[63] Both maintained that science was indeed in harmony with religion but only so long as it did not pretend to answer questions outside its natural limits. Edward Hitchcock, perhaps because he was actively engaged in scientific research, was more explicit and direct in using Genesis as a text to elaborate the *Religious Bearings of Man's Creation.* Reminiscent of earlier attempts to correlate the Mosaic account with contemporary scientific research, Hitchcock took refuge from evolutionary theory by noting the prevalent uncertainty about the time sequence of fossils and the debate over such matters of process as the drift or glacial theory. Hitchcock was less defensive and more positive in approach as he argued that "infidelity has long since claimed the testimony of science as on her side; and I fear too often the expounders of revealed theology have half admitted the claim and felt that the less they had to do with natural religion, the better."[64] Hitchcock argued metaphorically that if nature seemed to speak against science, it was ventriloquism; in fact, he found a unity, harmony, and mutual collaboration between science and religion.[65]

While a majority of the Association concurred that publication of sermons by the AAAS was somehow inappropriate, a few men realized that this decision involved a fundamental question regarding the appropriate role for scientists in dealing with general social or philosophical questions. The group dubbed the "young Americans" by the press and older scientists were, like the ambitious Josiah Dwight Whitney, men who might respect the scientific attainments of their scientific elders while being impatient with personal philosophical-religious aides. Whitney wrote wryly to Edward Desor after the meeting, "Agassiz treated us to an embryological demonstration of the existence of a personal God. . . . Also the first thing on the programme was to assemble at a church in the city and have religious exercises!"[66] The decision not to

63. Quoted at the beginning of the addresses cited in n. 61.
64. Hitchcock, *Religious Bearings*, p. 51.
65. *Ibid.*, p. 52.
66. Brewster, *Life and Letters*, p. 168. None of Agassiz's presentations, doubtless extemporaneous, were published that year.

print the sermons therefore registered a rejection of any implied responsibility to relate religion and science, just as the young men of an earlier decade had declined to assume roles as circuit lecturers on science.

Henry, Agassiz, and Peirce, who proved to be adamant against all kinds of pseudo-science, were bellwethers in keeping before the association the idea of divine providence operating as a vague but sustaining force.[67] Occasional allusions to religion slipped into the printed *Proceedings*, but usually such comments were simply eloquent oratory added to the conclusion of an otherwise scientific address. Joseph Henry's comments from the podium, for example, were recorded only by the public press, which noted with satisfaction that men of science had not lost religion.[68] The persistence of such verbal assurances indicates the immediacy of the subterranean issue of the relationship between science and religion.[69] An enforced informality indicates the ambiguity of men aware of their private and public dilemma but unable to formulate a solution. They retained extensive arguments for theological journals but also felt compelled to assert their beliefs before the popular but scientifically inclined audiences of the AAAS.

The reunion atmosphere of the AAAS meetings invited discussion on many of the concerns relating to men of science. Sometimes these matters prompted the association to take formal action, but equally often they were simply topics of mutual interest and occupied members during postsession discussions. Thus such matters as the Albany university scheme, the New York and London world fairs, and even scientific curricula in the colleges filtered through without formal action. Resolutions and memorials were reserved for matters of government sponsorship and specific research projects rather than for the more loosely coordinated and indirectly related projects and concerns of a few members.

The short-lived American Association for the Advancement of Education, primarily interested in modernizing collegiate programs as well as in the idea of a national, graduate-level university, was cautious in regard to the Albany proposal. The AAAS was even more so; it allowed ample opportunity for informal discussion at the 1851 meeting but took

67. AAAS, *Proceedings*, VII (1853), xvii–xx.

68. *Daily Cincinnati Commercial*, 12 May 1851; *Cleveland Plain Dealer*, 29 July 1853.

69. Often such statements were footnotes to a scientific paper or a discussion and were recorded only in newspaper accounts of the meeting. Occasionally they did appear in printed minutes, for example, *Proceedings*, III (1850), 147; IV (1850), 273; XII (1859), 9–10.

no action. This predisposition was indicated already in 1849 when Daniel Breed of New York submitted a resolution reported out by the standing committee, "that the Association authorize the standing committee to prepare and submit to the public, at the meeting in March next, a circular letter setting forth the great importance to the interests of American education and of true civilization, of a more general admission of the Physical and Natural Sciences into the systems of school and collegiate instruction throughout our country."[70] After an extended discussion the resolution was tabled for future consideration, but it was never revived; linkages between government, education, and private institutions were yet undefined.

Historians of the national university movement agree, however, that at the Albany meeting, during informal gatherings, the plan for a national university received its last major thrust, enabling its leaders to take action during the following winter.[71] Thus the AAAS meeting, despite its own restraint on the matter, served as a stimulus to local sponsors and provided a forum for the discussion of a strategy and a program for the proposal.[72] For example, during private gatherings Agassiz stated he would participate only if the faculty had guaranteed salaries rather than being dependent on student subscription; other men of science involved in the proposal agreed to accept appointment offers only as a group.[73] The entire topic of the Albany university is conspicuously absent in the published record, aside from a vague reference in Bache's address pointing out the institutional needs of science in America. The lack of resolutions of support suggest that the AAAS found the issue peripheral to its major purposes or perhaps politically explosive.[74]

Only in one instance did the members of the association consider for a moment their hopeful assumptions concerning the international community of science. The Altona Observatory, located in Schleswig-Holstein, duchies disputed between Denmark and the English Hanoverian kings, was headed by Heinrich Schumacher, who was also editor of

70. *Ibid.*, II (1849), 357.
71. See Norton MSS, YUA, and especially Norton's diary for 20–26 Aug. 1851. Interestingly, Norton was dismayed by the lack of a strong religious commitment among the scientists: "There is enough sound interest in its success but little of what may be considered a truly religious spirit. The men selected are mostly of very doubtfull orthodoxy though unexceptionable in other respects."
72. See Thomas C. Johnson, Jr., *Scientific Interests in the Old South* (New York, 1936), for the effect of the 1850 AAAS meeting on Charleston's scientific activity.
73. J. P. Norton diary, 23 Aug. 1851, Norton MSS, YUA.
74. Silverman and Beach, "A National University," pp. 106–167.

Astronomische Nachrichten, a truly international astronomical journal. Concern for the observatory during a time of war aroused astronomers throughout Europe as well as in the United States. As Edward Everett pointed out, the isssue was being discussed just as a preliminary treaty was being concluded.[75] Yet the decision to publish letters relating the concern of the AAAS membership demonstrated that American men of science were ready to assert some role in international science. Born in the nationalism so evident in mid-nineteenth-century America, hopes for an international spirit of competition and cooperation went for the first time beyond simply practical agreements concerning mutual meteorological investigation.

Sometimes topics of general contemporary interest were spotlighted by individual papers. At the first meeting S. W. Roberts spoke on the current, often polemical topic of priorities in selecting railroad construction routes.[76] Subsequent meetings heard about G. Borden's new preservative for protein flour and several speculative papers on disease theory.[77] The diminishing frequency of such papers and the relative lack of discussion indicate that the membership generally found such presentations outside the mainstream of their own scientific interests.

Occasionally, too, immediate concerns caught up and deflected the entire association's attention from the standard fare of research papers. In 1849, as in 1832, cholera threatened the United States, having swept through Europe in the previous year.[78] Worried observers suggested that the Boston meeting of the AAAS, scheduled for August of that year, should perhaps be postponed.[79] Although the meeting was held, personal anxiety of attending members was evident in several discussions relating to the epidemic during the meeting.[80] The diffuse comments mirrored, in fact, a nationwide proliferation of treatises and articles. While moral causes were no longer presented as an adequate reason for individual cases of cholera, no reasonable explanation of the origin or

75. AAAS, *Proceedings*, II (1849), 65–68; *Boston Daily Evening Traveller*, 16 Aug. 1849.

76. *Proceedings*, I (1848), 36–38.

77. *Ibid.*, III (1850), 36–38.

78. Medical, theological, and public response to the threat is outlined in Charles E. Rosenberg, *The Cholera Years: The United States in 1832, 1849, and 1866* (Chicago, 1962), pp. 101–172.

79. R. W. Gibbs to William Redfield, Columbia, S.C., 20 July 1849, Redfield MSS, BLY.

80. Alarm over the cholera threat limited attendance. See Edward Everett to William Whewell, Cambridge, Mass., 3 Sept. 1849, Whewell MSS, Trinity College, Cambridge, Engand.

spread of the disease had been found.[81] The theories presented to the AAAS, like their counterparts in medical journals, reached widely for some cause-effect relationship that might help prevent the dreaded scourge. Dr. Hare outlined rather empirical research which indicated that electricity was produced only with difficulty in places with cholera; while he stated his own skepticism about the apparent correlation, he urged that his facts not be overlooked.[82] Similarly, Increase A. Lapham of Wisconsin sent a paper on "medical geology," which was read by Charles T. Jackson, noting a relationship between particular substrata of rock and a region's evident susceptibility to cholera.[83] As scientists and without medical training, many members felt obligated to investigate situations and posit theories which might clarify facts about the environment of disease. They came before the association on their own initiative rather than that of the organization.

In matters of "public policy" the AAAS acted positively. Reports submitted by committees indicate that the scientific leaders of the association, many of whom were directly linked to state and federal government agencies, recognized that the AAAS provided an excellent supporting agency to reassure legislators that certain projects were approved by men of science. Local societies on the state level frequently appointed petitioning committees, and the AAGN had extended this practice to the federal government. Also, the geologists were anxious to enhance their knowledge of geology in states which had no surveys or whose reports had not been fully published. By 1852, when the governor of New York submitted a memorial to his legislature, it might claim

> The American Association for the Advancement of Science has hitherto had the good fortune to receive from the public bodies to which it has addressed itself, the most favorable regard. It is perfectly aware of the grave responsibility imposed upon it by these marks of respect and consideration, and is therefore careful not to utter sentiments and opinions, whether speaking through its committees or otherwise, which cannot bear the most rigid scrutiny. Under this sense of duty and responsibility the present committee is now acting.[84]

81. Rosenberg, *Cholera Years*, p. 164.
82. AAAS, *Proceedings*, II (1849), 406–407; similar arguments are elaborated in Rosenberg, *Cholera Years*, pp. 165–166.
83. AAAS, *Proceedings*, II (1849), 201–206, 406–408. Jackson's letter aroused open opposition from John C. Warren and Agassiz, the former arguing instead that cholera seemed to occur in areas where there was moisture to decompose animal and vegetable substances, as in Jackson's example of Sandusky, Ohio.
84. Washington Hunt, *Communication from the Governor, Transmitting a*

The association moved cautiously and circumspectly in such matters, seeking simply to "give direction" and a "stronger impulse" to scientific inquiry, as its membership suggested appropriate projects to support. Appointment of committeemen with a known competency was modified to include, for their political help, leading citizens interested in specific regional projects.[85]

The Washington group of scientists quickly recognized the AAAS's potential as spokesman for science and established the annual convention as a podium. Joseph Henry moved first, seeking to persuade the members to support his plans for the Smithsonian Institution and generally to popularize and to promote the concept of a research institute. Strongly contending forces had sought to direct the use of the Smithsonian bequest, and those seeking a museum or a library remained active. The AAAS potentially provided the force of opinion necessary to persuade the "timid politicians" who hesitated to commit themselves in some positive way to Henry's policy.[86]

Both the general meeting and the separate sections were utilized to circulate the goals and to promote the activities of the Coast Survey and the Naval Observatory. In 1849 a committee was established to "evaluate" the Coast Survey, and it was retained from year to year, making reports whenever Bache felt there was some opposition to his annual appropriation in Congress.[87] Similarly, when the Nautical Almanac was founded and placed under the supervision of Charles H. Davis, it presented a prospectus of its goals to the association and sought formal approval of its hope for establishing an American prime meridian.[88] The naturalists also established an ongoing committee to memorialize the

---

Memorial of the Committee of the American Association for the Advancement of Science, on the Subject of a Geographical Survey of the State (Albany, N.Y., 1852), p. 7.

85. A. D. Bache to B. Peirce, Deerfield, Mass., 23 July 1849, Peirce MSS, HLM. Bache considered the implication of committee appointments and suggested the alternatives: if the appointments were to be given to recognized men, "you then endorse activity and merit instead of reposing on the conservative bed of position. If your object is to propitiate then subdivide your committee so as to use as many names as possible. . . . If on the other hand you are strong enough to do without propitiation and want work—have a committee of fewer members and all working men without regard to position."

86. Howard S. Miller, *Dollars for Research: Science and Its Patrons in Nineteenth-Century America* (Seattle, Wash., 1970), pp. 14–23.

87. The committees extended from 1849 through 1853 and from 1857 to 1858, and a lengthy report was published in 1860.

88. AAAS, *Proceedings*, II (1849), 122.

federal government so that scientists would be included on all exploring expeditions, and it was called into action with each new proposal for a federal expedition.

When Judge John Kane was appointed chairman of a committee to report on the Coast Survey near the end of the decade, he circulated a letter to his associates on the committee, giving them wide range for their inquiry and suggesting that the object of the report was "to submit to the Association, and through it to the community, a popular view of the subject and course of the Survey, and of the influence it is exerting and may continue to exert."[89] Kane died before the project was completed, and the task fell to Frederick A. P. Barnard. Because the assignment sent out by Kane was broadly construed, Barnard had a wealth of information and comments through which to work. Yet, as he wrote to his friend Julius Hilgard, "I would not give two straws for all the materials furnished," adding, "except that which came from the Coast Survey Office itself."[90] The resulting report is, in fact, the longest and the most comprehensive submitted to the association during the 1850s, and it bears the unmistakable imprint of Bache's own organization.[91] It was separately published in the fall of 1858 and appeared in the AAAS *Proceedings* in 1860.[92] The latter's stamp of importance was evidently of use to Bache in both validating and documenting the survey's work.[93] Such direct use of the association was not untypical, as suggested by the consistent support for the Nautical Almanac and the Naval Observatory. Similarly, William H. Emory of the Topographical Corps of Engineers attended the annual meetings to reinforce his support among the scientists.[94] Thus, while some of the petitioning committees dealt with new geological or oceanographic surveys to which scientific men might be

89. A copy of the circular sent to the twenty-man committee, dated 21 Dec. 1857, is found in Gibbes MSS, LC.

90. Cited in John Fulton, *Memoirs of Frederick A. P. Barnard . . . Tenth President of Columbia College in the City of New York* (New York, 1896). Barnard found the survey report very helpful, and he probably synthesized the views of committee members into its general outline.

91. J. E. Hilgard to L. R. Gibbes, Washington, 10 Jan., 26 Feb. 1859, Gibbes MSS, LC.

92. *Report on the History and Progress of the American Coast Survey up to the Year 1858* (n.p., 1858), Toner Collection, Rare Book Room, LC; AAAS, *Proceedings*, XIV (1860), 27–150.

93. Bache pointed out to Virginia Congressman Muscoi R. H. Garnett, 26 Feb. 1859, Bache MSS, LC, "If you can take time to look over the accompanying Report to the Committee of the American Association for the Advancement of Science . . . it contains one of the most concise accounts of the survey. . . ."

94. William H. Goetzmann, *Army Exploration in the American West, 1803–1863* (New Haven, Conn., 1959), p. 130.

profitably attached, many simply encouraged projects already underway.

The committee structure, a potential arbitrator for the scientific community, served more often as a prompted spokesman for association leadership which resolved ambiguities on its own terms. A private libel suit against Louis Agassiz focused the attention of the scientific community on the need to establish some arbitrator. James T. Foster, a schoolteacher from near Albany, designed a geological chart for use in the New York school system; it was quickly denounced in the public press by James Hall and Agassiz as inaccurate humbug and, moreover, without the precise New York nomenclature. Foster brought suit against the two and simultaneously solicited the aid of Ebenezer Emmons for corrections. The latter not only updated the map but also inserted his debated Taconic system, which incensed Hall. The immediate problem seemed trivial, if irritating, to the men involved, but the question of review and evaluation of a scientific product was central. Leading scientists rallied and appeared to testify on behalf of Agassiz, including Josiah D. Whitney, James D. Dana, Eben Horsford, and Joseph Henry. When the jury after only a few minutes' deliberation ruled against Foster and for Agassiz, the case against Hall was dropped and scientific review seemed vindicated.[95] Success for the resourceful scientists had come only with a well-prompted defense attorney and considerable time from leading men of science. As a result these men were persuaded that such matters of science were inappropriate in the public press or in the courts before either a judge or a jury which lacked scientific competence. The association never took any action on the matter, but members were privately urged to help Agassiz withstand the court costs since the principle was of concern to them all.[96] The matter of establishing an appropriate arbitrator remained a dilemma to the democratic-minded members of the AAAS, who were divided about the merits of their

95. The most thorough account of the controversy is in Clarke, *James Hall*, pp. 214–216; also see Hall MSS, NYSL. J. D. Whitney to W. D. Whitney, Boston, 1 Apr. 1851, cited in Brewster, *Life and Letters*, pp. 115–116. Whitney jubilantly related, "The more they cross-questioned us, the more the truth would come out. The Judge took the highest ground possible of the right to criticize."

96. Eben Horsford to J. Henry, Cambridge, 5 May 1851, Henry MSS, SIA: Horsford wrote, "In this matter, Mr. Agassiz has stood in the breach for the scientific men of the whole country, defending the right of scientific criticism." L. Agassiz to J. D. Dana, n.p., 20 Mar. 1851, Dana MSS, BLY: Agassiz observed, "I consider it as a valuable contribution to the progress of science in this country to have shown the mischief the charlatans do everywhere they are allowed to lift their heads."

organization to fulfill that function. The need for review was evident, but the value of a resolution was not clear.

Because the validity of a scientific claim was not always self-evident, committees were appointed to consider the merit of new techniques or apparatus. In these committees and on the occasion of its meetings, the AAAS allowed for the vindication and self-justification of beleaguered scientists. The pronouncement of these committees was not the final judgment—each new theory or instrument ultimately had to stand in the public market place of ideas—but it might clarify the issues involved. In addition, fundamental evaluation could authenticate the claims of the proposal systematically. For example, when astronomers debated over the priority and the accuracy of O. M. Mitchel's new method of determining right declensions, Mitchel's friend Benjamin Peirce wrote a letter of support. But Peirce also recommended that the matter be brought before the association in order

> to discuss this question fully, and adopt a course which shall settle its worth beyond all future controversy. Would it not be a good plan to propose a committee of reference to be selected by the President of the Association after free and open correspondence with all parties interested? [John] Locke will probably decline all connection with it— and perhaps Bond also—but, even in this case, the very act of declining will be an argument against their claims, and need not prevent a committee from reporting upon the published facts, which are the only ones to be considered in reference to parties who decline the conference invitation [word uncertain].[97]

At the next meeting an appointed committee reported favorably and even recommended that another committee be appointed to memorialize Congress to secure an appropriation for Mitchel.[98] Such an exercise of power by certain AAAS leaders for the benefit of a favored individual fed a growing resentment of other powerful members. The report on Charles Spencer's microscope was a simple sign of approval for that rural New Yorker's efforts to advance microscope technology. In this case Jacob W. Bailey of West Point, a microscopist previously impressed with the work of Spencer, offered a report effusive in praise.[99] Similarly,

97. B. Peirce to O. M. Mitchel, Cambridge, (?) Aug. 1851, Mitchel MSS, Cincinnati Historical and Philosophical Society Library. George Bond resigned from the AAAS that year, apparently in protest.

98. AAAS, *Proceedings*, VI (1851), 395. *New York Times*, 5 Aug. 1853, reported, "This committee was appointed, as it now appears, merely as a compliment. Professor Mitchel disclaims any desire to receive Government aid, and asks only the approbation of men of science and the confidence of the public."

99. AAAS, *Proceedings*, VI (1851), 397–398.

committees formed to consider the establishment of specialized journals in zoology and astronomy usually registered approval for projects underway.

During the first twelve years the AAAS did not deviate from basic precedents of the AAGN and its own first meetings. Promotion of science through committee activity was one such function, and in 1850 a memorial to the Wisconsin legislature claimed,

> The high position of its representatives is acknowledged and respected in the older communities in the seats of learning which have long held supremacy in philosophical investigation. The scientific character of the Union is an object of anxious interest with all, and the Association which is here represented has been established to aid in bringing together and combining the labours of individuals who are widely scattered, into an institution that will represent the whole.[100]

Internal debate and a changing political climate did, however, modify the committee operation. An analysis of the committees appointed from 1848 through 1860 indicates a discernible shift from the first to the second half of the period under discussion. Table 2 shows that while com-

TABLE 2

AAAS Major Committee Assignments

This table summarizes the committees assigned during the 13-year period from 1848 to 1860, not including the research or summary reports assigned to individuals. Sixty committees, considering only those newly appointed but not those renewed or simply adding new members, were counted and tabulated under four headings:

a) Administrative—dealing with procedural or constitutional matters of the association.

b) Research—assigned to answer quantitative questions of a scientific nature, for example, the astronomical observations of an eclipse.

c) Inquiry—assigned to consider and report on an instrument or theory.

d) Recommendation—usually memorializing a legislative or other group on the advisability of a particular project, for example, a state survey.

The year 1854 was divisional because there were seven meetings before the Washington meeting and seven more to 1860; in addition, analysis showed that the committee total was almost equally divided numerically at that point.

	Administrative	Research	Inquiry	Recommendation	Total
To 1854	4 7%[a]	4 7%	9 15%	14 23%	31
1854 on	15 25%	3 5%	7 12%	4 7%	29

[a] Percentage of total committees appointed.

100. Copy of the circular in Lapham MSS, WHS.

mittees dealing directly with matters of science (research and inquiry) were relatively stable, there was a reversal of emphasis regarding administration and recommendations. In fact, the number of committees appointed to evaluate and recommend projects declined even as the number of administrative appointments increased. The AAAS thus withdrew from its lobbying function and concentrated on the bureaucratic workings of the organization itself. One explanation for the withdrawal was that economic difficulties after the mid-1850s and political preoccupation with the Union constrained science even before the outbreak of war. In addition, old goals were accomplished but few innovative plans were underway. There is no evidence that the membership judged the promotional activity ineffective; Bache's maintenance of the committee on the Coast Survey argues pointedly to the contrary. Rather, an important restraint was the association's own decision to be more circumspect in both evaluating and recommending projects.

The decline in total number of evaluating committees, which were virtually eliminated by 1860, seems related to disagreement within the association over the function of such appointments. After his trial in Albany, Agassiz felt more strongly than ever that science should establish its own tribunals, commenting to Dana, "We are too much afraid of controversies in this country. For my own part I see not the least objection to them. . . . It will lead also to a freer criticism of the works published or publishing."[101] William Redfield concurred that open and free arbitration and discussion among scientists were necessary but argued that the value of an idea or an instrument would be best attested to by common consent rather than through an official pronouncement; the effect of a positive committee report meant little if a discovery or invention could not stand on its own. His proposed amendment to limit the constitution on this matter dropped from sight.[102] But, significantly, an amendment supporting the practice was not accepted either. Ap-

101. L. Agassiz to J. D. Dana, Cambridge, 15 Oct. 1853, Dana MSS, BLY.
102. William Redfield to J. Henry, New York, 4 Sept. 1854, Henry MSS, SIA. Redfield declined to serve on the committee to investigate Basnett's theory of storms, "chiefly on the ground that I deem it altogether improper for the Association to assume through its committees or otherwise the function of deciding on the merits, or demerits, of any theories or views in science which may be put forward by any persons.

"It is enough that such theories and views, with any evidence by which they may be supported, be communicated to the Association, and through its publications or otherwise to the world of science. This last and only tribunal will neither ask nor respect the official dictum or judgment of any organized association or society."

parently meant for submission at the Cleveland meeting, the following resolutions were a summary rationale of those members seeking broad, definitive powers for the AAAS:

> Resolved that it is the appropriate & peculiar duty of the American Association for the Advancement of science, to consider & report on all public undertakings of a purely scientific nature, to which its attention may be called by Memoir read before it, & to give a favorable or unfavorable opinion by committees or otherwise on the same. Resolved that it is the further duty of this Association to consider reports on all instruments (using the term in its most extensive sense) intended for patents when the same shall be brought either in person or by memoir only, shall be exhibited to the notice of the Association, but the committees who may from time to time, be charged with these matters are hereby positively enjoined and instructed in each and every case, first of all to consider whether the instruments or means so tendered for examination and approval possesses [*sic*] any claim to novelty or any marked improvement over others previously in use—and in case such shall be the fact in their opinion, to withhold any report on the same.[103]

The self-limitation imposed by not assuming this responsibility was tied to the fear that such committees could be too easily misused. In subsequent years the issue was gradually eliminated as fewer controversial committees like that on Mitchel's method were appointed, and committees of inquiry estimated the feasibility of a project rather than offering evaluations of completed works.

The continuing appointment of committees in general suggests that they served some function. At the least, they expressed the concerns of leading scientists and seconded efforts for research on the state and national levels. The scope and intensity of activity demonstrated science's stand on other issues as well. The government did not attempt to establish a formal policy for science, and the committees provided a means for *ad hoc* decisions.

The single instance of an individually sponsored proposal came before the association when Chester Dewey in 1856 posed a problem in conservation. He observed that the giant fir trees of California, "*Abels Donglassei*" (*Abies douglassei*) were in danger of extermination by the westward movement. He reported that a stand of perhaps twenty-five remained and that "an expression of interest in their preservation by the Association . . . would save the residue of these giant inhabitants of our

103. A draft of resolutions is in the Bache MSS, SIA, undated but in the folder for Aug. 1856.

primitive forests from destruction."[104] By resolution Joseph Henry was appointed to correspond with authorities in California and in Washington on the deforestation; no report was subsequently given.

Although the association took no formal interest in either of the great exhibitions in London in 1851 or in New York in 1853, it began inquiry into a problem sharpened by these international demonstrations, that of uniform weights and measures.[105] The topic was under discussion in Britain in the 1840s and brought to the attention of the AAAS by Henry Taylor.[106] When at the subsequent fairs prize juries attempted to compare articles, they found that differences in coinage, weights, and measurements made parallel examinations difficult; uniform standards of measurement seemed essential for international interaction.[107] The AAAS committee, formed at the recommendation of Benjamin Peirce, submitted no report, although committee member John H. Alexander independently published a pamphlet suggesting the need for American action.[108] In 1855 Bache, whose responsibility for weights and measures on the Coast Survey sensitized him to the problem, presented a brief summary of the committee's activity. Pressure of time prevented discussion; a recommendation made at the meeting to divide the committee, separating coinage from the matter of weights and measures, did not pass.[109] Finally in 1860 Bache again submitted a progress report, but it was basically a review of a published report on weights and measures by Alfred B. Taylor of Philadelphia, which had been submitted to the American Pharmaceutical Association.[110] Although the AAAS took no

104. AAAS, *Proceedings*, X (1856), 239. This tantalizingly early suggestion of interest in conservation stands alone in early AAAS records and was highly approved by David Wells in his *Annual of Scientific Discovery for 1849*, p. iv, as a necessary step "for obtaining the protection of the government" against "cupidity of private individuals."

105. The important role of exhibitions in awakening an interest in a uniform, international system is the theme in the *Report* from the Royal Commission on International Coinage (London, 1868) and is continued in two accounts by Edward F. Cox: "The Metric System: A Quarter-Century of Acceptance (1851–1876)," *Osiris*, VIII (1958), 358–379, and "The International Institute: First Organized Opposition to the Metric System," *Ohio Historical Quarterly*, LXVIII (Jan. 1959), 54–83.

106. Taylor to the AAAS Committee, "to consider the practicability of introducing a universal system of scales and standards for scientific measurement," London, 15 May 1850, AAAS Library.

107. See n. 106 above.

108. John H. Alexander, *International Coinage for Great Britain and the United States: A Note Inscribed to the Hon. James A. Pearce* (Baltimore, 1855).

109. AAAS, *Proceedings*, VIII (1854), xii.

110. *Ibid.*, XIII (1860), 245.

further action, the discussion was informative for members. It undoubtedly consolidated opinion for a more direct, if unsuccessful, movement relating to adoption of the metric system after the Civil War.[111]

The AAAS was the principal agent in the decade before the Civil War for representing and directing the scientific concerns of the nation. It was popularly accepted as the scientific congress and was well received by urban centers at which annual meetings were held. The compromise of scientific goals and popularization was uneasy, however. Even as public newspapers dismissed some reports as "too scientific," Jacob W. Bailey declined to attend the Albany meeting, arguing, "It would be pleasant to meet many who will be there, but the meetings have become too big, everybody is in a hurry, and there is little real enjoyment. As for the science, there is a precious deal of big talk for Buncombe, not a little of 'can nee, can thee' among certain would be bigbugs, and very little of that comes up to the proposed object of the meetings viz: the Advancement of Science."[112] Bailey's dismissal was too broad, but it demonstrated that the Congress of Savants walked an uncertain middle line.

In its increasingly cautious response to complex social-scientific matters, the association reflected the consensus of the most active scientists, a self-conscious group that emerged in part from the discussions and actions at the annual meetings. Gradual withdrawal from controversial topics was not uniformly accepted, but individual episodes culminated in an unspoken agreement that the annual meetings provide an arena for specific and limited scientific debate but not for broad or polemical topics. Similarly, the gradual shift in committee activity was related to the question of the appropriate function of the association in relation to matters of public policy. As the next chapter argues, modification in both realms was related to the emergence of an elite within the association and to imprecise interpretations of what a professional scientist should be.

111. Cox, "International Institute," p. 59.
112. Jacob W. Bailey to unknown, West Point, 10 Aug. 1856, Bailey typescript of "Selected Letters," New York Public Library, New York.

VI

The Development of a Professional
Outlook in the AAAS

Because it worked to encourage an effective communication among scientific practitioners and to establish mutually acceptable standards for research and publication, the AAAS was, at the least, an understudy for the role of professional organization. This goal of the association was not explicit in 1848, nor were leading founders agreed on specific standards and supervisory procedures. The impetus for greater control over the training and quality of research scientists arose from distrust about amateur activity and the hope of raising the international status of American science. Reinforcement for a professional hierachy came from personal and social factors, including the need for self-identity and encouragement among the full-time practitioners. The professionalization occurring not only in America but also in Europe likewise promoted concern for a disciplined organization. Engineers, doctors, educators, and others experienced similar stirrings so that the methods and vocabulary for scientific professionalism were markedly similar to those of other emerging professional groups. Not unaware that an expanded and initially undifferentiated membership was antithetical to the concept of professionalism, the scientific leadership moved to consolidate power among themselves in the association. The explication and application of the standards initiated by them resulted in heated debate between scientists and amateurs, for certification has the negative function of exclusion. In a struggle between the older eclecticism and the implementation of checkpoints and more rigorous evaluation of competency, the latter inevitably won. As the following chapter demonstrates, however, the growing pains of professionalism, while undoubtedly necessary and educative, marred the earlier harmony of the association, which served as forum for the debate.

Persons eager to improve the status of American science privately and publicly echoed Agassiz's denunciation of America's presumed dependence on Europe. It was therefore hardly surprising that at the first AAAS meeting in 1848 Robert Hare called for "independence of thought in matters of science," wryly observing, "If France deigns to notice, damning us with faint praise, we are overflowing with delight and promises of self-reliance all vanish."[1] The new association began with the intention of representing scientific men both at home and abroad. Reputation, normally a cumulative value accruing from a long-term effort, seemed most appropriately sought first in the American community. The young AAAS made few international overtures, aside from those toward neighboring Canada, and concentrated on institutionalizing American science. Recognition from abroad would then be based on the visible progress of a community of American scientists.

The implicit need to distinguish and control the standards of scientific research directly related to awareness of the same professional needs which initiated the organization of the American Medical Association, the Society of Engineers, and the American Association for the Advancement of Education. Each of these groups reflected new professional outlooks and a contemporary trend toward bureaucratization and centralization. In addition, each had formal links to science, although little evidence exists that scientists directly or consciously paralleled any of the other professional modes. Professional aspirations in all disciplines grew from a desire for better communication with colleagues, a uniform code of ethical standards, adequate financial and psychic compensation, and higher status. Europeans had similar broad goals despite differing vocabulary and immediate concerns.[2] In America the impetus to establish standards in medicine and law was long-standing, yet the accomplishments to mid-century were meager and confined to the state level. Daniel Calhoun contends, perhaps too strongly, that

1. *Philadelphia North American and United States Gazette*, 26 Sept. 1848. Ironically this complaint was registered by Hare just as young and aspiring chemists found it necessary to study chemistry in Europe, especially in Germany, because there were no adequate postgraduate universities in America.

2. For an excellent discussion of similar European developments, see Everett Mendelsohn, "The Emergence of Science as a Profession in Nineteenth-Century Europe," in *The Management of Science* (Boston, 1964), pp. 3–47. Mendelsohn implies that voluntary societies were less important than universities and industrial schools in establishing professional standards for Germany and Britain.

"America had gained little from the difficulties that the learned professions had experienced during the previous century, except a certain caution, a certain provincial decency and inspirationalism."[3] He suggests that the rising tide of professionalism at the beginning of the century had been temporarily retarded by the democratic impulse of the Jacksonian period.[4]

At least one contemporary observer agreed that a major inhibition to professional development was the nature of American society. After agricultural chemist James Johnston visited Canada and the United States, he reported in *Notes on North America*, "A circumstance which early strikes the European traveller in the United States, is the comparatively small consideration in which professional men are held, and the small salaries they in general receive."[5] Johnston attributed this low prestige to the relatively higher degree of general education in the young nation and especially to the theoretical and practical equality of its citizenry. Jacksonian America appeared substantially more egalitarian than the class society of England. Johnston concluded that "every man you meet thinks himself capable of giving an opinion upon questions of the most difficult kind."[6] The American scientific men with whom Johnston conversed undoubtedly reinforced such impressions. His critical observation suggested the frustration of men whose immediate environment offered little incentive to enforce standards and failed to provide what they considered adequate monetary or psychic compensation for their creative efforts. Collegiate professors with a full-time teaching load or doctors with an extensive practice frequently limited their time in the laboratory or in field research, where financial and social rewards were minimal.[7] Isolation, scarcity of research facilities,

3. Daniel H. Calhoun, *Professional Lives in America: Structure and Aspiration, 1750–1850* (Cambridge, Mass., 1965), p. x.

4. *Ibid.*, pp. 191–196. Hofstadter and Metzger make a similar point in *Development of Academic Freedom*, pp. 245–246.

5. James F. W. Johnston, *Notes on North America: Agricultural, Economical and Social*, I (Boston, 1851), 136. There is little information about scientific salaries. A useful study of professional engineers for the mid-nineteenth century suggests that salaries showed increasing variance and thus higher income for the leading practitioners during this period. See Mark Aldrich, "Earnings of American Civil Engineers, 1820–1859," *Journal of Economic History*, XXXI (June 1971), 407–491.

6. Johnston, *Notes on North America*, p. 135.

7. B. A. Gould to Gibbes, Cambridge, 12 Nov. 1849, Gibbes MSS, LC: "The duties of American professors are usually so severe that they have felt themselves for the most part unable to devote their leisure hours to research and abstruse investigation. They have needed all the time for relaxation, which they could spare for [*sic*] teaching."

and a myth of independent creativity hampered the cooperation necessary to professionalization of science.

The incentive to become professional was largely internal to the discipline because the public offered little encouragement. Indeed, the excesses of the peripheral reform movement that tended toward individualistic and unorthodox practice, especially in medicine, renewed interest in standardizing practitioners' training and techniques. The experience of the American Medical Association as a national body suggests a professional reaction to an undifferentiated society. The medical practitioners established a distinct occupational status relatively early; in the eighteenth century such subgroups as the pharmacists were already clearly distinguishable in the formulation of an organizational hierarchy. Gradually, through medical colleges and local societies of doctors, medical professional criteria were established. Because medical theory was in transition and standards not uniformly accepted, however, differentiating between legitimate practitioners with creditable theory and techniques and "quacks" was a persistent problem throughout most of the nineteenth century.[8] Neither state licensing nor academic degrees were standardized throughout the nation. Founded in 1848, the American Medical Association, resulting largely from regional efforts to establish comprehensive standards for training and practice, became spokesman for medical men nationally.[9]

Educators involved in reform on the collegiate level concentrated on curricular changes in the period before the Civil War; professionalization came later.[10] This trend reflected the fact that the criteria for entrance into collegiate teaching were built into the system; in many liberal arts colleges the faculty was selected primarily from alumni.[11] The short-lived American Association for the Advancement of Education

8. For example, see Martin Kaufman, "Homeopathy and the American Medical Profession, 1820–1860" (Ph. D. dissertation, Tulane University, 1969).

9. Joseph Kett, *The Formation of the American Medical Profession: The Role of Institutions, 1780–1860* (New Haven, Conn., 1968), p. 170. Kett's study is an extension of the important work of Richard Shryock on American medicine. Doctors formed a considerably larger group than the scientists; Kett estimates 60,000 orthodox practitioners when the AMA constitution was adopted in 1848.

10. Hofstadter and Metzger, *Development of Academic Freedom*, pp. 230, 245–253.

11. This seems true at least for professors who were leaders of the AAAS, 41 percent of whom taught at their alma mater. This percentage is especially striking because so many of the younger professors took degrees at established eastern schools and then moved to newly founded colleges farther west. This percentage also accounts only for the period between 1848 and 1860, which is a small portion of most faculty careers.

(AAAEd), founded in 1850, thus concentrated on making collegiate curricula broader based and more practical. Like the AAAS, the AAAEd had members actively involved in promoting the idea of a national university, and both organizations recognized that a school for advanced training in science was needed in America, as evidenced by the number of students who went abroad for postgraduate study. The relationship between the AAAEd and the AAAS membership was close, and the annual meetings were arranged so that persons could attend both.[12] Promotion of far-reaching curricular change and institutional research programs proved premature, however. As a result, the National Education Association (NEA), which represented practicing elementary and secondary teachers rather than reformers in higher education, superseded the AAAEd in 1857.[13] Both organizations indicated the growing group consciousness of the educators, although neither was yet professional.[14]

About this same time engineers were establishing an occupational identity. Specialized training created a relatively small group of civil engineers acquainted with one another through studying at the U.S. Military Academy at West Point or neighboring Rensselaer Polytechnic Institute or by working for the Army Corps of Engineers on railroad and canal surveys. Until about 1850, in fact, the military engineers trained by the Army filled a technical vacuum not only in the Army but also as advisers on state or private engineering projects.[15] Although a tight-knit group, they did not gather into any formal organization. Mechanical engineering emerged about mid-century as a distinct branch of engineering and was the first to create a formal organization.[16] Monte Calvert, studying the new mechanical engineers as professionals, de-

12. Alonzo Potter to Joseph Henry, Philadelphia, 16 Aug. 1852, and excerpts from Joseph Henry's "Locked Book" for 23 Dec. 1854, both in Henry MSS, SIA.
13. Edgar B. Wesley, *The NEA: The First Hundred Years: The Building of the Teaching Profession* (New York, 1957), pp. 20–26.
14. Albertina Abramos, "The Policy of the National Education Association toward Federal Aid to Education (1857–1953)" (Ph.D. dissertation, University of Michigan, 1954), p. 28. Abramos suggests that the early goals of the NEA were humanitarian rather than professional.
15. Daniel H. Calhoun, *The American Civil Engineer: Origins and Conflict* (Cambridge, Mass., 1960), pp. 182–188. In addition, see Goetzmann, *Army Exploration*; Raymond H. Merritt, *Engineering in American Society, 1850–1875* (Lexington, Ky., 1960); Clark C. Spence, *Mining Engineers and the American West: The Lace-Boot Brigade, 1849–1933* (New Haven, Conn., 1970).
16. Monte A. Calvert, *The Mechanical Engineer in America, 1830–1910: Professional Cultures in Conflict* (Baltimore, 1967), pp. 107–138.

scribed several characteristics of their outlook: the development of a systematic, technical knowledge base, recognition of the need for prolonged and specialized training, recognition of the need for socialization and control of both old and young practitioners, a service orientation, development of self-consciousness about status and role, and predominance of colleague over client orientation among practitioners.[17]

Like the engineers, the men of science demonstrated a tendency toward all these characteristics during the decade of the 1850s, although none was uniformly accepted or explicitly adopted by the AAAS membership before the Civil War. The physical sciences intrigued young engineers studying along the Hudson River and at other New England colleges. Engineers, astronomers, and mathematicians were often casually acquainted and shared similar educational and even occupational experiences. Similarly, postgraduate medical training expanded passing acquaintance with natural history into a systematic study of biology and zoology. Scientists like Joseph Henry, Benjamin Peirce, and Benjamin A. Gould attended the AAAEd meetings with proposals for an expanded science curriculum. The link of professional awareness was not explicit but certainly operated among these intellectually compatible doctors, engineers, educators, and scientists.[18]

One major difficulty which confronted the mid-century scientists was that of effectively applying standards necessary to establish the goals suggested by Calvert as indicators of professionalism. Men of science retained the anxiety of the 1830s regarding their international reputation and were especially anxious that publications reaching Europe from the United States be of high quality. This required control at the source. In addition, they wanted to ensure that government support be granted only to scientists of proven ability; the ideal would be an agency to select and promote such men. A primary concern was therefore to differentiate the levels of participants and establish the relative rank and status of each. Unlike the medical, legal, and engineering professions, whose activities impinged directly on a public which might see the effects of incompetent medical practice, poor legal counsel, or a faulty bridge, the scientific community found its incentive for promoting professional standards primarily through internal discontent. Applied scientists like agricultural chemist John P. Norton found the

17. *Ibid.*, pp. xv–xvi.
18. The only direct link was informally forged when in 1856 the AAAS voted to send a delegation to the AMA. See AAAS, *Proceedings*, X (1856), 237.

problems less difficult to identify but knew of no practical way to keep "false science" from providing worthless soil analyses to credulous farmers.[19]

Sociologist of science Robert Merton has suggested that modern scientific ethos has four basic premises:[20] (1) universality—faith that natural phenomena are everywhere the same; (2) communism or communality—principle of freely shared knowledge; (3) disinterestedness—science should not exploit findings for personal gain; and (4) organized skepticism—responsibility of each scientist to assess the goodness of the work of others. While the ideal rarely matches reality, the AAAS leadership implicitly accepted the first but only partially subscribed to the latter two principles. Eager and optimistic about science as a way of knowing, they assumed its universality. Moreover, they refused to "patent" ideas, feeling that general truths of nature could be used maximumly.[21] Struggling to survive financially and interested in applications, the scientists did use their findings for what might be personal or society's gain. Criticism probed many comfortable relationships between research scientists, but the tendency to put questions before enthusiasm was certainly increasing.

As initiator of and agent for the emerging professional aspirations of men of science, the AAAS became the arena for a debate over the amateur tradition versus professionalism. By the mid-1850s it was commonly accepted that the AAAS was in fact the national scientific congress.[22] The membership, which nominally reached the unwieldy number of nearly 800 by 1851, included more than simply the leading scientists. A heterogeneous mixture of class, occupation, and scientific interest made consensus difficult. In the absence of another spokesman, however, the general scientific community considered pronouncements of the AAAS authoritative.[23] The issue of certification was complex

19. Rossiter, *Emergence of Agricultural Science*, p. 121.

20. Norman Storer, "The Sociology of Science," in Talcott Parsons, ed., *American Sociology: Perspectives, Problems, Methods* (New York, 1968), p. 203.

21. An excellent discussion of this issue is found in Nathan Reingold, ed., *The Papers of Joseph Henry*, I (Washington, D.C., 1972), pp. 367–372.

22. Scientists as well as the popular press used the image of "the science of America in Congress assembled" or some variation for the AAAS. For example, see F. A. P. Barnard to E. C. Herrick, Oxford, Miss., 27 Nov. 1858, Herrick MSS, BLY; J. W. Bailey to [William Bailey], [West Point, N.Y.], 10 June 1856, Bailey MSS, NYSL; *New York Times*, 3 Aug. 1860.

23. Two articles suggest the importance of the AAAS in the establishment of a professional outlook within the American scientific community: George Daniels, "The Process of Professionalization in American Science: The Emergent Period,

and remains so today. The amateur tradition was long-standing and had incorporated a democratic belief that any diligent individual might make some contribution to even specialized knowledge. It argued in Baconian fashion that all men were potential participants in the efforts to advance human knowledge, in this case the exact or observational sciences. In general the American public supported this view of science as open enterprise—it fitted their vision of self-culture and progressive technology. The new professional aspirants responded with a system which denigrated or eliminated amateur activity by insisting on a critical review of all research efforts. Decision-making powers came to reside in a leadership whose standards were often vague. Observers and men excluded from participation, uncertain about new procedure or authority, became hostile when the new policies were arbitrarily applied.

Professionalization appeared first in scientific publications, which were increasingly edited by leading specialists rather than casually produced by a local society's secretary. In contrast to medicine, where the development of a national, professional organization resulted from local activity, the new trend in science was initiated by national leaders within a national organization. The *American Journal of Arts and Sciences* and the Smithsonian Institution's *Contributions to Knowledge* were the first to institute regular review for articles to be published in the 1840s.[24] The AAAS provided the first opportunity to discuss the broader implication of such professional tactics.

At each annual meeting after 1851 the standing committee called for the titles and abstracts of papers proposed for presentation. They then adjourned the meeting briefly to decide which papers should be presented in general sessions and to arrange sections and subsections appropriate for the papers as submitted by title. Often they specifically assigned papers to be presented at each sectional meeting. The list of papers was then printed in a daily program or in a local newspaper. Prior submission offered an initial checkpoint, but during the early years of the AAAS most reviewing came after presentation and discussion of the paper took place but before publication. While decision-making was assigned either to a specific publication committee or to the stand-

1820–1830," *Isis*, LVIII (Summer 1967), 151–166, and Holmsfeld, "From Amateurs to Professionals."

24. S. F. Baird to Gibbes, Washington, 25 Dec. 1850, Gibbes MSS, LC; Joseph Henry to Charles Ellet, Washington, 8 June 1849, Ellet MSS, Transportation Library, University of Michigan, Ann Arbor.

ing committee, in practice the permanent secretaries, Spencer F. Baird and then Joseph Lovering, exercised the power of discrimination. Recognizing that the AAAS *Proceedings*, distributed at home and abroad, were taken as representative of American science, the standing committee or its leading members continuously urged more careful review. The introduction of such regular review of papers before publication by the AAAS marked its emergence as spokesman for the new professionals of the scientific community. The efforts of these professionals appeared arbitrary to some members, and inevitably protests from excluded scientists were heard. A reaction to the more rigorous policy was surprising only in its delay. The visible influence of the standing committee became an issue among members who felt that the AAAS was a tribunal, a democratic congress of scientists which should debate major issues in full assembly. Exclusion of papers without such a hearing seemed a violation of democratic principles. The policy was still vague and general. While acknowledging specialization through sectional meetings, no formal policy demanded that specialists critique a colleague's paper to judge both the data and theory proposed.

In the early years of the AAAS scientific leaders informally censured questionable presentations by general but direct critical commentary. The comments might deal with method of research but most typically questioned the logic of conclusions. Such papers were usually included in the *Proceedings* at least by title. Discussion was an integral part of the association's activity, and critiques by able commentators might be included in the *Proceedings* along with the original paper to indicate what points were questioned and by whom the paper had been criticized. For example, Louis Agassiz responded in 1848 to a scientific note sent to the AAAS by Samuel J. Parker.[25] Complimenting the absent Mr. Parker on his zeal for science, Agassiz observed that "new" species were too frequently "discovered" before a researcher carefully examined the literature for known species. His reprimand was pointed but unoffensive. Many leading scientists exercised a gentle critical technique, but some, such as Benjamin Peirce, were sharp and abrasive. Their condescension and sarcasm compounded hostility toward the increasingly strict reviewing procedures.

In an attempt to avoid open confrontation at meetings, the AAAS leaders tried to avoid frivolous or sensational presentations whose scientific value was dubious. When someone suggested a paper relating a recent court case on poisoning, James Hall spoke forcefully against

25. AAAS, *Proceedings*, IV (1850), 348.

it. He recommended that the association not review the trial and evidence because this could "possibly place the Association in a position on every account undesirable."[26] Expert testimony by chemists in the courtroom was far more appropriate. Aging Robert Hare responded that all presentations should be allowed and that "the speakers alone were responsible not the Association."[27] The membership apparently did not discuss the issue and the paper was not presented. Hare's view was becoming outdated, and organizations were expected to exercise some responsibility for presentations and publications produced under their auspices.

A paper presented by Cleveland's Jehu Brainerd in 1853 forced the first public discussion of the procedures of evaluation and review that might result in exclusion of a paper from publication. With other western amateurs, Brainerd eagerly awaited the first regular meeting held west of the Allegheny Mountains and dutifully registered his paper. Immediately after its reading, however, James Hall rejected Brainerd's highly speculative paper on pebble formation and publicly urged that it not be published.[28] No one but Brainerd countered Hall's abrasive critique at the meeting.[29] The paper was printed in full in a subsequently suppressed issue of the Cleveland *Proceedings*, but when a revised Cambridge edition appeared in 1855, the offending article was omitted and not even listed by title.[30] At the Providence meeting that year Brainerd raised the issue of exclusion and requested that his paper be published. The matter, assigned to a committee consisting of James D. Dana and Benjamin Peirce, was dropped when the members reported that Brainerd's paper was inferior, "containing erroneous reasoning and unsound views, which, if sanctioned by this body would injure the cause of science and bring the Association into disrepute."[31] The AAAS upheld the committee's decision.[32] Brainerd vainly tried to defend both his theory and his right to a hearing, arguing that such exclusion would

26. *Washington National Intelligencer*, 27 Apr. 1854.
27. *Ibid.*
28. Holmfeld, "From Amateurs to Professionals," pp. 28–29.
29. *New York Times*, 1 Aug. 1853. Hall refused to respond to Brainerd's questions, stating somewhat arrogantly that the subject was not of sufficient importance to occupy his time or that of the association. Brainerd had submitted an earlier paper which was not challenged. See AAAS, *Proceedings*, V (1851), 222.
30. A copy of the suppressed *Proceedings* is in the Academy of Natural Sciences of Philadelphia Library; see Holmfeld, "From Amateurs to Professionals," for an extended comparison of the suppressed version with the authorized edition.
31. *Providence Journal*, 21 Aug. 1855.
32. AAAS, *Proceedings*, IX (1855), 305–307.

discourage any person whose ideas were different or new.[33] The AAAS ruling sustained the principle of totally excluding papers reviewed negatively by an established authority in a field.[34] Yet the policy was not clear. That same year A. Bigelow's paper "On the Drift" was interrupted during presentation, and the author was asked to take his seat. The *Providence Journal* recorded its puzzlement in a critical tone: "A hint is said to be sometimes as good as a kick. In this instance the rule appears to have been reversed. The idea seems generally to prevail that a paper which has once been received by the committee and announced on the printed programme, should at least be heard. To stop a man in the midst of his paper, on the ground of its being elementary, seems hardly fair according to the ordinary standards of judging and deciding."[35] In this case the right to exclude the paper was not at issue, but the time and technique of so doing seemed inconsistently applied.

The Cleveland meeting at which Brainerd delivered his controversial paper marked a turning point for professionals in the association. As John Holmfeld has argued, the *Proceedings* for 1853, first published locally, were republished in Cambridge two years later because they did not meet the AAAS leadership's standards.[36] Equally important, the "provincial meeting" had produced few sound research results. Waldo I. Burnett wrote later to Baird,

> By the way will not something be done by which committees will have more censorial power in reference to the Comm^s. as in the British Association? Truly some papers were presented at Cleveland—containing no more Science than a country Lyceum's communication and not half as well written—The object of the Association is to advance science in all directions, but its constitution is so democratic that, unless there can be some lateral conservatives kept up against the pseudo-science that is flocking in, the Association will fail of its object—to advance true science—and will go down. . . .[37]

Quietly the implementation of better controls began the following year. Two later dissidents affected by the new procedures were more force-

33. *New York Times*, 22, 24 Aug. 1855.

34. Undoubtedly in response to this challenge, the association voted more explicit directions concerning papers presented and read. It agreed that most papers presented would be read, but only those considered worthy would be printed. Their resolutions declared, "1. That papers read at this meeting be referred to the Standing Committee, to determine in reference to their publication. 2. That the papers not accepted be returned to their authors." AAAS, *Proceedings*, IX (1855), 282, 279.

35. *Providence Journal*, 20 Aug. 1855.

36. Holmfeld, "From Amateurs to Professionals," pp. 33–36.

37. Burnett to Baird, Petersburg, Va., 25 Apr. 1854, Baird OFF, SIA.

ful than Brainerd had been and refused to allow the matter to rest within the AAAS, which they believed to be controlled by a "Washington-Cambridge clique" whose purposes were less professional than personal. Both John Warner of Smithville, Pennsylvania, and Charles H. Winslow of Troy, New York, claimed that they were mistreated by Harvard mathematician Benjamin Peirce, with the implicit sanction of the AAAS. The purported antagonist, Peirce, was possessed by a desire to lead American science, and his arrogance scarcely masked his impatience with those who refused to acknowledge his scientific pre-eminence. Peirce's two priority disputes with George Bond, astronomer of the Harvard Observatory, and his accusation that the planet Neptune was technically discovered "by mistake," marked Peirce as a man eager to establish his importance and authority.[38] Tall and lean with a massive shock of hair and dark, deep-set eyes, Peirce was an imposing figure at the AAAS meetings, where his sharp wit and bitting sarcasm were notorious. Others tolerated his temper because of his skillful humor as well as his stature as a mathematician. An elitist predilection, however, made him unsympathetic to the efforts of amateurs, and his brusque handling of Warner and Winslow added to criticism of the "inner clique" of AAAS leaders with whom he associated. Neither of the two amateurs, in the end, hampered the establishment of a firm reviewing policy, but they forced a closer scrutiny of the procedures involved.

The Warner-Winslow controversy developed from two distinct grievances which coalesced in a joint attempt to exert pressure on the AAAS leadership through the public press. John Warner, an amateur mathematician from rural Pennsylvania, had Pythagorean ideas about numerical relationships and the generation of curves.[39] After reading several European authors on morphology—the sciences of forms—he agreed that the mathematics of curves had important implications for zoology and botany as well as for the solar system.[40] He decided to

38. Lenzen, *Benjamin Peirce*, p. 625. Lenzen has an excellent, although undocumented, analysis of Peirce's work. The personal interpretation of Peirce, however, is this author's own, based on familiarity with the Peirce MSS, HLH.

39. Warner's only other work was his *New Theorems, Tables, and Diagrams for the Computation of Earthworks* (Philadelphia, 1863), and he was a founding member of the Pottsville Scientific Association. See Pottsville Scientific Association, *Bulletin for January and February, 1855* (1855), p. 10. Daniel Kirkwood in an open letter described Warner as a "mathematician of highly respected attainments, an original thinker, and a high-minded honorable gentleman." Bloomington, Ind., 31 Dec. 1856, Warner MSS, APS. A note in the files of the American Philosophical Society Library indicates that Warner was a "Pennsylvania inventor and engineer" and that he died in 1873.

40. Morphology was of growing interest in Europe because it coincided with

"introduce" the topic into the United States by local publication.[41] His initial work on the classification of curves, submitted to the Smithsonian, brought mixed results.[42] Warner took the advice of one Smithsonian reviewer to expand his argument with more specific examples; then, although not a member, he decided to submit a paper to the AAAS. Louis Agassiz received it in 1855, and Warner anticipated that his findings would be presented at the Providence meeting that year. When the *Proceedings* from the meeting appeared in 1856, Warner found no reference to his paper but did discover a paper on a similar subject by Benjamin Peirce.[43] He then began an extended effort to retrieve his paper as proof of his own initial efforts in the field.[44]

the ideas of German *Naturphilosophie*. That it never was commonly studied in the United States is shown in Walter B. Hendrickson's "Naturphilosophie in the United States," *Ithaca: Proceedings of the Tenth International Congress for the History of Science . . .*, II (1962), 977–980.

41. His ideas were eventually published as *Studies in Organic Morphology: An Abstract of Lectures Delivered before the Pottsville Scientific Association in 1855 and 1856* (Philadelphia, 1857).

42. Warner to Joseph Henry, Pottsville, 26 July 1852, Henry MSS, SIA; Henry to Warner, Smithsonian Institution, 1, 19 Dec. 1853, and William G. Peck to Warner, n.p., 25 May 1853, both in Warner MSS, APS. The anonymous reviewer suggested that the graphical discussion of conical sections was not new, although Warner's suggestion of making conic sections a basis for classifying curves was. The reviewer added that "when the author has obtained a more profound knowledge than he now lays claim to he will find other fields more promising of rich harvest."

43. Warner outlined his active response to this discovery in "Professor Peirce and His Use of the Labors of Others" in the *Boston Daily Journal*, 28 Oct. 1857. He also had a fellow Pottsville Scientific Association member, P. W. Schaeffer, substantiate his own claim in an open letter dated 7 Nov. 1857, Warner MSS, APS. Warner admitted that the notices of Peirce's work on morphology "contain nothing which conclusively proves to my mind that his investigations in this department preceded any knowledge of my paper or were continued independently of it. But be this as it may, I cannot feel satisfied with his course in continuing to treat this subject whilst my paper was in his possession, without communicating with me or offering any explanation when called upon."

The precise reason for Warner's irritation seems to be in the loss of his paper and the failure of acknowledgement. In the Warner MSS is an outline of events surrounding the controversy which suggests that Warner knew of early work on the morphology of cells by W. I. Burnett and on elastic curves by Peirce presented at the Cambridge meeting but believed his work was more detailed and theoretical.

44. When he was unsuccessful in trying to get his paper back, Warner asked his Boston friend W. W. Whitcomb to call on Agassiz in person. Whitcomb reported that Agassiz had quite a cluttered office and that the paper might even be lost. He called on Peirce, to whom Agassiz said he had given the paper. Peirce first declared positively that he had never seen it and then acknowledged a faint recollection. Peirce sent Whitcomb to Lovering, who checked the report and

To Warner, who had worked without access to basic source materials, the Peirce article appeared to be plagiarism—of idea if not detail—and the act was compounded by deceit because Peirce even refused to acknowledge having seen his work. Warner believed that his papers, sent first to Henry and then to Agassiz, were pioneering efforts subsequently utilized unethically by Peirce.[45] The controversy only obliquely related to the AAAS because Peirce reported there on what Warner considered his conceptualization of a new and, in America, relatively unexplored area of science. Warner considered taking the matter to court but was dissuaded by Horace Greeley, who pointed out the difficulty of proving plagiarism and alternately recommended that Warner simply publish his findings.[46] Because he believed that the AAAS was dominated by an aristocratic clique, Warner resisted suggestions that he take the issue to a regular meeting.[47] Initially he solicited proof of his early effort from friends and verification of his mathematical ability. Then a press notice of another dissident, Charles Winslow, inspired him to turn to the public press.

At the time of the Montreal meeting in 1857 Warner read that Charles Winslow also charged Peirce with plagiarism.[48] Warner wrote Winslow, and the two became allies in a struggle to gain "justice" from the scientific establishment. In the case of Winslow the AAAS was more directly involved, and the two men reinforced their preconception that a vicious and self-serving elite dominated the association.

Charles H. Winslow, a physician in Troy, New York, had become interested in the problem of physical forces and the structure of earth forms. He expressed his thoughts in a preliminary essay in 1854 en-

---

found no record. Whitcomb concluded, "My opinion is that Prof. A. in his haste and as is customary with him laid it aside with other papers and that is the last of it." Whitcomb to Schaeffer, Boston, 7, 24 Mar. 1857, Warner MSS, APS. Schaeffer himself finally decided that "the paper alluded to has been rejected by the American Society or that its merit was not great enough to enforce attention." Schaeffer to [Whitcomb], Pottsville, Mar. 1857, draft copy, Warner MSS, APS.

45. See n. 43 above.

46. Warner to Greeley, Pottsville, 30 Jan. 1858, and Greeley to Warner, New York, 13 Feb. 1858, both in Warner MSS, APS.

47. This proposal by J. W. Yardley was rejected by Warner, who noted that the Pottsville publication made another paper on the subject unnecessary. Yardley to Warner, n.p., 10 May 1858, Warner to Yardley, Pottsville, 13 May 1858, William P. Foulke to Warner, Philadelphia, 3 June 1858, and Warner to Foulke, Pottsville, 6 June 1858, all in Warner MSS, APS.

48. *Rochester Union and Advertiser*, 17 Aug. 1857.

titled "Cosmography."[49] Subsequently traveling in the Pacific, Winslow developed a theory of continent formation as related to solar activity and made a trip to the Sandwich Islands to gather further evidence.[50] In 1856 he returned just in time for the gala Albany meeting of the AAAS and presented his results. Although the newspapers reported almost no discussion on the presentation, positive or negative, the copy Winslow presented to Lovering was not published in the *Proceedings* for that year or even mentioned by title. Winslow, although believing that an erroneous theory was harmless, acknowledged the validity of scientific reviewing, even after his paper was rejected.[51] But the decision-making power, he maintained, must be systematic and entrusted to men dedicated first to the advancement of science; the power to make decisions should not be arbitrarily wielded.

Winslow charged that in suppressing his paper Peirce and his colleagues in the AAAS had acted arbitrarily. In the late spring of 1857 Lovering informed Winslow that his paper had been rejected for publication. Winslow requested its return but received no reply. When the *Proceedings* actually appeared without mentioning his paper even by title, Winslow again inquired of Lovering about it. The latter replied that it had been rejected because the standing committee at Albany found it "inadequate."[52] On checking with a local AAAS member who had served on the standing committee, Winslow discovered that his paper had not been discussed by that group as a whole and concluded that in fact Lovering and perhaps Benjamin Peirce were responsible for the exclusion of his paper.[53] Winslow suspected the latter because Peirce had elaborated a somewhat related theory on the shape of the

49. *Cosmography; or, Philosophical Views of the Universe* (Boston, 1854). Winslow (1811–77) was apparently a physician. Born in Nantucket, he was living in Troy, N.Y., during the early stages of the controversy, but later moved to West Newton, Mass. See *Herringshaw's Encyclopedia.*

50. *Boston Daily Journal*, 27 Oct. 1857. In an open letter to Peirce dated 5 Aug. 1857, reproduced in the *Journal*, Winslow traced the controversy's history.

51. Winslow somewhat glibly introduced his *Cosmography* with the assertion that "the views, however, if altogether erroneous, will do no harm to mankind; but, if correct, they will . . . impart vigor to scientific inquiry. . . ." P. iv. He added later, in the midst of the controversy, that it was entirely proper to review articles for publication if "every gentleman connected with the Scientific interests of our country were fair & honest." What he objected to was the activities of a clique of the leading scientists in Cambridge and Washington, bound first "to each others interests & defense at all times and under all circumstances." Winslow to Warner, Troy, 2 Oct. 1857, Warner MSS, APS.

52. *Boston Daily Journal*, 27 Oct. 1857. Lovering apparently explained that the paper contained "nothing new or valuable" or "leading to positive results."

53. Winslow to Warner, Troy, N.Y., 21 Nov. 1857, Warner MSS, APS.

continents and was currently enjoying high praise for his discovery, which Winslow found suspiciously similar to his own ideas.[54] After considerable delay Winslow received his paper, minus its final section on continental shape. As a result, in August 1857 Winslow submitted an open letter to the public newspapers suggesting duplicity on the part of Peirce and then went to the Montreal meeting of the AAAS to vindicate his claims. Witty but evasive, Peirce refused to discuss the matter and simply countered the comments and queries of Winslow by claiming that the latter's ideas were a "conglomeration of mental house vageries [sic]."[55] Relatively unknown and virtually excluded from any further challenge within the AAAS, Winslow extended his attacks in the press.[56]

This activity brought him to the attention of David A. Wells as well as the frustrated Warner. Wells, a graduate of Lawrence Scientific School under Eben Horsford, combined his academic training with an interest in journalism. In 1850 the two interests were first linked in the *Annual of Scientific Discovery*. Wells began each yearly volume with an account of the AAAS, and as debate over the constitution became heated in 1855 and 1856, he took the side of dissident members, convinced that the outcome of the vote would determine the very existence of the organization.[57] When he heard of Winslow's difficulty, he volunteered to publish the disputed paper, adding that a similar paper had been published by the BAAS without any suggestion of nonoriginality or speculation.[58]

Similarly, Wells supported Daniel Vaughan, a western mathematician and astronomer who had submitted three papers at the Montreal meeting, all of which were rejected.[59] Wells again deliberately challenged

54. *Christian Examiner*, LXIII (July 1857), 151–52; *Boston Daily Journal*, 26 Oct. 1857.

55. *Boston Daily Journal*, 27 Oct. 1857.

56. For example, see *Boston Daily Traveller*, 21 Aug. 1857; *Boston Daily Courier*, 18–21 Aug. 1857.

57. In the *Annual . . . for 1854*, p. 5, Wells predicted, "Upon the action which may be taken in regard to the constitution, and the future government of the Association, its harmony and prosperity will essentially depend." The annual volumes of Wells offer a helpful index to the most outstanding scientific and technological advances for a given year; although not directly engaging in polemics, they did furnish an opposition press for men thwarted by the AAAS or the *American Journal of Science*. Unfortunately the Wells MSS, LC, emphasize the period after 1864, and none of the early letters deal with Wells's editorial opinions on the AAAS.

58. *Annual . . . for 1858*, pp. 364–370.

59. Vaughan (1818–79) was interested in astronomy, mathematics, and chemistry. Teaching in Kentucky and Ohio and out of the mainstream of science, he was

the assessment of the AAAS by publishing one of the rejected papers with a note of European approval.[60] Vaughan himself introduced yet another tactic in response to his exclusion. In 1858, unable to attend the Baltimore meeting, he sent a circular to AAAS members with the abstracts of three papers. He prefaced these simply with a comment that private circulation of new findings was necessary because there was "so little security in the AAAS."[61]

The newspaper controversy undoubtedly damaged the reputation of Peirce, already a controversial figure in Cambridge. Yet the tactic of going outside the community of science to court popular opinion ultimately rebounded against the critics. The early Foster map case against Agassiz and Hall persuaded nearly all scientific leaders that the courts were not appropriate arbiters of scientific disputes, and the emerging professionals also suggested that popular opinion was totally unqualified to judge scientific matters. All arbiters external to the scientific community itself were rejected.[62] Warner, Winslow, Vaughan, and Brainerd were provincials outside the East Coast scientific mainstream; they had no leverage in the scientific "establishment," the AAAS, and therefore felt forced to appeal to public opinion.

During the fall and winter of 1857 Warner and Winslow maintained a barrage of editorials and open letters. The press coverage, anti-intellectual in tone, was generally sympathetic to the dissidents and hostile toward the Cambridge professors.[63] Yet Winslow soon recognized that they were defeated. When Winslow's friend Thomas Hill supported Peirce in an unsigned editorial, Winslow knew that the Cambridge scientists were united.[64] Similarly, early in the controversy J. Peter Lesley had written to his friend Warner that "it is unsafe in every way and useless to *show up* great men," instead suggesting, "They succeed in the end in doing that for themselves and save us the trouble." Lesley

apparently a quiet, gentle man characterized by Daniel Kirkwood as an "odd genius." *DAB*; Kirkwood to Warner, Bloomington, Ind., 8 Apr. 1858, Warner MSS, APS.

60. *Annual . . . for 1857*, p. 372.

61. Copy of the circular is included with a letter from Vaughan to Warner, Cincinnati, 4 May 1858, Warner MSS, APS. To Warner he observed, " I regret that original inquiries meet with so little liberality from our American Association and so little security for the fruits of labors that the members are compelled to resort to this plan of giving their views to the public."

62. See Chapter V.

63. Warner to W. F. Miskey, Pottsville, 9 June 1858, Warner MSS, APS. Warner was generally pleased with the coverage, noting recent editorials in the *Pottsville Miner's Journal*, the *Springfield Republican*, the *Troy Whig*, and others.

64. Winslow to Warner, Troy, 19 Sept. 1857, Warner MSS, APS.

urged that Warner publish his ideas as original.[65] Later, however, he discussed the matter with Peirce's friend John LeConte, with the result that he brusquely recommended that Warner cease publishing "innuendoes" in the press and take the matter to a scientific society.[66] Men of science agreed that scientific disputes should not be fed into the "scandal-loving maw" of the public press.[67] In any case, as Lesley argued to Warner, "a victory over public opinion as such is worth nothing."[68] With former friends in opposition, Warner and Winslow retreated from public battle.

Although Warner and Winslow realized no satisfaction from their efforts, the issue they raised intensified debate in the AAAS about its exercise of power and contributed to a more general questioning of the leadership of certain men. At the least, many AAAS members could agree that Vaughan had been treated less well than propriety demanded and that the Cambridge scientists mishandled the papers of Warner and Winslow.[69] Winslow found support in such men as Edward Youmans, Benjamin F. Green, and Charles Wilkes, all of whom agreed that the association under current leadership was a negative force for advancing science. They accused the association, as directed by Henry, Bache, and other Washington and Cambridge scientists, of being the "machine of a clique for their self-aggrandizement," used to "keep them before the political men of the country as the greatest scientific men of the republic for the purpose of continuing their situations and securing government pay."[70] There was a fleeting attempt in Albany in 1856 to found a counter organization and one at the Baltimore meeting in 1858 to force a change in leadership. Too many AAAS members seemed unaware of the problems of Warner and Winslow and apathetic about the concentration of power to bring significant change. When the small gatherings of dissidents concluded that nothing could be done, a few followed the examples of Matthew F. Maury and Wells in deciding simply not to attend the meetings or to withdraw from the association entirely.[71]

Both the nature of the work of the overt dissidents and their tactics

65. Lesley to Warner, Philadelphia, 21 Feb. 1857, Warner MSS, APS.
66. *Ibid.,* 9, 26 June 1858. After a talk with LeConte, Lesley stated that Warner had no case, while admitting the "integrity and originality" of his work.
67. *Ibid.,* 26 June 1858. Peirce to John L. LeConte, [Cambridge], 23 May 1858, LeConte MSS, APS.
68. Lesley to Warner, Philadelphia, 9 June 1858, Warner MSS, APS.
69. Kirkwood to Warner, Bloomington, Apr. 1858, Warner MSS, APS.
70. Winslow to Warner, Troy, 13 Oct., 25 Dec. 1857, Warner MSS, APS.
71. Wells to Warner, New York, 25 June 1858, Warner MSS, APS.

of taking the controversy beyond the scientific community practically precluded their success. The association leadership, considered a clique by its opponents, offered scant response, depending on a growing professional spirit to vindicate in theory and in practice a stricter reviewing policy. An anonymous letter written by Thomas Hill, John LeConte, and Chauncy Wright partly answered the charges made by Warner and the others but primarily emphasized the impropriety of handling the problem anywhere but in a scientific tribunal. Science closed ranks. A common ongoing concern was the establishment of high standards for papers presented. Although the AAAS leadership maintained that the issue was one of scientific purity or integrity, the distinction which they made between true and speculative science was not so clear as they assumed. As Warner noted, Peirce himself indulged in speculative and theological orations. Inclusion of certain papers suggested that the leaders in power were as concerned about group solidarity as about professional standards.

Eventually the complainants retreated to requesting only the return of their original papers. The reasons for the delay in returning the papers remain a mystery. Peirce's own testimony, recorded second-hand, that he had seen the papers in question suggests that he carelessly mislaid them or in a fit of temper against amateurism decided to ignore the problem. The resulting bitterness from these encounters was deep and long-lasting. In 1867 Winslow pointedly dedicated his *Force and Nature* to the "British Association for the Advancement of Science and to the scientific students of the astronomical sciences in America."[72] Indignation had a moral edge, for when Hill protested that Peirce was too righteous to have utilized the work of someone else, Winslow merely commented, "I don't believe any wings have grown out of these Cambridge managers of the AAAS since they first acquired direction of its affairs."[73] If they failed in their purpose of bringing "the reign of scientific despotism to a close,"[74] the challengers forced some rethinking about the manner in which professional standards and designation of authority should be established. At subsequent meetings individuals were generally treated with more patience and candor.

In situations where members were in fairly common agreement on the quality of a presentation, the situation could be almost humorous. At the Newport meeting in 1860 Dr. Barratt, a physician from Con-

72. (London, 1869).
73. Winslow to Warner, Troy, 19 Sept. 1857, Warner MSS, APS.
74. *Ibid.*

necticut who had earned some distinction in botany, asked to read a paper of dubious merit on the tridactyl. The standing committee assigned him a presentation time in the natural history section coincident with that for the report of the astronomical expedition returning from Labrador in the other section. The latter was of such current interest that only two persons went to hear Dr. Barratt, and when they realized the situation they quickly retired from the room.[75] Such a tactic might have prevented an awkward confrontation. However, Barratt held his place, and when the session resumed he began to present his paper. Eventually suppressed laughter and criticism about his conclusions from Lewis Gibbes and William Rogers forced him to take a seat. No notice of this paper appeared in the AAAS *Proceedings*. Such experiences strengthened the policy of reviewing papers not only for publication but also for presentation.

Yet a more direct policy could not prevent bitterness on the part of a would-be author. When Robert Lachlan's proposal for a joint meteorological observation system was not even placed on the agenda for discussion at Montreal in 1857, he decided to renew his application. Lachlan checked in advance with James Hall about the proper procedure for submission to ensure that he would receive a hearing at Baltimore in 1858.[76] He did, but after Joseph Henry voiced skepticism about the proposal at the meeting, the report was handed to the standing committee for consideration and subsequently entered only by title in the *Proceedings*.[77] Lachlan resored to the tactic of Warner and Vaughan, privately publishing his *Paper and Resolutions* in Cincinnati. There he commented that when a reading of the Baltimore *Proceedings* indicated that the "Standing Committee of that Institution had adhered to their hastily-formed resoluion of limiting all notices of a Paper and Resolutions ... [he] decided to appeal against so mortifying a decision to the impartial tribunal of public opinion."[78] Once again the public tribunal remained eloquently silent, and the AAAS was unaffected.

The general policy of excluding all nonscientific or quasiscientific matters was sometimes modified, however, for men who had established reputations. In the case of Robert Hare, distinguished for his early re-

75. *New York Times*, 11 Aug. 1860. Deference for age and reputation was clearly operative.
76. R. Lachlan to Hall, Cincinnati, 10 Apr. 1858, Hall MSS, NYSL.
77. AAAS, *Proceedings*, XII (1858), 283.
78. Robert Lachlan, *A Paper and Resolutions in Advocacy of the Establishment of a Uniform System of Meteorological Observations, throughout the Whole American Continent* (Cincinnati, 1859).

search in chemistry, the AAAS confronted a peculiar problem as the aging Philadelphian became persuaded of the truths in the spiritualist movement. His 1848 presentation to the AAAS, "Definitions and Discrimination Respecting Matter, Void, Space and Nihility," was basically a plea for spiritualism.[79] His preoccupation with the topic persisted. When he introduced the subject in conversation after a scientific meeting in Boston, A. A. Gould commented that it seemed to indicate the "break-up" of a powerful mind.[80] Hare's old friend Benjamin Silliman was deeply concerned that this mystical religion was not Christian, and other men of science were gentle with Hare, who still spoke on occasion with expertise on scientific topics.[81] Although Joseph Henry prevented Hare from presenting a paper on spiritualism in 1854 before the AAAS,[82] Hare's continued insistence for a hearing gained him permission to speak at Albany in 1856 with no record of his address printed.[83] Like other spiritualists, Hare was convinced that spiritualism could be adequately demonstrated on scientific terms. In Cambridge in 1857 H. F. Gardner persuaded Agassiz, Peirce, Horsford, and Gould to "investigate" spiritualism in a specially arranged session, and the four used the opportunity to denounce the entire enterprise as a "stupendous delusion."[84] When Hare went to the Montreal meeting that year and again asked for a hearing, he was verbally attacked by a member of the AAAS, defended with deference by Peirce, but not allowed to speak again on the subject.[85]

79. AAAS, *Proceedings*, I (1848), 76–78.

80. A. A. Gould to Benjamin Silliman, Boston, 29 Dec. 1854, Gratz Autograph Collection, HSP.

81. Edgar F. Smith, *The Life of Robert Hare: An American Chemist, 1781–1858* (Philadelphia, 1917), pp. 500–504.

82. Joseph Henry to J. D. Dana, Smithsonian Institution, 24 May 1855, Dana MSS, BLY; Henry to Hare, Smithsonian Institution, 3 Feb., 7 July 1857, Hare MSS, APS.

83. Smith, *Life of Robert Hare*, p. 483. Handwritten minutes of the AAAS record Chester Dewey's motion that Hare be allowed the use of the hall after the AAAS meeting adjourned.

84. H. F. Gardner to Hare, Boston, 23 July 1857, Hare MSS, APS. The *Boston Courier* had offered a $500 reward to anyone who could prove spiritualism scientifically, which Gardner refused to contend for on principle. He did, however, also hope to establish the legitimacy of spiritualism through scientific approbation. Also see Hare to unknown, n.p., n.d., Hare MSS, APS.

85. *Boston Daily Journal*, 27 Oct. 1857. Thereafter the subject did not appear before the AAAS except when a Mr. Pierce of Flushing, Long Island, startled the Springfield meeting with a discourse on spiritualism which he claimed was preamble to his request for a motion to appoint a committee of three to examine whether spiritualism ought or ought not be considered a science. "In two minutes the subject had been tabled and sobriety restored." *New York Times*, 9 Aug. 1859; such events were never recorded in the *Proceedings*.

Amateurs and onlookers were increasingly excluded from participation at AAAS meetings and in subsequent years turned to local societies and private publication for a hearing.

The new scientific professionalism remained incomplete and the review policy loose and informal until after the Civil War, with the decision-making power still the preserve of the standing committee and exercised most often by the permanent secretary. Undoubtedly convenience dictated this pattern, as it was easier to decide the merits of a paper with colleagues over tea in Cambridge than under the pressure of a heavy agenda and in a gathering of fifteen men at the close of a busy annual meeting. The questions raised by Warner, Winslow, and Vaughan provoked both private and public discussion on professional rights and responsibilities, which in turn reinforced and clarified the professional outlook of the organization. A recognition of the need for codification of professional principles resulted in the association's request that Joseph Henry write a code of ethics.[86] Perhaps equally important for the mid-century AAAS, the dissidents' complaints brought into public view a leadership elite which, despite honest intentions to advance science, seemed too often to serve its own purposes. By 1860 the American scientific community had only touched on the modern concept of professionalism. The establishment of an organization did not necessarily indicate more amateurs but did make those persons more visible.

In other ways the AAAS did provide mutuality and an information network. The probability of critical comments undoubtedly persuaded authors to be thorough and sure in their presentations; the majority of papers represented the very best work being done in the country at that time. Informally the association provided a kind of marketplace for jobs, an opportunity for exchange and criticism of ideas, and a platform from which to expand support for science. In retrospect, the uncertain steps toward review were essential to the establishment of standards. Unfortunately the scientific leaders did not uniformly apply to their closest peers the same standards imposed on the larger scientific community. This failure to be precise and impartial, as Chapter VII demonstrates, nearly destroyed the AAAS.

86. AAAS, *Proceedings*, XIV (1860).

VII

The Dilemma of Leadership

Inspired by Alexander Dallas Bache's address in 1851, James Dwight Dana wrote later that year, "However much motives may be impugned, it is of the utmost importance that those who know what true science is should strive to keep the Association in the right path."[1] Dana's assumption that a few men best understood the essence of true science and had the necessary administrative and professional ability to direct its path was indicative of the spirit which created a leadership elite in the AAAS. Paradoxically, this elite symbolized professional unity at the same time that it bred a bitter struggle within the young association. In the early years of the association, as in the AAGN, democratically minded scientists somewhat naively assumed that the association could function with minimal forethought and planning.

Yet, in discussing the new Smithsonian Institution, T. R. Beck had warned Joseph Henry,

> I will in conclusion only hint at a danger which unless early and constantly guarded against, may render your scheme unpopular and in a measure impair its usefulness. It is the possibility of the selection of particular persons—or of associations in different places, who may appear to assume control in any particular department of science—in other words, the formation of predominant cliques. These are the curse of most of our distinguished societies at home and abroad—and in this country the danger is greater, from the fewness of men well grounded in science and the disparity that exists between these claiming to be adept.[2]

In fact, the highly visible and inclusive AAAS created a leadership vacuum when it assumed the broad role of national scientific spokes-

1. J. D. Dana to A. D. Bache, New Haven, 6 Sept. 1851, Rhees MSS, HHL.
2. Beck to Henry, Albany, 29 Nov. 1847, Rhees MSS, HHL.

man and elected to represent professional standards in the presentation and publication of papers. A group of like-minded scientists, conscious of their own reputation and of the new national organization's potential, rapidly joined together to fill that vacuum and direct AAAS activities by logical rhetoric and the power of office. Not surprisingly, the elite's presumptions and its manipulative tactics created a loyal opposition within the association. Because professionals comprised the new elite, the struggle of amateur versus scientist seemed to many to be the issue. The concentration of power in the hands of a few leading scientists also aroused their professional peers, who resented the accumulation of authority and questioned its procedures.

THE BACHE FORMULA FOR LEADERSHIP

An elite was perhaps inevitable and not incompatible with the democratic composition of the AAAS. As Richard Merritt suggests for western political processes in general:

> It is they [elites] who at once shape and reflect the values, attitudes, and beliefs of a society, who set the standards of taste and elegance, who are most instrumental in molding the public institutions that are the framework for our lives. It is they who, individually or collectively, formulate or articulate demands within the political system, who maintain the flow of communications within the system, and who enact, interpret, and enforce the rules that govern us.[3]

Such elites are implicit because a political democracy assumes that there are "positions of power in the society . . . open in principle to everyone, that there is competition for power, and that the holders of power at any time are accountable to the electorate."[4] But elites are not confined to political processes, for they are evident in business, the arts, and intellectual life as well.[5] Politically and socially the era of Jacksonian democracy was not without the elitist tendencies which became apparent among the scientists. The scientists obviously enjoyed the hospitality afforded by civic leaders at the annual AAAS meetings. Some,

3. Richard L. Merritt, *Systematic Approaches to Comparative Politics* (Chicago, 1969), chapter 4, p. 1.
4. T. B. Bottomore, *Elites and Society* (New York, 1964), p. 10.
5. See "Elites," in *International Encyclopedia of the Social Sciences*, ed. David Sills (New York, 1968), pp. 26–29. Suzanne Keller points out that such elites are more vaguely defined than those in politics and diplomacy and hence are also more controversial.

too, moved among the best circles at home. Bache, a great-grandson of Benjamin Franklin, was accepted by the high society of Philadelphia and, not surprisingly, he enjoyed the intellectual fellowship there. In the 1850s elite groups, formal or informal, existed in most major American cities, and leading men of science, by virtue of accomplishment if not birth, frequently were guests if not actual members of such groups. In Cambridge Agassiz and Peirce were members of the famous Saturday Club of Oliver Wendell Holmes.[6] The Century Club of New York included Oliver Wolcott Gibbs and Lewis M. Rutherfurd, while nearly all American Philosophical Society members could join the Saturday night Wistar parties in Philadelphia.[7] All such groups were known for their congenial atmosphere and prestigious associates. The leading scientists, as members and as guests at such gatherings, accepted without question the community leadership of these men. Such social familiarity could bring a working relationship which for the energetic, ambitious Bache was the ultimate justification for participation. He was similarly impressed when he joined the more exclusive members of the British Royal Society at their club, whose aristocratic tendencies were part of the accepted scientific hierarchy of Britain.[8]

The elite identified by unhappy dissidents John Warner and Daniel Vaughan as the "Washington-Cambridge clique" emerged among several leading AAAS scientists because, as individuals, they agreed together on the need to preserve the integrity and enhance the reputation of American science. They exercised authority primarily because they had established reputations in their own right through affiliation with leading national institutions and agencies and by virtue of their own scientific activity. At informal gatherings, frequently in conjunction with AAAS meetings, their common goals for American science, their opinion of its current efforts, and their incisive and often witty repartee drew them into a spirit of camaraderie. Out of an endless drive to secure appropriations for research came a casual agreement to call themselves the

6. Edward W. Emerson, *The Early Years of the Saturday Club, 1855–1870* (Boston, 1917).

7. *The Century, 1847–1946* (New York, 1947), pp. 3–24. There is no study of the informal Wistar Club, although a delightful anonymous satire suggests its activities: *Philadelphia: or, Glances at Lawyers, Physicians, First-Circle, Wistar Parties, &c. &c.* (Philadelphia, 1826).

8. Archibald Geike, *Annals of the Royal Society Club: The Record of a London Dining-Club in the Eighteenth and Nineteenth Centuries* (London, 1917), p. 323. Charles Wilkes and Henry D. Rogers were among the American guests on other occasions.

Lazzaroni, or "scientific beggars."[9] Their clearest point of consensus was that American science could progress only if adequate financial support were given to scientific research. They were far less concerned with the popularization of science than with its advance. Members of the self-proclaimed Lazzaroni were empire builders, men committed to promoting American science through research and institutions for its support.

All discussion of the elitist Lazzaroni reflects the elusive nature of the confidential fellowship.[10] Historians have often noted the central role played by the group, known first as the Florentine Academy and then the Lazzaroni, in projecting the Albany university scheme, directing affairs of the Dudley Observatory, and forming the National Academy of Sciences. In addition, the men gave mutual support to one another's institutions. This attitude was most evident for the Smithsonian Institution and the Coast Survey when they were under fire in the 1850s and for the Lawrence Scientific School under Agassiz. Some analysts have also noted that the key Lazzaroni members were active, if sometimes disgruntled, participants in the AAAS. No one, however, has

9. In Italian *Lazzaro* referred to the lower class of Naples, and *Lazzaroni* gradually came to mean persons of humble origin or low caste who were restless or in revolt. *The New American Encyclopedia: A Popular Dictionary of General Knowledge* (New York, 1858-63) suggests that the name was imprecise in detail yet did parody the elitist scientists in a curious way: "They still annually elect their chief, the capo lazzaro, the election taking place in the open air, and being determined rather by clamor than by choice." The author of the piece is unlisted, although both B. A. Gould and C. C. Felton contributed articles to the encyclopedia.

10. Earlier references to the Lazzaroni are in Edward Lurie, "Nineteenth-Century American Science: Insights from Four Manuscripts," Rockefeller Institute, *Review*, II, no. 1 (Feb. 1964), 11-19, and Storr, *Beginnings of Graduate Education*, pp. 67-74, 82-84. More recent analysis has concentrated directly on the men involved and their goals. Richard Olson, who generously allowed me to read his seminar paper written at Harvard University under I. Bernard Cohen, found the group ultimately ineffective because of its intellectual snobbery and elitist goals; his "The Scientific Lazzaroni: A Case Study in the Social History of American Science" identifies six certain and four probable members of the Lazzaroni. Mark Beach's "Was There a Scientific Lazzaroni?" in Daniels, ed., *Nineteenth-Century American Science*, suggests that the Lazzaroni were basically a social group and stresses their lack of mutual agreement on most major issues. The variety of interpretations reflects the incomplete and inconclusive nature of evidence scattered throughout various manuscript collections. A recent exhibition and catalogue, *The Lazzaroni: Science and Scientists in the Mid-Nineteenth Century*, produced by the National Portrait Gallery (Washington, D.C., 1972), offers a brief look at the Lazzaroni activities and presents an excellent series of portraits of the members.

suggested that perhaps the fellowship owed its very existence to the young association and that a broad concern for viable scientific institutions was the crucial—and perhaps only—point of consensus, aside from a love of stimulating conversation and fine food.

Bache was the first member to speculate at length on the future of the emerging AAAS. His experience as head of the Coast Survey taught him that "organization here [in the United States], for good or for evil, is the means to the end. While science is without organization it is without power...."[11] The young, amorphous AAAS needed guidance, and the politically and administratively astute Bache sought to direct it toward the "good" end of professional power to advance science, against the efforts of "enemies or injudicious friends" with less high-minded goals.[12] In response to Bache's comments on the nature of the new association in 1851, Dana self-consciously asserted, "There are a few men in the country upon whom its scientific reputation has seemed to me to rest."[13] Such men found that as their confidence increased, so did "an oppressive sense of responsibility." Responding to this responsibility in the same letter, the associate editor of Silliman's *Journal* indicated his willingness to assume an official position in the AAAS as one who knew "true science." Such men respected Bache as both scientist and administrator and supported his evident leadership. It had been a slow and irregular evolution from a vague notion that scientific societies were essential to developing the cultural life of the nation to recognition that organization could be an agency for advancing scientific research and promoting professional standards. Bache, familiar with the continuum, became spokesman for the final transformation.

Bache, with Joseph Henry, was the leading scientific administrator of his generation. Although capable of popular lectures and persuasive argument before a large audience, he personally preferred repartee with fellow scientists in small gatherings. While in Philadelphia in the 1830s he had been active at the Franklin Institute and at the American Philosophical Society, experiences from which he derived his later assumptions about professional science. Once in Washington, he acknowledged that despite his important and fulfilling role as head of the Coast Survey, he missed most the familiar evening sessions with fellow scientists in a kind of scientific club.[14] There a group consisting usually of Joseph

---

11. AAAS, *Proceedings*, VI (1851), lii–liii.
12. *Ibid.*
13. J. D. Dana to Bache, New Haven, 6 Sept. 1851, Rhees MSS, HHL.
14. Smithsonian Institution, *Annual Report* (Washington, D.C., 1859), p. 109.

Henry, newly arrived at Princeton, John Torrey of New York, and Philadelphians James Espy, Sears C. Walker, and Henry D. Rogers enjoyed open discussion to the end of "promoting research" and "scrutinizing the labors of its members."[15] Once actively an officer in the AAAS, Bache vigorously assumed responsibility by urging Lewis R. Gibbes to do ever more to produce the Charleston proceedings rapidly and in a scholarly form. Then, realizing the need for better cohesion in the association, he encouraged the establishment of permanent officers to whom the details of administration might be assigned. Finally, with the acumen of an experienced administrator, he himself worked to build the standing committee into a controlling agent capable of establishing professional standards within the organization.

It is tempting to superimpose a nonexistent cohesion on the activities of the self-defined Lazzaroni because the individual members took themselves so seriously. Information on the Lazzaroni is dependent on very sparse manuscript materials; neither formal minutes nor a constitution exists. Careful study of such sources reveals that even the annual meetings were neither regular nor well attended and that the influence of the group derived far more from the individual status of members than from their group affiliation.[16] This led Mark Beach to suggest that, in fact, the group was simply social. The Lazzaroni were important, however, as one lens through which professional aspirations were focused for the larger scientific community,[17] most visibly through the agency of the AAAS.

Bache's address as retiring president in 1851 marked a watershed for the association.[18] Before that date ordinary members and ambitious professionals had subordinated their personal expectations to the primary purpose of establishing the AAAS as a viable working organization. That year, however, had brought out a restiveness within the group of leading scientists and a tension between the leadership and the membership, which increasingly resented the paternalism of Bache, Dana, and others. The ensuing debates continued throughout the 1850s and con-

Also see Henry to Torrey, Princeton, 20 Dec. 1834, 23 Feb. 1835, Torrey MSS, NYBG.

15. *Ibid.*

16. Among the most useful collections are the Rhees MSS, HHL; Peirce MSS, HLH; Bache MSS, LC and SIA; Joseph LeConte MSS, University of California at Berkeley (not used by author); and Frazer MSS, APS.

17. Robert V. Bruce, "Universities and the Rise of the Professions: Nineteenth Century American Scientists" (paper presented at the American Historical Association meeting, Dec. 1970).

18. AAAS, *Proceedings*, V (1851), xli–xlix.

tributed to the formulation of a new consensus with regard to science in America. Bache's address provided the rallying cry and a tentative program for like-minded professionals.

Although belatedly acknowledging the formation of the AAAS, Bache devoted himself to its successful development and was less easily daunted than his fellows by its slow progress. In 1839 he had been a member of the American Philosophical Society committee which defeated Warren's proposal for an American Institution for the Promotion of Science. In 1851 he noted that it had "prudently" been left to the rather well established geologists to create a nucleus for national organization.[19] By 1851, more than a decade after the formation of the AAGN, the AAAS brought together cultivators of science who were by then "numerous, zealous, and not closely gathered in one community."[20] Bache admitted that genius was not fostered by associations but insisted that they created a climate of openness by removing opposition to new theory and by stimulating "second rank" scientists to a higher level of knowledge. In this sense, Bache argued, associations fostered group research and also appealed to the strongest and best motives of man's nature.[21] By inference, the most recent and best research was presented at professional meetings.

The entire address made it clear that Bache accepted the contemporary self-analysis which found American science empirical and greatly influenced by popular demands for the technical application of physics, mathematics, and natural science. In fact, empiricism was "emboldened by the absence of accredited tribunals to try its claims . . . [and] proffered boldly its pretentions to public notice, calling itself by the respected name of science."[22] As a result he felt that empirical studies and basic research lacked support and rightful prestige. Alluding to examples of quack medicine, Bache suggested that science itself experienced a "modified charlatanism" when men having ability in one area of science claimed to be arbiters in others. Incompetent tribunals could not curb the distortion of science, and in them "notoriety [was] often dearly purchased by the sacrifice of some portion of real reputation."[23] The abuses to which Bache alluded seemed all too clear to him in American societies, but he anticipated that the AAAS, well directed, might create a new model of scientific organization. Essentially a con-

19. *Ibid.*, p. xlii.
20. *Ibid.*, p. xli.
21. *Ibid.*, p. xlii.
22. *Ibid.*, pp. xliii–xliv.
23. *Ibid.*, p. xliv.

servative, he wanted the specialization necessary for modern science to be tempered by opportunity for interdisciplinary discussion. This, too, could be accomplished in the broadly based association.

Throughout his address Bache emphasized parameters for the future activities of the association. He felt it was important to bring together members of local institutions and stimulate their activity by holding meetings in various places. Moreover, as a scientific congress the nascent association could answer questions for legislatures which belonged properly to scientific rather than political bodies. Bache approved without qualification this advisory role: "Thus far no single recommendation made by the American Association has fallen powerless . . . they have both done positive good and avoided positive evil."[24] However, Bache recognized that the association lacked funding and working spirit to carry out extensive observations like those of the BAAS and rightly limited its function to directing operations sponsored by others. Trying to involve the full membership in all decision-making processes of the association was overambitious, he suggested, and he proposed that time be set aside for appointed committees to deliberate on special topics during the meetings. Correspondence seemed a slow and unsatisfactory alternative to personal discussion. Bache insisted that the association was the most authoritative scientific body in America, responsible for circulating new ideas and establishing priorities by its opinions on questions of scientific research.

A subtle but major point in his address indicated that Bache did not find the AAAS a final national organization. He argued that practical application flowed from the principles of science and that perhaps the world should bestow "rewards for principles, instead of for application." Yet the intermittent and voluntary character of the AAAS rendered it unable to assist the neglected area of "abstract research." The United States required another institution to supplement the AAAS, one similar to the French Academy of Sciences, which provided financial support for research and also published scientific papers. Bache reiterated the complaints of professors that they were overworked and lacked time and equipment for research. This want of direct support was the greatest obstacle to the progress of American science. He specifically proposed "an institution of science, supplementary to the existing ones, to guide public action in reference to scientific matters,"[25] and envisioned an institute representing men of various states who would meet "only at

24. *Ibid.*, p. lii.
25. *Ibid.*, p. xlviii.

particular times, and for special purposes," a permanent consulting body undertaking specific research at the request of the federal government. While he hoped that the primary function of the group would be basic research, Bache's definition of "abstract" was not precisely that of "pure" research, for he concluded that "there are few applications of science which do not bear on the interests of commerce and navigation, naval or military concerns, the customs, the lighthouses, the public lands, post-offices and post-roads, either directly or remotely. If all examination is refused, the good is confounded with the bad, and the Government may lose a most important advantage. If a decision is left to influence, or to imperfect knowledge, the worst consequences follow."[26] Scientists already affiliated with the government had assumed large responsibilities and could not alone meet the imperative for such organization.

Several times in the address Bache chided scientists for their lack of enthusiasm in establishing scientific institutions. Bache argued the need for more opportunities for scientists isolated geographically to interact on a personal level. In his opinion working committees previously assigned had been sluggish about completing projects or observations. Further, younger men seemed deliberately to avoid the laborious details of administration. Critical yet optimistic, his vision of what common purpose and zeal might do for American science and his candid assessment of the contemporary AAAS obviously inspired several attending members, including James Dwight Dana.[27] From 1851 on, Bache was the acknowledged spokesman for professional science and the mentor for a coming generation of scientists. During the rest of the 1851 meeting members discussed the future of the association and reassessed its constitution.

Private conferences concluded that the level of American science would be raised only if there were institutions for research and better training for researchers, topics which undoubtedly fueled discussion of the proposed Albany university. Backed by local business and professional men and by some leading scientists, a university had been chartered by the New York legislature in April 1851. James Hall was a prime mover and received strong backing from Louis Agassiz of Harvard and agricultural chemist John Pitkin Norton of Yale. Bache had not directly endorsed the university scheme in his address, nor did the AAAS for-

26. *Ibid.*, p. l.
27. John P. Norton, for example, regarded it as a "noble address." Diary, 20 Aug. 1851, Norton MSS, BLY.

mally record its support. But Hall, Agassiz, and Norton were rapidly joined by Benjamin Peirce, who wrote still another proposal for the venture. Because many key scientists attending the August meeting seemed approving, plans were soon announced for a course of lectures in Albany in February to promote the university proposal and simultaneously a series of meetings to discuss it.[28]

In the winter of 1851–52 a new name, the Florentine Academy, appeared in letters between Bache and Peirce, undoubtedly an outgrowth of the cordial evening fellowship of Bache, Peirce, Agassiz, John LeConte, and his cousin John L. LeConte at Albany.[29] The support of these prominent men persuaded Norton that the chartered university was close to formulation. He suggested beginning with an agricultural college and offered to give an initial course of lectures. Others invited to discuss the plan that winter were Bache, Peirce, Agassiz, Dana, and Oliver Wolcott Gibbs. The scientists provided support only temporarily, however, for when they confronted the obstacle of a local legislature and a proposed fee system, again linking salary primarily to teaching, they balked.[30] Withdrawal of prominent scientific support left the Dudley Observatory and a law school, but the dream of a university of graduate education in science vanished.[31] Although Peirce and Gibbs maintained an ongoing interest in the project, by the late summer of 1853 Peirce suggested that their annual dinner be a "final continuation of the Florentine," since hope for both the university and a national institute was fading.[32] Bache, in turn, acknowledged that "the last attempts at Albany were worse than failure."[33] The Florentine Academy seemed intrinsically related to the university movement, yet its intimate members were broadly interested in advancing science.[34] The new-

28. Silverman and Beach, "A National University," pp. 708–710.

29. Lenzen, *Benjamin Peirce*, pp. 44–45. In a letter dated 15 May 1852, Peirce suggested to Bache in regard to the university that "as president of the Florentine Academy, you should call an early meeting for its discussion." Peirce MSS, HLH. Beach suggests that the first reference to the group was in 1850, but this author was unable to find his citation.

30. W. Gibbs to B. Peirce, New York, 15 Feb. 1853, Peirce MSS, HLH.

31. Alonzo Potter considered it abandoned when he suggested to Joseph Henry that perhaps the plan could be implemented in Philadelphia. Philadelphia, 1 July 1853, Henry MSS, SIA. In 1856 the university idea was revived but found minimal support. See Henry Tappan to B. Peirce, University of Michigan, 25 Apr., 19 June 1856, Peirce MSS, HLH; Benjamin A. Gould, "An American University," *American Journal of Education*, II (Sept. 1856), 29ff.

32. B. Peirce to A. D. Bache, [Cambridge], 30 Aug., 25 Sept. 1853, Peirce MSS, HLH.

33. Bache to (?), n.p., 9 Sept. 1853, Bache MSS, APS.

34. Richard Olson stresses this linkage and notes that Samuel Ruggles, a New

found spirit of fellowship and mutual purpose enjoyed for nearly two years by its members was not easily abandoned.

The term "Lazzaroni" appeared first in a letter from Peirce to Bache in late 1853.[35] Whether the Lazzaroni was a deliberate extension of the old Florentine Academy or simply a new group consisting of most of the same participants—Bache, Peirce, Agassiz, John LeConte, and Gibbs—is not clear.[36] Institutional concerns were different, for the university plan had receded from discussion. Until 1855 the letters exchanged among Lazzaroni mentioned only vague hopes for science without specific recommendations. The Lazzaroni, although distinguished from the Florentine Academy by a somewhat defensive tone, nevertheless retained the witty interchange which had developed among the Florentines. The newer fellowship emerged just as its organizers found their mutual assumptions about American science and their direction of the AAAS challenged by men who reacted to their sometimes patronizing and exclusive activities.[37] Correspondingly resentful that their authority was questioned, Lazzaroni fellows reinforced each other in the rightness of their causes and determined to work more closely to direct national science. Their power was grounded in the federal agencies they influenced, the public recognition they individually enjoyed, the colleges, societies, and peers which respected them, and their temporary group cohesion.

Peirce's retiring address from the AAAS in 1854 is indicative of his euphemistic style and of the overtly supportive role the Lazzaroni played for each other. A separately printed copy of the address, apparently that of Cornelius C. Felton, has a special titlepage: *Ben Yamen's Song of Geometry, Sung by the Academy, at the Accession of Her Majesty the Queen, Degraded into Prose by Benjamin the Florentine.*[38] The volume is dedicated to the "Queen of the Florentines" by Benjamin from "Function Grove."[39] The game was a serious one, however. Amidst rhapsodic lines on the geometric nature of the universe, Peirce gave extended praise to the work of Agassiz, Bache, and Henry. He called

---

York lawyer interested in education, was considered a confidant of the Florentine Academy but not of the Lazzaroni.

35. B. Peirce to A. D. Bache, 30 Aug. 1853, Peirce MSS, HLH. It is probable that the group originated in the spring of 1852 when key members met to discuss the Albany university; Peirce's letter refers to the group as an established fact.

36. Most of the analysts mentioned in n. 10 above assume that there was simply a change of name.

37. A. D. Bache to B. Peirce, Denmont, 16 Oct. 1855, Peirce MSS, HLH.

38. Copy in Widener Library, Harvard University.

39. Mrs. John LeConte was apparently the Queen.

again for an institution of such scientists, "supplementary to existing institutions, to guide public action in reference to scientific matters."[40] Peirce sought a general society in which "scientific influence should predominate," unlimited by either demanding patrons or an apathetic public. His conclusion swept into vague optimism: "If American genius is not fettered by the chains of necessity, and helplessly exposed to the results of envious mediocrity, but is generously nourished in the bosom of liberty, it will joyfully expand its full wings, and soar with the eagle to the conquest of the skies." After the spring of 1855 mention of the Lazzaroni became more frequent, if incidental, in the correspondence among Peirce, LeConte, Bache, and John F. Frazer, as well as the younger Wolcott Gibbs, Benjamin A. Gould, and Fairman Rogers. Wolcott Gibbs returned from his European studies eager to participate in American science and anxious to promote European standards of research at home. He was active in promoting the first university scheme in 1851 and associated with Samuel Ruggles's efforts to the same end later in the decade. Bache solicited aid from such men in his 1851 address and found a responsive audience. Gould, an astronomer who also boasted a European Ph.D., was a protégé of Peirce. In a letter introducing Gould to Bache, Peirce commented that Gould had a "burning likeness to the engine which will carry him from Massachusetts to Maine—he is all fire and fume to the enemy upon whom he advances while the friends whom he leaves behind see nothing of him but what is tender."[41] The younger man was quickly incorporated into the Lazzaroni fellowship and became its most intense member, taking the initiative for the first important meeting of the group. Similarly Fairman Rogers, a student of John F. Frazer, met the latter's mentor, Bache, at a dinner party. He was subsequently invited on a Coast Survey expedition to Florida and wrote back happily to LeConte that he was collecting many good jokes for the "Philadelphia members of the Lazzaroni."[42]

Cambridge, however, remained an intellectual headquarters, with members there acting in consultation with Bache. Gould wrote to James Dwight Dana outlining their proposal, which was simply that "once in each year (winter as farthest removed from the annual meeting at the Association) some of us should eat *one outrageously good dinner* to-

---

40. Peirce, *Ben Yamen's Song*, p. 16. Subsequent references are to this copy.
41. [B. Peirce to A. D. Bache], Function Grove, 28 Oct. 1853, Rhees MSS, HHL.
42. F. Rogers to J. LeConte, Key Biscayne, Fla., 20 Apr. 1855, LeConte MSS, APS. Fairman Rogers was apparently only an "associate member" and never appeared on any of the Lazzaroni rosters.

gether."[43] The first annual meeting was planned for Washington during the third week of January, with the place thereafter to rotate. Gould named the nine persons invited to join—Agassiz, Joseph Henry, Peirce, Dana, Gibbs, Gould, Bache, Frazer, and Cornelius C. Felton—suggesting that others be admitted only by unanimous vote. Gould, at least, saw the Lazzaroni reunion as a supplement to the annual AAAS meetings, suggesting, "It will do as much as the Am. Assn. to stimulate and encourage us all and it will taste better."[44] Notably absent from his list were John and Joseph LeConte, Ruggles, J. D. Whitney, and Fairman Rogers, individuals frequently mentioned as members of the Lazzaroni.

The group gathered under Bache's guidance was competent, energetic, arrogant, and grounded in the belief that theirs was a generation seeking science for science's sake.[45] Each had evidenced a young and somewhat precocious interest in science, and their individual enthusiasms were sparked by the fertile mind of Bache. Charles Peirce, the philosopher son of Benjamin Peirce, remembered and was himself shaped by the "steadily burning enthusiasm of the scientific generation of Darwin" who had met with his father in Cambridge. In his end-of-the-century assessment Peirce argued that the great men of the nineteenth century were in fact characterized by their "devotion to the pursuit of truth."[46] Manuscripts reveal a jealous and often petty exclusiveness in daily affairs, but Peirce accurately captured their dedicated spirit. They were high-minded in assessment of themselves and of science, but their idealism clouded their perception of the work of outsiders and led them to question the motives of those with whom they differed. Predominantly physical scientists, the Lazzaroni differed from the geologists. The latter were also eager to raise their research to European standards, to verify new theoretical conceptions, and to justify their efforts to a basically utilitarian public. The Lazzaroni were at once more harshly realistic about the limits of science in popular culture and more idealistic in their hopes concerning scientific organization than many fellow professionals. Skeptical of elaborate schemes for international research and most impatient about limited resources, they sought institutions to facilitate important

43. B. Gould to J. D. Dana, Cambridge, 22 Nov. 1856, Dana MSS, BLY.
44. *Ibid.*
45. Merritt, *Systematic Approaches*, p. 26. Merritt found that leaders typically have supportive superegos and strong self-assurance—two characteristics conspicuous in the leading Lazzaroni.
46. Charles Peirce, "The Century's Great Men in Science," Smithsonian Institution, *Annual Report . . . for 1900* (Washington, D.C., 1901), pp. 693–699.

research efforts in America. But they wanted these institutions to be free from naive scrutiny by the general public.

Counterbalancing their serious concern for American science was the Lazzaroni's comic wit and love of good fellowship. Gould wrote for Peirce and Bache to invite Dana to join the Lazzaroni for "one good feast," indicating that congenial company and superb dinners formed an important purpose in their meeting. Among themselves they wrote of good wine, carefully prepared oysters, and dark lager beer. In Philadelphia and Boston leading citizens entertained the scientific leaders during the AAAS meetings, and that entertainment became the Lazzaroni model. Sociability was important. Frazer urged Bache to select a time for the annual meeting that would not interfere with "Wistar Party Day—which ought to be much more important to the Lazzaroni than the Am. Phil. Soc."[47] Moreover, fellowship outside science proved useful as the Lazzaroni endeavored to make scientific institutions more exclusively their domain, purportedly to secure independence for research. Special dinner guests like Cornelius C. Felton, Smithsonian regent and later president of Harvard, Senator James Pearce of Maryland, Congressman Charles Upham of Massachusetts, and businessman H. H. Hawley of Albany were strategically selected. Thus Benjamin Peirce, concerned about an investigation of the Smithsonian, recommended that Lazzaroni members in Washington invite the Massachusetts congressman, chairman of the House Investigating Committee, to dinner. Peirce admitted that he did not "like or respect Upham," yet "I fancy that he would not resist the cheering influence of good fellowship and oyster supping, and I cannot believe it possible that, if he were once to listen, in a serious spirit, to the argument against the exclusive adoption of the library scheme, he would return to it again."[48]

The assertion of scientific fellowship was timely. The personal hegemonies of the leading Lazzaroni were being threatened in various ways, and within the AAAS there was a rising tide of dissenters against the Washington-Cambridge clique. Henry's proposals for the Smithsonian were still questioned in some legislative circles. Gould was antagonizing Albany citizens and Dudley trustees alike by his financial and administrative requirements and his autocratic handling of the observatory. A subsequent controversy involving Christian Peters spilled over to affect the Coast Survey as well. Bache had appointed Peters to do longitudinal

47. J. Frazer to A. D. Bache, Philadelphia, 23 Dec. 1857, Bache MSS, LC.
48. B. Peirce to A. D. Bache, [Cambridge], 17 Jan. 1855, Peirce MSS, HLH.

observations in conjunction with the Dudley Observatory while on the Coast Survey payroll. Proving unacceptable to Gould, Peters was dismissed and returned to Washington to rally influential political and scientific backing against his former sponsor. In the process he raised angry questions about the power wielded by a small group of scientists in the federal service.[49] Discussion among those Lazzaroni who met in Washington in early 1857, perhaps its most crucial meeting aside from that in 1863, concentrated on these difficulties. The groups concluded that the integrity of the scientific community was at stake and agreed to stand firm together rather than compromise principles. When the Dudley Observatory trustees appointed a scientific council, it supported Gould. This was hardly surprising: three of the four council members —Peirce, Bache, and Henry—had agreed in advance. It was just such working harmony that aroused outside interest in the close and friendly group of leading AAAS scientists.

The Lazzaroni developed through the AAAS and remained actively interested in association activities. All Lazzaroni were members of the association, and the key Lazzaroni served as officers and were thus a critical voice on the standing committee. The winter meetings (usually in January) were intended to supplement the AAAS's summer sessions, and plans for forthcoming meetings were discussed. When Agassiz, a leading scientist, threatened withdrawal from the association, the other Lazzaroni argued that they needed a unified leadership that could withstand internal challenges as well as direct association policy. The nearly autonomous actions of the standing committee, guided by the plans of the Lazzaroni, came under attack, and dissidents proposed constitutional limits on that committee as early as 1853.

AUTHORITY IN THE AAAS

In 1854 Agassiz deliberately avoided the Washington meeting, and James Dwight Dana returned to New Haven to report skeptically, "I do not look for much real good from the Association."[50] Growing discord reinforced Bache's decision to draw leading scientists into working harmony on association affairs. Shortly after the 1854 meeting he pondered the future of the AAAS: "As now advised there is more hope in me than there was a year ago & yet I see more clearly the danger from breakers ahead. Should even a few 'leading men' (Rogers) stay away

49. Miller, *Dollars for Research*, pp. 45–47, is the best summary of the controversy.
50. J. D. Dana to J. Hall, New Haven, 24 May 1854, Hall MSS, NYSL.

from the meeting the whole tone of things would at once change."[51] Peirce concurred in the association's potential because in retrospect its *"position* is remembered and its negatives forgotten."[52] This serious concern for the future of the association was counterbalanced by the ability to laugh, paternalistically and sometimes sarcastically, at the foibles of the association. Among the Lazzaroni the AAAS became the "American asses," the "asses of science," and even the "Amazing Asses Adverse to Science."[53] The Lazzaroni tempered ultraserious purposes with wit.

Occasionally the concern bordered on personal possessiveness. In 1856 young Gould, appointed to the Albany local committee, became responsible for extending invitations to European scientists. Because several steamship companies offered free passage to a limited number of visiting European scientists, Gould drew up a hierarchial list of men to invite. James Hall, as head of the committee, was incensed with the results, claiming Gould allowed "personality and prejudice" to affect his judgment. He promptly sent out another circular to supplement Gould's.[54] The latter felt, however, that the action was justified, if arbitrary; Robert Dale Owen was excluded for having treated Jeffries Wyman "outrageously," and Murchison was "down on the list" because of his controversy with Sedgwick in England.[55] Such attitudes increased antagonism among leading scientists. Hall was not a Lazzaroni confidant and readily saw that such actions were not totally disinterested.

After 1853 questions relating to the constitution and to the exercise of power by the standing committee confronted the AAAS annually. Scientists were learning political strategy. Bache headed a group which sought greater centralization of power and a more incisive role as spokesman for the AAAS. Aligned against Bache were scientists more inclined toward an open society and those antagonistic to the manipulation of the association by the leading physical scientists. Debates over constitutional revision revealed the specific issues. The composition of the standing committee and the scope of its prerogatives seemed critical

51. A. D. Bache to B. Peirce, Washington, 5 May 1854, Peirce MSS, HLH.
52. B. Peirce to A. D. Bache, [Cambridge], 8 May 1854, Peirce MSS, HLH.
53. *Ibid.* Also see A. Gould to J. F. Frazer, Cambridge, 22 Nov. 1856, Frazer MSS, APS.
54. J. Hall to Dr. Armsby, Washington, 2 May 1856, draft copy, Hall MSS, NYSL. Hall pointed out some prominent French and German scientists who were not included and some "3rd rate" men who were.
55. B. A. Gould to J. Hall, Cambridge, 21 Apr. 1856, Hall MSS, NYSL. Hall contended that the Murchison-Sedgwick debate was no concern of the local committee and that both men must be invited.

to both sides. Significantly, Lazzaroni members appeared on the roster of every standing committee in the 1850s. After Bache, Agassiz and Peirce and then Dana were presidents of the AAAS, followed in turn by Wolcott Gibbs and Benjamin A. Gould as general secretaries. Each of these positions meant not only an ex officio position on the standing committee for the year in office but also for the subsequent year. John Fries Frazer and Cornelius C. Felton rarely if ever attended association meetings, but all the other Lazzaroni were standing committee members at least three times. If there was a strategy, it was to elect as at-large members at least two Lazzaroni, two or more like-minded friends such as the LeContes, Benjamin Silliman, Jr., or Stephen Alexander (Henry's brother-in-law), leaving the two remaining seats for easily influenced local luminaries. For reasons explained later in the chapter, William Rogers was not nominated for a major office or a standing committee position from 1851 to 1860, and nomination had come to mean election.

Still, the Lazzaroni never constituted a majority of the full standing committee. Bache realized, as perhaps his Cambridge fellows did not, that measures still had to be carried in the general meetings by weight of argument rather than by authority alone, and he used logic, wit, and idealism to defend specific proposals. The rest of the Lazzaroni were less successful than Bache in these tactics, for not all of them had the personality to be effective in debate.

As suggested earlier, 1851 marked a turning point, organizationally, for the association. In that year Baird assumed duties as permanent secretary and began to centralize bookkeeping operations. Thereafter, except for the suppressed Cleveland volume, the annual *Proceedings* were arranged systematically by type of science rather than reflecting the actual order of events at a particular meeting. Encouraged by Bache, the standing committee assumed responsibility for organizing the program by checking authors' abstracts and arranging the selected papers into specific order for presentation.[56] Members generally accepted this basic review system as it was gradually implemented. Order and efficiency in themselves demanded a systematic schedule, and the standing committee, composed of past and present officers as well as six at-large officers, seemed a logical agent for exercising control.[57] Activities such as Robert

56. See Chapter VI.
57. AAAS, *Proceedings*, IV (1850), 159–164. The need for centralization was the underlying theme of Bache's report on the semiannual Charleston meeting, for he observed that "some permanent form of organization is needed" to interlock the annual meetings. The standing committee for Charleston had considered a rotating secretariat, with perhaps one permanent officer.

Hare's persistent efforts to address the AAAS on spiritualism reinforced the membership's consciousness of the need for careful management; the nuisance of such speakers was evident to scientists and amateurs alike.[58]

Although the scientist-amateur discussion polarized some factions in the AAAS, the constitutional debates from 1851 through 1859 were more deeply divisive for the association. Information is difficult to trace in detail because the *Proceedings* remained prudently silent concerning the discussions, and the more candid public press mirrored the tension rather than recording the substance of debates. Often arguments were couched in theoretical terms, yet their roots arose from the exercise of authority within the association. Many regular members did not comprehend the nature of the struggle and were quite willing to postpone constitutional questions from year to year. More active opponents discussed the issues privately after the public meetings. The major concerns were two—the power and composition of the standing committee and, directly related, the organization and independence of the section meetings. Actual leadership and its exercise of control were in question; constitutional debates essentially mirrored a power struggle.

Beginning in 1851 members considered in open meetings the assigned function of the standing committee, the residual functions of the sections, and the composition of membership. The key protagonists were Bache and William Barton Rogers, leader of the "strict constructionists."[59] These two prominent scientists, representing the physical and natural sciences respectively, gathered supporters less along scientific than ideological lines. Neither was radical, and they held certain basic premises in common. Both accepted with unconcealed regret the increasing specialization of science and stressed the need for open discussion of scientific problems and the review of questionable papers. Rogers was as vociferous as Bache in denouncing false science. After a tour of the Great Lakes region he had observed, "Credulous, incompetent & in some instances not impartial [observers], ranked among the scientific by popular acclaim, throng along the shores of the Lake & have left their hammer-marks on the least vein ... that gleams obscurely from the dark surface of the bare rock. . . . Need I add that the sanguine dreamers

58. When frustrated by the association, Hare published the results of his experiments on the "spiritoscope" in *Experimental Investigation of the Spirit Manifestations, Demonstrating the Existence of Spirits and Their Communication with Mortals* (n.p., 1855?).

59. Peirce so labeled the dissidents in a letter to Baird, Harvard University, 13 Mar. 1853, Peirce MSS, HLH.

who follow their march listen to the quackery with a perfect faith that every such trend must lead infallibly to the coffer eldorado."[60] But that observation did not lead him to cynicism. Rogers and Bache stood fairly close as emerging professionals in promoting government sponsorship for science, advanced scientific education, a code of scientific ethics, and some distinction between scientists and amateurs. The debate was, however, more than a struggle between power blocs seeking to control the association. Both men sought to have only research scientists in office and a professional organization; but while Bache wanted to centralize control and limit the accessibility to office, Rogers insisted that the process be open and competitive. The manner in which to establish supervision and exercise control became a source of bitter conflict. A split between the leading scientists reached beyond the early years of the AAAS into the Civil War period and similarly disrupted the National Academy of Sciences.

Alexander D. Bache had little patience with well-meaning amateurs and preferred to direct them from a position of authority. Although he disliked the heterogeneous nature of the AAAS, Bache recognized its positive potential in securing support for science. He personally felt justified in and even responsible for consolidating the fledgling organization into a body useful for science. The character of his own reports and his general comments before the association indicate that he did not find the sessions conducive to intensive debate. His own papers were expositions designed to demonstrate the importance of the Coast Survey practically and scientifically. In short, the AAAS provided a podium for promoting his survey and other projects. Those few men conversant with geophysics he arranged to meet less formally for discussion. These same colleagues also assumed leadership positions in the AAAS so that together they could guide American science in "the proper course."

William B. Rogers, a leading member of the AAGN, had made strenuous efforts to keep that organization well established and reputable. He did not, however, as did some in the natural sciences, denigrate the importance of field observers and instead maintained their right to present findings to the organization. Recognizing the distinction between regular scientific researchers and men interested simply in hearing the findings of others, he did support his brother Henry's proposal to form two "classes" of membership in 1849.[61] In this scheme associate members

60. Rogers to George S. Hillard, steamer between Cleveland and Buffalo, 1 Oct. 1845, Rogers MSS, University of Virginia Archives, Charlottesville (hereafter UVA).

61. AAAS, *Proceedings*, II (1849), 179.

were to enjoy the meetings without direct participation. Regular members would present papers, suggest new areas for research, and share the prestige attached to holding office, as in the AAGN. This distinction would have left power still rather broadly distributed among the scientific researchers, but the resolution was not passed. When he perceived power concentrating in the hands of a very few men, Rogers realized that the new association, far more than the old, was being manipulated by certain men for individual purposes. The other issues, therefore, were clouded by personality. Unwilling to bind the association narrowly, Rogers attempted to keep the constitution open and flexible so that in principle as well as in fact the association would be protected from oligarchical control. Ambitious and with a following, he resented the accumulation of power by another bloc. His challenge was limited in success, but the debates forced his peers to specify their goals and demonstrate their methods of assessing others. Rogers's standing as a geologist meant that his opposition could not be ignored.

An abortive attempt in the 1850s to found a geological society pinpointed two particularly discontented groups, the geologists and certain western scientists. As the AAAS expanded, leadership subtly shifted away from its geological founders toward men in the physical and geophysical sciences. The deliberate effort to balance the presidency between the physical and natural sciences and the separate sectional meetings masked this transformation. Nonetheless the geologists found themselves outside the inner leadership circle and without the working dynamic that had earlier characterized the AAGN. In general, western scientists were concentrated in astronomy, meteorology, geology, and the natural sciences, for which raw data were readily available, although the establishment of colleges increased the number of physical scientists beyond the Hudson River by mid-century. Even within their specialized fields, however, the western men frequently felt excluded from the scientific community; the existence of a national AAAS made distinctions increasingly obvious. Western naturalists had been basically data gatherers for other collectors who had better resources for the research work of classification and analysis. Down into the nineteenth century frontier scientists continued to share their new findings with Europeans as well, sometimes to the anger of naturalists in Boston and Philadelphia.[62] The feeder system proved unsatisfactory when donors found themselves denigrated as amateurs and their "discoveries" analyzed by and credited

62. Joseph Leidy to J. Hall, Philadelphia, 20 Apr. 1853, and J. Hall to J. Henry, Ceres, Pa., 24 Apr. 1853, typed copy, both in Hall MSS, NYSL.

to eastern correspondents.[63] Long-standing western resentment was a contributing factor to the movement, which originated in Ohio, to found a distinct geological society.

There is very little evidence on the origins and nature of the movement to form a geological society because the effort was deliberately obscured by its cautious founders. Undoubtedly William B. Rogers's questioning of the AAAS leadership in 1851 created an opportunity to consider changing the direction of the association back to the earlier intimacy and incisive debates of the AAGN. In November 1851 Josiah D. Whitney and John Foster, then working on the Lake Superior Survey, met with Henry Rogers and Edward Desor to discuss the Pottsville anthracite coal basin in Pennsylvania. During their meeting they discussed establishing an exclusively geological society which would have a library, a permanent research collection, and eventually a journal devoted "to the science of geology."[64] Reflecting John Locke's early complaint of the high cost of urban meetings, they proposed that their smaller meetings should be in places of general interest to members and "not necessarily in a large & expensive city."[65]

Whitney subsequently suggested to James Hall that the formation of a society comparable to those in England and Germany be discussed when geologists met to defend Agassiz at his Albany trial.[66] There is no record of such discussion, however. Postponement of the Cleveland meeting until 1853 compounded the westerners' irritation about eastern attitudes toward them.[67] Baird polled leading members before announc-

63. Numerous letters hint at the reciprocity of suspicion. John Locke, for example, observed that when Ephraim Squires visited him in Ohio, "In the interview he honoured me by remarking that he had come to me because I was a 'smatterer' in these affairs. As he was an Editor from the East, the salient East, and then only a yearling in the West I was amused." Locke to J. Henry, Cincinnati, 19 Feb. 1848, Henry MSS, SIA. Many easterners would have concurred with Rogers when he came back from the mining region of the Great Lakes and wrote, "Exact research is only *beginning* in that region. Credulous, incompetent & in some instances not impartial observers [are] ranked among the scientific by popular acclaim." W. B. Rogers to George S. Hillard, steamer between Cleveland and Buffalo, 1 Oct. 1845, Rogers MSS, UVA.

64. J. D. Whitney to J. Hall, Philadelphia, 7 Nov. 1851, Hall MSS, NYSL; also see Brewster, *Life and Letters*, pp. 120–121.

65. J. D. Whitney to J. Hall, Philadelphia, 7 Nov. 1851, Hall MSS, NYSL.

66. Clarke, *James Hall*, p. 225; J. D. Whitney to J. Hall, 23 Dec. 1851, Hall MSS, NYSL. The two had apparently discussed such a project earlier, perhaps while together on the Lake Superior Survey.

67. J. D. Whitney to J. Hall, Copper Falls Mine, 23 Aug. 1852, Hall MSS, NYSL. Commenting on the postponement, Whitney wrote, "I am sorry that anything has been done to increase the ill-will of the western men...."

ing the decision for postponement; the decision was reasonable, for even if the threat from cholera was minor, as the Cleveland sponsors claimed, fear would have inhibited attendance.[68] Yet the Cleveland members felt that their recommendation had been ignored and were unhappy with the selection of August for the 1853 meeting, a month when many local citizens took country holidays.[69] For reasons related to both western frustration and the geologists' belief that they could recapture earlier group spirit, Charles Whittlesey, a geologist of limited experience, renewed the movement in the winter of 1852.[70]

Whittlesey had worked with Whitney and Foster in 1850 and garnered the tentative support of William Rogers, William Redfield, and James D. Dana for a geological society.[71] None of these men was willing to assume the initiative, however. In early 1853 Whittlesey complied with Redfield's suggestion of quietly circulating a proposal to determine interest.[72] Conscious that some scientists would construe the new society as a challenge to the AAAS, Whittlesey inserted a specific disclaimer of any such intention, simply observing, "It was not proposed that in case a society is formed the members should sever other connexion" with the AAAS.[73] He argued positively that this proposal resulted from recognition that "the science of geology is so progressive & important that it is constituted so much of well-observed and well-

68. S. F. Baird to W. B. Rogers, Carlisle, Pa., 23 July 1852, Rogers MSS, MITA; Baird to J. Henry, Carlisle, Pa., 24 July 1852, Henry MSS, SIA; L. Agassiz to Baird, Cambridge, 27 July 1852, Baird OFF, SIA.

69. When word of the postponement arrived, the *Daily True Democrat* for 10 Aug. 1852 indignantly reported, "...at this season Cleveland was never more healthy." Also see J. P. Kirtland to Baird, Cuyuhoga, Ohio, 3 June 1853, Baird OFF, SIA.

70. For skepticism about Whittlesey's scientific qualifications, see Dana to Silliman Jr., New Haven, 9 Dec. 1852, Dana MSS, BLY; J. Hall to A. Gould, Albany, 29 Apr. 1849, Gould MSS, HLH.

71. Whittlesey to W. Redfield, Cleveland, 11 Dec. 1852, and Redfield to Whittlesey, New York, 8, 18 Dec. 1852, both in Redfield MSS, BLY; Whittlesey to I. A. Lapham, Cleveland, 2 Dec. 1852, Lapham MSS, WHS. Dana apparently had heard of the movement but knew none of the particulars. See Dana to Baird, [New Haven], 6, 14 Dec. 1853, Baird PP, SIA.

72. Whittlesey to Redfield, Cleveland, 23 Dec. 1852, Redfield MSS, BLY. Whittlesey sent the draft of a circular agreed upon by James Hall and Samuel St. John. Subsequent quotations are from this draft copy. No printed copy has been found, although Hall alludes to one in a letter to L. Agassiz, Albany, 9 Feb. 1853, draft copy, Hall MSS, NYSL.

73. James Dwight Dana interpreted it somewhat differently: "I suppose it is started by the discontented of the Amer. Assoc., as it looks to discussion for mutual enlightenment as its main end." Dana to Silliman Jr., New Haven, 9 Dec. 1852, Dana MSS, BLY.

established facts & is also now becoming so independent of European dissatisfactions that more frequent & less formal meetings of geologists had become necessary or at least desirable." Once again the spirit of scientific nationalism engendered the assertion that migratory meetings at places of special geological interest and "a more full & rapid publication" of reports would make "conclusions more satisfactory at home and more respected abroad."

Whittlesey concluded by asking for comments and recommendations because "our impression is that the number of active geologists is such that an organization they constitute can be maintained in a creditable manner." Louis Agassiz disingenuously replied that he knew little about the movement.[74] By the 1853 AAAS meeting the movement seems to have died, perhaps because the few leading geologists attending the Cleveland meeting were not interested.[75] Western discontent had not dissipated. In fact, Alexander Winchell of Michigan felt that the "investigators of the Northwest feel that they have interest in common which cannot be adequately subserved through the medium of any organization the center of whose influence is located in some remote section of the country."[76] But the indecisive support from established geologists rendered the project not feasible. Rogers and the others wanted reform of the AAAS, not an alternative to it.

The geologists within the AAAS remained discontent with the leadership of the physical scientists. By mid-decade the association was "endured rather than controlled" by the old members of the AAGN.[77] Peirce asserted that the "earthen venet is afraid of the iron retort."[78] In fact, the men in the earth and natural sciences section preferred their sectional meetings to the joint meetings in which they were clearly second-class participants, and the tensions remained throughout the decade. Table 3 demonstrates that while the earth and biological sci-

74. L. Agassiz to J. Hall, [Cambridge], 3 Feb. 1853, and Hall to Agassiz, Albany, 9 Feb. 1853, draft copy, both in Hall MSS, NYSL. Agassiz attributed the movement to an "intrigue" of Rogers for the "design of undermining the American Association." Hall replied that it was more attributable to dissatisfactions among the people of Cleveland, although not denying that Rogers might be a participant.

75. Alexander Winchell to Baird, Selma, Ala., 12 Oct. 1853, Baird OFF, SIA.

76. No heading or signature, [Spring 1855?], Box 1, Winchell MSS, Michigan Historical Collections, Ann Arbor.

77. J. W. Bailey to J. Hall, West Point, 20 Dec. 1856, Hall MSS, NYSL. In Edward Hitchcock to Hall, Amherst, 29 Mar. 1854, Hall MSS, NYSL, Hitchcock had belatedly found out about the proposal for a renewed geological association and "sympathized with such an object."

78. B. Peirce to A. D. Bache, [Cambridge], 8 May 1854, Peirce MSS, HLH.

TABLE 3

Papers Presented to AAAS by Type of Science

Papers listed only by title in the AAAS *Proceedings* are included in this table because such listings could mean either that the author exercised his prerogative not to submit a paper or that the paper was deliberately omitted by the standing committee. In general, the ratio between papers included in full and those merely mentioned by title was similar in each of the categories listed below for a given year, with the exception of 1855, when nearly all papers in physical science were printed. Chemistry is included as a distinct category because the association itself was unclear about this science's appropriate place, shifting it between the other two categories until 1855, after which it was placed with physical science. The number of papers in each scientific category is recorded below as a percentage of the total number of papers presented in a given year.

Meeting	Type of Science		
	Biological/Earth	Physical	Chemical
1848	53%	30%	17%
1849	61	24	15
1850a	70	30	—
1850b	40	38	22
1851a	66	27	7
1851b	25	39	36
1853	38	57	5
1854	28	62	10
1855	51	44	5
1856	40	57	3
1857	46	44	10
1858	36	48	17
1859	46	44	10
1860	34	49	17
Total Number	531	529	150
Cumulative %	44%	44%.	12%

ences dominated the early meetings, at least in sheer number of presentations, by 1854 the physical sciences usually were in the plurality. There are several possible explanations for the shift. The geological sciences declined as state surveys were halted or completed at the same time that the physical sciences became more firmly established in such agencies as the Naval Observatory, the Coast Survey, and the Smithsonian Institution. Many of the older geologists like Hitchcock and Silliman did not regularly attend association meetings, and their places were taken by physical scientists. The data are rough and must be treated cautiously, yet the shift in papers presented is clear and suggests a similar shift in the locus of power.

Bache, active leader of the standing committees in 1850 and 1851, consolidated the power of that body as the central administrative unit of the AAAS.[79] Aside from the arrangements made by the local committee, the standing committee established the daily program of papers and resolutions to be considered. With the creation of permanent officers who were also on the standing committee, a new continuity was created for that body. Down to 1851 the publication of the *Proceedings* had been left to the general secretary and the local committee, but at Cincinnati that spring the standing committee appointed a committee on publication, not unlike that of the *American Journal of Science*, with men of established reputation revising papers in their area of competency.[80]

Agassiz became president at the Albany meeting in 1851, and, like Bache, he assumed a dominant role in guiding AAAS affairs during the week-long meeting. Before the meeting he had suggested to Bache that the standing committee meet early to determine its nomination of at-large members as well as the succeeding president.[81] Evidently Bache agreed that Peirce was a logical candidate in order to maintain the unofficial policy of alternating the presidency between the physical and natural sciences; Peirce was subsequently nominated and elected as president. Such preplanning became the norm among the Bache group. When that fact became more evident, other AAAS members revolted.

In particular, some members resented the exercise of power in relation to forming and directing the distinct section meetings. The constitution gave the standing committee power to organize sections by subdividing proposed reports into appropriate scientific groupings; the increasing number of papers could not all be presented in open session during the single week's meeting. Once sections were formed, their

79. Bache, elected president for the year beginning with the annual meeting of 1850, presided in Henry's absence over the Charleston meeting and then served his own term at the New Haven meeting in 1850 and the semiannual Cincinnati meeting in 1851.

80. AAAS, *Proceedings*, V (1851), 178. Three groups were appointed to examine all submitted papers "with authority to present abstracts of the communications instead of full papers, and to omit the publication of any paper which they do not approve." The groups were Peirce, Henry, and Bache for mathematics and physics; J. W. Foster, Hall, and Dana for geology and chemistry; and Agassiz, Torrey, and Leidy for zoology and botany.

81. L. Agassiz to A. D. Bache, Cambridge, 2 Aug. 1851, Rhees MSS, HHL.

elected chairmen would become members of the standing committee, and each section would organize its own daily program. Absence of sections thus concentrated power in the standing committee, keeping it more compact and responsible for reviewing all papers.[82] When the standing committee for 1851 dallied in forming the sections, Ormsby M. Mitchel challenged it. Henry D. Rogers seconded Mitchel's remarks, observing that there had been "a usurpation by the Standing Committee of the duties assigned to the sections."[83] Rogers did not impugn the motives of the committee but did argue that to organize the presentation of papers before delegating sections was unconstitutional; comparing the situation to that of the federal government, he suggested that the "states' rights" of the sections had been violated.

Agassiz stepped down from the chair to reply and deftly avoided the issue of the power of the standing committee. He stated that the delay in creating sections resulted from the limited number of papers initially submitted and proceeded to suggest that in actual fact the standing committee had assumed a heavy burden by making all the arrangements. Because the sections, once established, had not created their own committees, the standing committee had had to arrange the program for the sections as well as for the general meetings. Paternalistically he added the undoubtedly real reason for the delay (as viewed by the Lazzaroni) and for the standing committee's action: the immature AAAS was not ready to exercise its own control, although "after an experience of two or three years the members would be left free to take up business in the sections as they saw fit."[84] He further claimed that the standing committee took no more authority than that voted to it by the AAAS in resolution. Quite openly the leadership thus proclaimed its peculiar capability for directing the association's business. Agassiz's explanation elicited qualified approval for the standing committee in a vote which stated that "considering the circumstances in which they were placed the members of the Association do entirely approve the action of the Standing Committee this year."[85] It was the last such blanket approval given.

82. B. A. Gould to J. Hall, Cambridge, 27 May 1856, Hall MSS, NYSL. This conscious tactic was expanded when the size of the association required formation of additional sections. Gould wrote, "The great good point of only two Sections ... [is that] we shall have only two chargeable members a day, enough & not too much, & the secretaries of each will be permanent."

83. *Albany Evening Journal*, 23 Aug. 1851.

84. *Ibid*.

85. AAAS, *Proceedings*, VI (1851), 402.

CONSTITUTIONAL DEBATE

The action of Rogers and Mitchel had some effect, however. The association also voted that the constitution of 1848 and all subsequent resolutions relating to governance of the association be printed for the membership. Two other resolutions which explicitly stated the mode of organizing sections were passed to ensure the authority at that level of the association. In addition, as a reaction against the standing committee's rearranging the order of papers on the program, three pertinent resolutions were suggested for consideration at the following annual meeting. These proposals basically made the order of submission of papers to the permanent secretary the order of presentation, with the suggestive hostile limitation that papers by members of the standing committee were liable to postponement by a decision within the sections themselves.[86]

Members were determined to make the constitution a more concrete expression of their expectations and were moved to action by the standing committee's emphatic exercise of power. The Rogers brothers, joined in this instance by Mitchel, John H. Coffin, and Alexis Caswell, did not challenge the system of reviewing papers intended for presentation but insisted that this function was the legitimate and appropriate prerogative of the sections themselves. In this case Bache and his fellow travelers, seemingly spokesmen for professionalism, actually resisted the final implications of specialization, which would decentralize the scientific community and allow specialists to determine the legitimacy of a paper. They obscured the problem by their discussion of amateurs—the growing opposition was headed by Rogers and Alexis Caswell, neither of whom was a charlatan.

When the meeting scheduled for 1852 was postponed because of the cholera threat, Agassiz hoped that the time lapse would "let some people come to their senses" and cease their opposition, which was beginning to undermine the society.[87] At the Cleveland meeting the following year there was limited debate over the issues of 1851. In accord with the standing committee's recommendation, a committee of ten was appointed: A. D. Bache, J. Lawrence Smith, John LeConte, Wolcott Gibbs, B. A. Gould, W. B. Rogers, J. D. Dana, Joseph Leidy, S. S. Haldeman, and A. A. Gould. It was to be "so constituted as to represent as equal as possible the several departments of science, with due

86. *Ibid.*, p. 403.
87. L. Agassiz to S. F. Baird, Cambridge, 27 July 1852, Baird OFF, SIA.

regard also to the interests of different institutions and geographical sections of the country."[88] They were to redraft the constitution, incorporating resolutions which had been passed in previous years relating to the function of the association.[89] The committee reported a long series of amendments, including not only the earlier resolutions but also relating to the selection of members, the overly elaborate arrangements made by local committees, and an enlargement of the standing committee.[90] To these Rogers recommended adding a rule "precluding all action in the way of recommendation or otherwise, either of instruments, books, researches, or other scientific, public, and private enterprise."[91]

As earlier mentioned, Rogers questioned the function of appointed committees on the grounds that they were liable to abuse of power by the standing committee which appointed them. The AAAS leaders contended that recommendation of specific enterprises had been successful in promoting science. In proposing such a limit Rogers was on less firm ground, for he was protesting an activity which he had encouraged in the 1840s in the instance of state and federal surveys. As to the rest of the constitution he believed that changes should be minor: "What is needed is not so much a change in its provisions as a better knowledge among the members of what these provisions are, and a more careful adherence to them."[92] He agreed that membership should be more selective. He also insisted that the constitution implied that at-large standing committee members were to be nominated and elected by the total membership.

The constitutional question created a permanent and outspoken division between the leaders in 1854. Rogers anticipated the meeting, hoping the issue would be discussed in Washington with a "full mustering of the Scientific workers from all parts of the Union," which would truly represent the diverse geographical, disciplinary, and occupational backgrounds of the membership.[93] However, the normally

88. Unpublished "Report of the standing committee on the resolutions of Dr. B. A. Gould, introduced July 20th 1853," AAAS Archives.
89. AAAS, *Proceedings*, VIII (1854), xi.
90. *New York Times*, 8 May 1854.
91. W. B. Rogers to A. D. Bache, Boston, 2 Apr. 1854, Rogers, ed., *Life and Letters*, I, 339; Rogers to J. Lovering, Yellow Springs, Pa., 30 June 1854, Rogers MSS, MITA.
92. W. B. Rogers to A. D. Bache, Boston, 2 Apr. 1854, Rogers, ed., *Life and Letters*, I, 339. At this point the two men's dealings were still fairly congenial and direct.
93. Rogers to Baird, n.p., 16 Oct. 1853, Baird PP, SIA.

quiescent Elwyn notified Baird, "The meeting seems to have been less harmonious than usual."[94] According to Rogers, the series of earlier resolutions were intended to clarify the provisions of the original constitution rather than to propose any substantive changes in the operation of the Association. The original constitution of 1848 stated that in addition to the officers for the current and preceding years, the standing committee should include six members elected "by ballot." This phraseology was indefinite, and by 1850 the standing committee itself assumed the right to nominate its six additional members; the nominees were inevitably voted into office by the general membership. As long as the nominees were selected to provide a balance of scientific interests and geography, the practical policy aroused little early skepticism. As the standing committee became more powerful, however, and its core membership assumed a repetitious profile, opposition arose from men like Rogers, who resented less the activities of the group than its potential autocratic power. When Rogers challenged the right of the standing committee to nominate its own membership and this right was then made the basis of a proposed amendment, he went into active opposition. No one denied the right of recognized leaders in science to hold the major AAAS offices and to publish their own papers. Even the authority to establish a daily program which might exclude "lesser" contributions had general approval. But the autocratic, self-perpetuating techniques fueled the newly fired torch of the opposition. Initial consolidation of power in the standing committee was the result of Bache's forethought, but its maintenance required a coterie of like-minded men.

The proposed amendments formulated by the committee were supposed to be printed and distributed among members by mail for consideration before the Providence meeting in 1855. They were not circulated until the meeting itself, however, and the question was postponed several times by Bache and others who argued that the scientific presentations had highest priority.[95] The committee on the constitution met twice and changed the original draft by adding specific powers for the standing committee.[96] When this new draft of the constitution

94. A. Elwyn to S. F. Baird, Philadelphia, 19 May 1854, Baird PP, SIA.
95. *New York Times*, 8 May 1854.
96. *Ibid.*, 24 Aug. 1855. A letter fragment from W. B. Rogers, probably written in late Aug. 1855, reported events. Printed copies of the old and new constitutions were distributed early in the meeting; the committee for the constitution met on Saturday and Monday and reported out a modified draft. Rogers felt that the committee, far from following the implied recommendations for change, had

was presented and recommended for adoption, pandemonium occurred. "It was as if a shell had suddenly dropped among combustibles," the *New York Times* reported, "with the fuse burning."[97] Debate was heated and bitterly personal. Ultimately, after two hours of haranguing, the members once again voted to table the subject until the next annual meeting.[98] The *Times* concluded that the "bugbear" was the extended power of the standing committee: while the Washington draft proposal had increased its membership to twenty-six, the Providence draft reduced it to eighteen, provided for a vice-president, and gave the committee more power—an ironic departure from the original proposal to limit the power of that group.[99] In detail, the new proposal provided for authority "to assign papers, arrange the business, suggest places and times of meeting, examine or exclude papers, appoint the local committee, nominate persons for membership, and decide on publication."[100]

Rogers, although ill, attended most sessions and worked feverishly to keep the "obnoxious" constitution from passage. His major support came from Mitchel, J. Lawrence Smith, and William Hackley. Their success in thus stopping "Bache, Peirce, Gould, Agassiz & c." seemed to Rogers "the best rebuke that had been administered to those who assert authority."[101] When in his absence the committee had modified the draft constitution to grant explicit authority to the standing committee, his sense of propriety was offended and, like Chester Dewey, he found the action not only irregular but immoral. Rogers left the meeting convinced that the general association membership finally recognized the clique and that the "despotic" exercise of power by the standing committee would be defeated the following year at Albany.

Rogers was ready there in 1856 when the standing committee nominated the six at-large members. Rogers, Robert Hare, and others immediately protested that the nomination was not consistent with the constitution, which required that standing committee members be elected by the association but did not intend "that the committee should be a

---

"recast it so as to restore in great part the obnoxious features of the Standing Committee as formerly organized & in other ways to retain existing influences." Rogers MSS, MITA.

97. *New York Times*, 24 Aug. 1855; *Providence Journal*, 21 Aug. 1855.

98. The initiative for postponement came from the Rogers faction, with one press account mentioning Mitchel and the other Alexis Caswell. Charles Wilkes MSS Journal, LC, recorded for 21 Aug. 1855: "Attended meeting constitution. Question came up postponed after dispute for a year. We gained our point."

99. *New York Times*, 24 Aug. 1855.

100. See n. 80 above.

101. AAAS, *Proceedings*, V (1851), 178.

self-electing and self-propagating corporation."[102] After much discussion the matter was tabled, and when taken up at the next general meeting the nominees were duly elected by the entire association.

Once again the constitutional question was postponed until late in the meeting. When finally brought forward on Monday morning, the majority report of the committee remained in favor of the Providence draft, with one change—that the sixth article specifically be changed from "open nomination" to read "nomination by the Standing Committee." The only dissenting vote was by Rogers, who submitted a minority report eliminating that change. The majority report was to go into effect in 1857. The final vote of the association decreed that at-large standing committee members were to be elected after open nomination, as were the sectional leaders. In addition, the sections were grantd authority for working out their own programs.[103] Rogers was jubilant because as the probable results of the vote became evident, Bache had gallantly proposed that the vote be considered unanimous—giving Rogers an opportunity to "thank the gentleman for coming over *to our assistance.*"[104] Now, under the new constitution, he felt the membership would be able "to maintain some control over the combination which has been so arrogantly claiming to regulate the Assoc. & with it the science of the country."[105] The victory was mixed, however, for the standing committee remained at eighteen members, with the rest of its extensive authority explicitly granted.[106] In addition, Rogers's hoped-for amendment limiting committee activity had been eliminated.

Many members did not understand the nature of the controversy and reacted to the evident disharmony among leading scientists. A press reporter observed, "Constitutional questions exercise the mind of the

102. *Albany Atlas and Argus,* 21 Aug. 1856.

103. W. B. Rogers to H. D. Rogers, Sunny Hills, 1 Sept. 1856, typescript copy, Rogers MSS, MITA: "This is therefore a return with a little more strictness of phraseology to the Old Constitution....The Coast Survey [Bache] and its alliances were at first quite sure of legislating their usurped powers and their defeat in this matter has been a serious blow."

104. W. B. Rogers to Lorin Blodgett, Cambridge, 1 Sept. 1856, Rogers MSS, MITA.

105. *Ibid.* Also see Brewster, *Life and Letters,* p. 168, where a Whitney letter reports, "I am glad to say that the democratic party carried the day against Bache, Henry, and the Cambridge clique in the matter of revisal of the constitution, which has been so often up for discussion, and as uniformly staved off by the Bache and Co., party...."

106. Bache to J. F. Frazer, Mt. Desert, Me., 5 Sept. [1856], Frazer MSS, APS. Bache claimed that the Lazzaroni "held their own" on the matter. Also see B. A. Gould to J. Hall, Cambridge, 27 May 1856, Hall MSS, NYSL.

Association. With many men, there are many minds. No great Expounder appears and the body is tossed to and fro by divers winds of doctrine."[107] One semi-serious response to the acrimonious debate was that since the association seemed unable to form its code peaceably, a convention with *no* rules simply meet yearly and organize all members equally.[108] A disenchanted Albany reporter, observing the weather during the week of the Albany meeting, wryly commented, "Since the savans have been among us we have had nothing but storms within & storms without. . . ."[109] The confusion of members uninitiated into the controversy is clear in a letter from George Swallow, who had apparently been chided by Hall for voting against Bache: "I think the Association has been well managed by those who have had control of it, but still when the proposition for the constitution came up without any agency of mine, I voted according to my judgment of what in the nature of things would be for the best interest of the Association. As to the previous difficulties on this I knew nothing of them as to their character, the parties interested, or their relation to the Constitutional question."[110] Aware that the "best part of the association for activity and talent was on the other side," Swallow independently voted on the issue rather than on personality.

In 1858 the standing committee reported a resolution, not unlike that of Rogers in 1849, that "none but members who are practically scientific" be elected officers.[111] When voted upon in 1859, however, the motion was defeated. Chester Dewey, joined by former Smithsonian librarian Charles Jewett, argued that such legislation was unnecessary. As the *Times* reporter observed, "It was no use to legislate to make oil rise to the top of water, neither was it necessary for the Association to take action on this subject; the scientific members would naturally be above the others without any rule for keeping the classes separate."[112] The almost defensive proposal suggested that the Bache contingent was no longer sure of its place. Yet the general membership seemed tired of constitutional proposals from either side. At the same 1859 meeting James Coffin proposed an amendment that "the titles of all papers read at the meeting be published as a record." Peirce responded that the as-

---

107. *New York Times*, 24 Aug. 1860.
108. *Ibid.; Annual of Scientific Discovery . . . for 1856*, p. 4.
109. *Albany Atlas and Argus*, 26 Aug. 1856.
110. Swallow to J. Hall, Chester, Ill., 2 Oct. 1856, Hall MSS, NYSL.
111. *Baltimore Sun*, 5 May 1858. Also see Dewey to J. Torrey, Rochester, 21 Sept., 1 Oct. 1859, Torrey MSS, NYBG.
112. *Springfield Daily Republican*, 10 Aug. 1859; *New York Times*, 12 Aug. 1859.

sociation should have an opportunity to review a paper after its reading. Although Coffin thought that "historically it would be found that the Standing Committee had seriously erred in judgment rejecting papers that the Association had approved," his motion was defeated as well.[113] Again the constitutional debates aroused the press: "Unquestionably there is less of harmony and less of usefulness in the Association because of the dominant influence of this so-called aristocracy."[114] Constitutional revision had not freed the association from domination by Bache and the other standing committee stalwarts. They continued to hold leadership positions and to regulate publication, although under closer scrutiny. Instead, the continuous debate bred both internal and external disillusion with the association. The savants demonstrated to the outside world that they were men, capable of pettiness, temper, and irrationality, as well as sociable men and outstanding scientists. The association had not resolved the implicit problem in most new groups—how to organize and consolidate without making the original leadership an entrenched oligarchy. Extravagant bids for power in the later years of the decade eroded the authority granted without question to men of established reputation in the early 1850s. By using the AAAS not only as a professional arbiter but also as a private mouthpiece, Bache's credibility was curtailed. Members were more skeptical about irregular procedures, as in the cases of Winslow and Vaughan, and resentful of the high-handed attitude of Peirce.

The discontented who grouped loosely around Rogers were a loyal opposition dedicated to preserving what they believed to be the original purposes of the association.[115] They warmed to outsider Matthew F. Maury, who observed, "I have no sympathy with those philosophers, who mounted on scientific stilts, stalk about the physical domains of the earth saying, 'Those who would follow us must mount stilts as high.' "[116] But the power he opposed was no longer so formidable.

Along with the decline of the AAAS the Lazzaroni declined, disintegrating from within rather than succumbing to outside attack. Concerned with financial and political problems and unimpressed with the quality of science at the AAAS meetings, leading scientists, including the

113. New York Times, 9 Aug. 1859.
114. Springfield Daily Republican, 9 Aug. 1859.
115. C. Wilkes to W. B. Rogers, Washington, 30 July 1859, Rogers MSS, MITA. Wilkes, unable to go to Springfield, urged Rogers to "keep them within the strict interpretations of the Constitution."
116. Towle, "Science, Commerce and the Navy," p. 498, quotes a letter from Maury to Lt. F. Julien, 18 July 1859.

Lazzaroni, attended less often. Bache came regularly, usually joined by Benjamin Gould, Benjamin Peirce, and Joseph Henry, but Wolcott Gibbs, John LeConte, and John F. Frazer felt less obligation to participate. Even the former became less creative in their reports to the association. Rogers reported in evident disgust, "We had as usual many scraps from the Coast Survey ... and innumerable unintelligible things from Peirce. Henry repeated his molecular theory given at the Cleveland meeting & presented sundry other scraps, for which he carried no credit from the knowing ones. ... Although they figure somewhat largely in the papers, Bache, Henry and Peirce did not make much impression. Agassiz did better."[117] The disintegration of the Lazzaroni was gradual. Josiah D. Whitney, an early confidant, by 1856 rejoiced in Bache's "defeat" on the constitution.[118] Dana managed politely to decline dinner invitations while maintaining open relations with Bache.[119] Joseph Henry became increasingly unhappy with the quarrelsome Gould and finally ceased supporting him in the Dudley Observatory affair. Similarly, Peirce broke with his early protégé and regarded only Bache as a trustworthy confidant.[120] By early 1858, when Bache peremptorily insisted as "chief" that Fairman Rogers was a Lazzaroni, Frazer informed him that election required unanimous approval by all members.[121] Thus, while the Lazzaroni displayed professional zeal and felt a mutual responsibility to lead science in America, they differed widely in conducting the specific activities of scientific institutions and ultimately splintered as a group. Dana remained inactive but observed, "I believe as much as ever that the honor of the Science of the country depends on its being controlled by the power that has controlled it; I wish that it were possible to make that power more felt in the details of its administration. My seeming indifference comes from a natural aversion to pushing into the front ranks of action."[122] Bache was deliberately more quiescent and conciliatory at the last two AAAS meetings and conceded that the Lazzaroni was a failure.[123]

Bache continued to use the association but less conspicuously. Shortly after Gibbs lost the Columbia post, he was invited to give a report on

117. W. B. Rogers to unknown, n.p., c. Aug. 1855, Rogers MSS, MITA.
118. Brewster, *Life and Letters*, p. 169.
119. Dana to Bache, New Haven, 2 Feb. 1856, 24 Dec. 1857, Bache MSS, LC; Dana to James Hall, New Haven, 12 June 1858, Rhees MSS, HHL.
120. B. Peirce to A. D. Bache, Cambridge, 23 Dec. 1857, Rhees MSS, HHL.
121. A. D. Bache to B. Peirce, Philadelphia, 30 Nov. [1857], Peirce MSS, HLH; J. F. Frazer to Bache, Philadelphia, 23 Dec. 1857, Bache MSS, LC.
122. Dana to Bache, New Haven, 11 May 1858, Bache MSS, LC.
123. Bache to Peirce, Washington, 12 Jan. 1860, Peirce MSS, HLH.

chemistry before the AAAS, and 300 copies were printed at association expense.[124] Similarly, Gould was nominated as vice-president months after his dimissal from the Dudley Observatory, an action intended as a vindication of his scientific ability.[125]

Just as the overconfident and sometimes abrasive Lazzaroni forgot that their positions were won by creditable leadership, the opposition to the "Washington-Cambridge clique" had naively assumed that constitutional explication would modify the Lazzaroni's control. Few critics understood the nature and composition of the group. They missed the quiet John LeConte and assumed Henry's brother-in-law Stephen Alexander's intimacy with the inner circle. Never known as the Lazzaroni to outsiders, the "mutual admiration society," a phrase created by David Wells, became the stock description for the hostile press after 1856. There was more than a hint of the group's most effective technique in Wells's observation:

> It may be very pleasant and agreeable for some individuals to discourse popular science by the hour to popular audiences; to indulge in fulsome adulations of one another; for one to designate the other as a second Kepler, and for another to rise in his place and "thank God that such men as Profs. S., Y., and Z., existed." These occurrences may be well enough in a mutual admiration society, but do not properly belong to the proceedings of an American Association assembled for the discussion of abstract science.[126]

The amorphous Lazzaroni proved to be more cohesive in their abstract discussion than in conducting practical affairs. The members of this group are significant in the history of science in America because they represented vividly if not consistently an emerging professional outlook and because their own enthusiasm inspired others to work to higher standards. For the AAAS they were, albeit indirectly, a guiding force which drew together the functions of the organization; they were also the source of a continuing and harsh division among the leadership and between the leadership and the membership. The Lazzaroni were an

124. AAAS, *Proceedings*, XIII (1859), 282. In 1863, largely through the efforts of Peirce and Agassiz, Gibbs got the Rumford Chair at Harvard.

125. AAAS, *Proceedings*, XIII (1859).

126. *Annual of Scientific Discovery ... for 1857*, p. iv. Bache answered the charge in his *Anniversary Address before the American Institute, of the City of New York at the Tabernacle, October 28th, 1856, during the Twenty-eighth Annual Fair* (New York, 1857), pp. 49–50. Henry also replied angrily to the continuing hostile press and was quoted in the *New York Times*, 12 Aug. 1859: "We did indeed abound in mutual congratulations. We delighted to congratulate each other when we brought truths to light that had never before been revealed to man."

intellectual elite whose snobbery and high-handed measures toward amateurs alienated even some professional fellow travelers. Their personal exclusiveness may have prevented the association from becoming more like the organization of working scientists envisioned by the Lazzaroni. Nonetheless, the Lazzaroni helped implement and clarify principles which motivated the scientists of their generation. It is impossible to distinguish carefully between the positive and negative influences of the "party in power" throughout the 1850s. On the one hand, the standing committee's firm handling of affairs ensured reputable and effective annual meetings, and the continuing participation of the nation's outstanding scientists enhanced the AAAS's prestige. On the other hand, the attitudes and tactics of certain leaders darkened the association's public image and divided scientific leaders. Looking back to the previous decade, Joseph Henry wrote to Louis Agassiz in 1864, "It is lamentable to think how much time, mental activity, and bodily strength have been expended among us in personal altercations which might have been devoted to the discovery of new truths; to the enlargement of bounds of knowledge, and the advancement of science."[127]

To demonstrate the specific nature of the AAAS, the following chapter outlines the characteristics of both active and inactive members. It suggests that many members, like Joel Howard, had little pretension about their scientific attainments or time to pursue their interests:

> I do not regard myself as able to do much in any way for "The Advancement of Science." I contracted a taste for scientific inquiry in younger life; but the duties of a laborious and absorbing profession have left me little leisure for its pursuits till I find myself not only distanced by my contemporaries but left out of sight by my own pupils.—It was by them that I was first invited to become a member of the "association." And it was a gratification to me to find that,—though I was left so far behind, I was not entirely forgotten.[128]

These persons were supportive but not persistent AAAS members. Others, the new professionals discussed in this chapter, took responsibility in the AAAS in part because their training, occupation, age, and geographical location coincided with the requisites for association leadership.

127. Henry to Agassiz, Washington, 13 Aug. 1864, Agassiz MSS, HLH.
128. Howard to S. F. Baird, in "Scientific Addresses," Baird OFF, SIA.

VIII

Profile of a Voluntary Scientific Community

Although loosely coordinated, informal, and replete with tensions, a discernible scientific community emerged in the United States with the establishment of the American Association for the Advancement of Science in 1848. The AAAS grew out of the self-conscious need for a symbolic and operative association of scientists to become the major arena for establishing professional aspirations as well as for meeting public expectations. Because the AAAS membership was essentially voluntary and its leadership elective, the association provides an index to the pre–Civil War generation of active scientists. A descriptive profile of the membership becomes, in essence, a profile of the active scientific community from 1848 to 1860.

New interest in collective biography and, simultaneously, in quantitative analysis has resulted in several studies relating to nineteenth-century American scientists. In fact, quantification, described by some as prosopography, is increasingly applied to discernible political, social, and professional groups.[1] The monographs relating to American science primarily consider the changes of a particular sample over an extended or transitional time period.[2] With the exception of Robert V. Bruce, all

1. Lawrence Stone, "Prosopography," in Felix Gilbert and Stephen Graubard, eds., *Historical Studies Today* (New York, 1971), pp. 107–140.
2. Five studies of American scientists are of interest. Donald Beaver, "The American Scientific Community, 1800–1860," is primarily concerned with historiographical methodology, and he systematically selected journals of the period from which a roster of 138 individuals was tabulated and analyzed. A second is Daniels, *American Science*. Daniels used statistics on the biographical information of 56 men, whose names were selected because of their high number of publications in 16 American scientific and medical journals between 1816 and 1845. A third is Clark Elliott, "The American Scientist, 1800–1863: His Origins, Career, and In-

of the authors have drawn their data groups, with varying degrees of sophistication, from the publication records suggested by one or more standard indices of publication.[3] Often they imply that their group is only a sample and that the actual community would be considerably larger; they readily admit that they are considering an elite group of the leading, published scientists.[4] The basic assumption, then, is the present-day view that a scientist is a person who publishes a scientific paper.[5]

In distinction from previous studies, this analysis of the AAAS began with the assumption that in a given scientific community leadership not only could be expressed through publication but also could be

terests" (Ph.D. dissertation, Case Western Reserve University, 1970). Elliott based his study on 502 men and women who wrote three or more articles cited in the *Royal Society Catalogue of Scientific Papers, 1800–1863*, and he therefore has a more comprehensive group of scientists. All of these concentrate on the first half of the century; two others analyze the Civil War period. The first is Amos J. Loveday, "A Statistical Study of the American Scientific Community, 1855–1870" (M.A. thesis, Ohio State University, 1970). Loveday used the *Royal Society Catalogue* to generate a list of 377 scientists. His analysis is related to that of Daniels by his effort to use the survey to analyze not only the men but also the climate of science; it is rather brief in discussion. The most recent is Robert V. Bruce, "A Statistical Profile of American Scientists, 1846–1876," in Daniels, ed., *Nineteenth-Century American Science*, pp. 63–94. Bruce selected 477 persons from *DAB* accounts indicating an interest in science for a period somewhat more comprehensive than Loveday's. While formats vary considerably, all of these studies include the data most readily available in standard biographies, that is, birth date and place of birth, areas of scientific interest, occupation, education, and geographic location. While Elliott and Bruce further attempt to study such items as motivation for entering science as a profession, rank in family, socioeconomic status of parents, and religious preference, the relative percentage of data in such categories is considerably lower and even the information available is frequently conjectural.

3. Daniels, *American Science*, pp. 6–33. Daniels simply uses 16 "national" journals of his own choosing. Beaver devotes a major portion of his dissertation to methodology and develops a sophisticated technique for ascertaining the journals most representative of his period, using the *Royal Society Catalogue* as well as Henry C. Bolton, *Catalogue of Scientific and Technical Periodicals (1666–1882), together with Chronological Tables and a Library Checklist*, Smithsonian Institution, *Miscellaneous Collections*, vol. XXIX (Washington, D.C., 1885), and Samuel H. Scudder, *Catalogue of Scientific Serials of All Countries, Including the Transactions of Learned Societies in the Natural, Physical and Mathematical Sciences, 1633–1876*, Library of Harvard University, *Special Publications*, vol. I (Cambridge, Mass., 1879).

4. Bruce is an exception, for he assumes that the selectivity basis of the *DAB* does produce a group who are the key figures in science; he states, "My list is presented not as a sample but as the whole concern."

5. Derek J. DeSolla Price has established this working definition in several places, including "Is Technology Historically Independent of Science? A Study in Statistical Historiography," *Technology and Culture*, VI (Fall 1965), 556.

established through some other activity—such as teaching or unpublished research communicated informally—recognized as significant by peers. Correlation of officers and publishing authors in the AAAS is, in fact, high, justifying for most scientists the hypothesis of earlier scholars. By the 1850s publication was already associated with peer recognition of achievement and thus with professional stature. The present analysis also differs from other studies because it is not concerned with tracing changing characteristics such as residence or occupation for individuals within a selected group but, rather, emphasizes developments for the organization as a whole. The 13-year upper limit on AAAS participation, with many members active for fewer years, is too small a portion of individual career patterns to show significant changes with time. The shift in activity and goals within the organization has been considered earlier, as a function of attitude (Chapter V) and of the scientific papers presented (Table 3).

The AAAS was the first truly national voluntary organization in the United States in which the leading persons in all fields of science took an active interest. Not all were willing to travel to the annual meetings or participate in the sometimes volatile sessions; yet nearly every eminent scientist was enrolled as a member, and most assumed leadership positions during the period under consideration.[6] This breadth and level of interest subsided after 1860 and was never regained during the following century of the AAAS.

SELECTION AND DEFINITION OF GROUPS ANALYZED

The first step in quantitatively studying the AAAS was to develop a working list of the membership for the relevant period. The published lists of members for each of the 14 volumes of AAAS *Proceedings* for the period (1848–60 and 1866) generated a cumulative file. The date of election was recorded for each name and a year-by-year record was maintained so that the date a member became inactive or was dropped from membership could be determined. If a person joined, then became inactive, and subsequently rejoined, the number of years inactive were

6. A comparison with the individuals selected for the previously mentioned quantitative studies suggests that the AAAS leadership list includes most of the same persons who were active for the period from 1848 to 1860. The AAAS was represented by 166 leaders and 116 general members in Elliott's group; most of those not members were either active before 1848, marginal producers, or geographically isolated. Seventy AAAS leaders coincide with Beaver's list; as Beaver himself points out (p. 182), 106 of a possible 124 persons (85.6%) joined the AAAS.

thus also recorded. A cumulative list of 2,068 members emerged, considerably longer than the year-by-year membership rosters—none of which had more than 1,100 members—might imply.[7]

The next step was to ascertain the extent of each member's participation. Ideally, a measure of participation (active interest) would include a record of the number of meetings attended by each individual. Such data, however, proved impossible to obtain because the AAAS archives have signature lists for only five meetings, and these do not always coincide with newspaper or private accounts concerning who attended a given meeting. From a page-by-page analysis of the *Proceedings* the specific offices held and papers presented were recorded for each contributing member. Such active members were designated "leaders." When collating this information, it became clear that a sizable number of individuals who presented papers and held offices were not members. Thus three distinct categories with a total of 2,205 persons were established: the leadership of 337 men, the general or noncontributing membership of 1,731 persons, and the participating nonmembership of 132 persons. In subsequent discussion the term "membership" will refer to the 2,068-person full membership unless otherwise designated.

Biographical Investigation

To collect the biographical information necessary to develop a profile of the membership, standard biographical dictionaries were first consulted.[8] The several standard works, however, provided no data for a number of individuals, limited sketches for many, and redundant information for others. After the initial search many persons remained identified by name only. Because the proportion of unavailable information was considerably higher for the regular membership, the biographical search beyond the standard dictionaries was abridged to include only leaders and active nonmembers.

7. See the Appendix for a list of these persons. The number of members peaked with the 1854 meeting.
8. The full list was compared to *Appleton's Cyclopedia of American Biography;* the *Dictionary of American Biography; Herringshaw's Encyclopedia of American Biography of the Nineteenth Century* (Chicago, 1905); Howard Kelley and Walter Burrage, eds., *American Medical Biographies* (Baltimore, 1920); and the *National Cyclopedia of American Biography.* The indices of several older histories of American science were also checked: Bernard Jaffe, *Men of Science in America* (New York, 1958); David Starr Jordan, *Leading American Men of Science* (New York, 1910); and E. L. Youmans, *Pioneers of Science in America* (New York, 1896).

Frequently more than one source was consulted to fill in basic bio-graphical data. When the AAAS member was included in a major dictionary, it was relatively easy, in general, to determine such informa-tion as dates of birth and death, place of birth, level of education, col-legiate institution with date of A.B., additional training for a profession, adult residence while a member, and the major fields and subfields of scientific interest if the person was a scientist or active amateur. Some of the more personal characteristics which might reveal subjective rea-sons for a person's scientific involvement proved more difficult to find. None of the biographical dictionaries, including the *Dictionary of American Biography*, provided adequate or consistent information on the important questions of socioeconomic status of the member's parents and his relative position in the family structure.[9] Similarly, the member's own economic holdings or his social status was difficult to ascertain. Data on participation in contemporary reform movements, or religious beliefs, or political affiliation were also recorded when found, but these data were rare in basic reference volumes. Only infrequently did evi-dence indicating a scientist's reasons for entering the field exist.[10]

As suggested by other group analyses, leading scientists were occupa-tionally and geographically quite mobile. However, the 13-year period considered here is too brief for these shifts to yield a significant trend. Therefore, for geographical analysis, the urban or rural location and the state in which a member spent the majority of his time during the period of his membership—often less than the full period—were recorded. A person's classified occupation was the one dominating his AAAS membership years. Since this spanned a brief interval in his total career, the listed occupation might not represent a person's major area of en-deavor, especially if he were young. The purpose of this research was to uncover the origins and living patterns of scientists in the decade or so under discussion and, more basically, to characterize the AAAS membership during its formative years.

9. P. M. G. Harris, "The Social Origins of American Leaders: The Demographic Foundations," *Perspectives in American History*, III (1969), 203.

10. Elliott, "American Scientist," p. 224. Elliott found suggestive information on motivation for only 40% of his group and had to include such imprecise categories as "inborn science abilities," "contact with nature or natural events," and "accidental circumstances." Not surprisingly, he found that the factors most frequently mentioned were family encouragement and acquaintance with an older scientist. A study of twentieth-century scientists emphasizes that interest in science was especially strong for persons in transition between acceptance of orthodox Protestantism and acceptance of a more secular world view. See Robert H. Knapp and H. B. Goodrich, *Origins of American Scientists* (Chicago, 1952).

The first and second areas of scientific interest were also of central importance and were considered fixed for the time of membership. Occasionally this characteristic was indicated by the individuals analyzed. In 1852 Spencer F. Baird, then the permanent secretary of the AAAS, sent out a questionnaire to AAAS members and others.[11] It primarily concerned occupation and areas of scientific interest. Whenever possible, this reference was used because it clearly stated a person's self-assessment. Otherwise, data in biographical sketches and publications during the period of membership were taken as indicators of scientific preference.

Differing levels of activity among the membership categories became evident in the initial stage of collating a membership file. Later, during the biographical investigation, it became clear that foreign members, in particular the men from British North America, were nearly impossible to describe in detail.[12] The decision to omit this group from some analyses was sustained by a survey of the length of membership which demonstrated that their interest, stimulated by the meeting in Montreal in 1857, was short-lived. The local committee composition was distinct from all other appointed committees in that nonmembers and otherwise noncontributing members were frequently included. Because of the peculiar composition of the local committee, it will be treated separately in subsequent analyses. The participants on the local committee, however, are distributed into one of the three basic categories given above for purposes of general discussion; for example, a person who held a post in addition to that of local committee was considered a leader, while a member who served only on a local committee was assigned membership status.

Categories of information for analysis were governed basically by the information available. In all 24 variables were considered.[13] The data

11. Bound as "Scientific Addresses," these questionnaires were sent out under Baird's name as secretary of the AAAS but seem to have also gone out to other Smithsonian correspondents, especially those doing meteorological observations. The volume does less than it promises, however, for many members apparently did not respond, and the volume itself shows mutilated areas where autograph seekers apparently removed signatures and even full replies. Among the most interesting questionnaires was one not bound in the volume but held by a collector—that of Henry David Thoreau, who coolly expressed his disinterest in such efforts. See Walter Harding, *Mr. Thoreau Declines an Invitation: Two Unpublished Papers by Thoreau: An Original Letter and an Annotated Questionnaire* (Richmond, Va., 1956).

12. The *Dictionary of Canadian Biography*, vol. I— (Toronto, 1966), has been completed only through the first two volumes, covering the years to 1800.

13. These were: the number of publications, the number of offices held, local

were recorded in numerical code, punched on computer cards, and then compiled into one- and two-dimensional frequency tables.[14] These tables were analyzed to determine which variables and pairs of variables showed patterns striking enough and had enough observations to justify closer scrutiny. At this point trends could be observed and information categories placed in juxtaposition for comparison.

RESULTS

The major purpose of this investigation is to provide a descriptive profile of the men and women in science in the immediate pre–Civil War, a time of transition for American science. The AAAS membership roster is an index which is more broadly based than the publication records used by several other historians. The most evident limitations on collective biographical studies are imposed by the inadequacy of biographical reporting, which leaves some interesting categories with data too limited for authoritative statements. Another difficulty is that statistics on the general population for the period are lacking for meaningful comparisons in several categories. Where available, such data are considered. Until similar profiles of other professional groups and of the general population are available, the following survey of scientific persons should be considered basically a descriptive profile, valid as such and useful for comparisons with other early professional associations, similar foreign organizations, and contemporary scientific groups.

The AAAS membership pattern was both larger and more unstable than historians have assumed for the pre–Civil War years.[15] The fluctuations in membership and attendance might have been expected of a new, untried organization. In the term of social scientist Robert Berkhofer, the AAAS was a "collective phenomenon" because individuals pooled their

---

committee participation, year joining AAAS, year leaving AAAS, years inactive, city of birth, state of birth, decade of birth, general educational level, place and date of college degree, occupation, suboccupational category, teaching college for professors, major scientific field, subfield of scientific interest, second major scientific field, city of residence while a member, state of residence, religion, political affiliation, membership in the National Academy of Sciences, and European travel or study.

14. A prepared SOUPAC program at the University of Illinois provided both a frequency count, complete with item percentages, and a correlation program to measure the linear dependency between the appropriate numerical variables.

15. Beaver projected that the total number of active scientists for the first half of the nineteenth century would be 2,000. While all 2,073 persons in this study might not be considered active scientists, they represent only the period from 1848 to 1860.

particular talents for the common purpose of promoting American science.[16] The identity of the association developed in response to the problems met in accomplishing that goal. Because there were certain ethical and professional principles implicit in the drive to create a viable scientific community, the leadership of the new organization took on a symbolic aura, at least until challenged. Means to the end of a better scientific community were not readily found, and the initially eager membership became restive during the searching years. Evidence of this tendency appears in the pattern of affiliation and the regional characteristics of the full AAAS membership, which are considered below.

Although the total number of members for any single year from 1848 to 1860 only went above 1,000 in 1854, the cumulative membership reached 2,068 persons, including 215 foreign and Canadian members. Table 4 describes the percentage of total members who, having joined in a designated year ("Joined"), left either by exclusion or drop-out in a specific year ("Left"). For example, the number of persons who had joined in 1848 and left in 1851 was 5.3% of the total membership. The percentage in each cell is based on the total membership to allow comparisons in both directions on the table. Turnover in the AAAS membership was rapid; only one-fourth of those joining the association from 1848 to 1860 were still active members when the organization reassembled in 1866.[17]

The drop-out rate for persons joining in 1848, 71% of those who joined in 1848 (or 15.3% of the total membership), can be accounted for in part by the fact that the list of founders included many local scientists who were not solicited for membership before being added to the roster. Thus the initial group also showed great persistence, with 20.0% (4.3% of the total membership) still active in 1860, despite the fact that 13.9% (3.0% of the total membership) of the persons on the 1848 list died by 1860. A significant drop in membership in 1851 resulted from Spencer F. Baird's efforts to weed out nominal members who were tardy about dues. Thereafter the rule of carrying nonpaying members for no more than two calendar years was exercised. This fact can be traced by noting the high values for 1853 Joined 1855 Left, 1854 Joined 1856 Left, and so forth. More than 44% did leave the AAAS within the first three years after they had joined. The member-

16. Robert F. Berkhofer, Jr., *A Behavioral Approach to Historical Analysis* (New York, 1969), pp. 75–77, 309. Berkhofer distinguishes three types of groups in the order of their cohesion and self-identity: aggregate, conjunctive, and collective.

17. To determine that a member remained active through 1860, the 1866 roster was used to verify membership continuance.

TABLE 4

Year-by-Year Participation Level of Total Membership

This cross-tabulation indicates the percentage of members, based on the 2,068 total, who having joined in year "Joined" left in year "Left." Percentage totals are not always 100% because each entry is rounded to the nearest tenth of a percent. In addition, the table indicates the number joining each year who remained active through 1860, the percentage of individuals joining each year, and the number joining in each year who were dropped by 1860.

Left	Joined												Total
	1848	1849	1850	1851	1853	1854	1855	1856	1857	1858	1859	1860	
1848–49													
1850	0.4												0.4
1851	5.3	1.0	0.6	0.2									7.1
1853	0.4	0.1	0.1	0.1	2.6								3.3
1854	5.3	1.0	5.3	5.4	1.5	0.1							18.6
1855	0.9	0.2	0.4	0.6	2.5	0.1	0.5						5.2
1856	0.2	0.1	0.1	0.2	0.5	1.2	0.1	1.4					3.8
1857	0.2	0.3	0.2	0.5	0.7	0.2	1.2	0.1	1.6				5.0
1858	0.2	0.1	0.2	0.4	0.7	0.1	0.2	1.7	0.0	0.1			3.7
1859	1.0	0.4	0.4	1.0	1.3	0.3	0.7	3.7	8.0	0.1			16.9
1860	0.4	0.2	0.5	0.4	0.5	0.4	0.3	1.4	1.2	2.5	0.1		7.9
Died[a]	3.0	0.2	0.2	0.1	0.1	0.1	0.1	0.1	0.1	0.1	0.0	0.0	4.1
Active in 1860	4.3	0.9	1.2	1.4	1.7	0.8	1.3	2.6	2.8	1.5	3.3	1.5	23.4
Total Percentage Joining	21.6	4.5	9.2	10.3	12.1	3.3	4.4	11.0	13.8	4.3	3.4	1.5	99.4
Total Percentage Dropped	14.3	3.4	7.8	8.8	10.3	2.4	3.0	8.3	10.8	2.7	0.1	0.0	71.9

[a] A member's death was indicated on the roster, often a year or more after the actual death date. This table indicates the actual year of death when known.

ship gains made at the intermediate meetings at Charleston in 1850 and at Cincinnati in 1851 were not long lasting. By 1854 more than half of the new members who had joined at these meetings left association membership. The dramatic drop in 1854, like that in 1851, was due to the attempt of the new permanent secretary, Joseph Lovering, to rid his books of nonpaying members. To that year membership had grown steadily, with the number joining in any year always greater than the number withdrawing or being dropped for nonpayment of dues. The meetings at Albany in 1856 and at Montreal in 1857 added a significant number of New York state residents to the roster, but most of these were dropped after another rigorous effort to collect dues in 1859. By that year, too, internal problems had brought a number of withdrawals.

Table 5 indicates a rather different joining-leaving pattern for the 337 leaders from that of the general membership. Most striking are the facts that the percentage of leaders still active in 1860 was nearly twice that of the membership and that 26.6% of the leadership group left within their first three years. Almost 50% of the leadership had joined by 1850, compared to 25% of the membership. The persistence of the leadership from the earliest years of the AAAS may be explained in part by the time it takes for any new member to gain enough peer confidence to be nominated for office. Also, publication suggests an on-going commitment to scientific research. The pattern of withdrawal for

TABLE 5

Year-by-Year Participation Level of Leadership

This cross tabulation, similar to Table 4 but indicating the leadership subgroup, outlines the percentage of leaders, based on the 337 total group, who having joined in year "Joined" left in year "Left."

Left	Joined												Total
	1848	1849	1850	1851	1853	1854	1855	1856	1857	1858	1859	1860	
1848													
1849													
1850													
1851	2.1	0.9											3.0
1853	0.0	0.0	0.0	0.0	0.9								0.9
1854	6.2	2.1	4.2	2.1	0.3								14.9
1855	2.4	0.6	1.2	1.2	0.0	0.6							6.0
1856	0.6	0.3	0.3	0.3	0.9								2.4
1857	0.6	0.6	0.0	1.5	0.0	0.0	0.3	0.0	0.3				2.8
1858	1.2	0.3	0.6	0.6									2.7
1859	1.2	1.2	1.2	0.3	0.6	0.3	0.3	1.2	2.7				9.0
1860	1.2	1.2	1.2	0.3	0.9	0.3	1.2	0.6	0.3	0.6	0.3		8.1
Died[a]	5.9	0.3	0.6										6.8
Active in 1860	16.6	3.8	4.1	2.9	2.3	2.7	1.2	3.8	1.7	1.2	1.8	0.9	43.0
Total Percentage Joining	38.0	11.3	13.4	9.2	5.9	3.9	3.0	5.6	5.0	1.8	2.1	0.9	100.1
Total Percentage Dropped	15.5	7.2	8.7	6.3	3.6	1.2	1.8	1.8	3.3	0.6	0.3	0.0	50.3

[a] All dates are corrected to actual year of death.

nonpayment of dues after two years is practically missing from the leadership table, except in the case of Montreal in 1857. Only 25 non-participating members dropped out for several years and later rejoined; most simply left. Of the 34 leaders who became inactive and then returned, only 15 were unlisted for three or more years. The leaders thus seem to have felt more responsible to and interested in the association.

A relatively complete regional comparison is possible because the membership rosters usually indicate city and state addresses for each member. Table 6 shows the total number from each of five major regions of the United States. The state recorded is that from the membership

TABLE 6

AAAS Membership by Region and Year Joining

The home states of the 1853 American members, grouped into five regions, and the countries of the 215 alien members, tabulated as "Foreign," are presented in a two dimensional frequency table with the year in which members joined the AAAS. If a member left and later rejoined the AAAS, only the initial year is indicated; 67 men are in this category.

Region	Joined												
	1848	1849	1850	1851	1853	1854	1855	1856	1857	1858	1859	1860	Total
Northeast	145	51	45	28	31	9	55	43	30	4	32	19	492
Massachusetts	98	44	14	16	19	6	10	20	12	2	22	8	271
Connecticut	28	1	29	9	3	3	4	8	8	1	7	3	104
Mid-Atlantic	221	20	39	106	92	46	24	128	63	65	12	6	822
Washington	29	4	6	6	13	16	2	5	3	5		1	90
Pennsylvania	65	4	5	13	20	7	1	7	6	10	1	1	140
New York	101	12	22	75	49	19	18	104	45	9	9	3	466
South	41	14	84	22	50	6	4	17	25	12	6	2	283
South Carolina	10	8	65	1	6	1			1	2			94
North-Central	25	7	16	56	74	6	6	20	17	3	18	2	250
Ohio	14	4	13	48	46	3	3	4	5		4	1	145
West	3				1					2			6
Foreign	15	4	7	2	5	2	2	20	154		3	1	215
Total	450	96	191	214	253	69	91	228	289	86	71	30	2,068

Northeast includes Connecticut, Maine, Massachusetts, New Hampshire, Rhode Island, and Vermont.

Mid-Atlantic includes Washington, D.C., Delaware, Maryland, New Jersey, Pennsylvania, and New York.

South includes Alabama, Arkansas, Florida, Georgia, Kentucky, Louisiana, Mississippi, Missouri, North Carolina, South Carolina, Tennessee, Virginia, and Texas.

North-Central includes Illinois, Iowa, Indiana, Michigan, Ohio, and Wisconsin.

West includes California.

roster unless additional research proved that the person had spent the majority of his or her years as an AAAS member in another place and the roster was therefore incorrect.[18] In addition, specific states which contributed a significant proportion of regional membership are designated for comparison. Two important trends are immediately evident. The first is that states in which the AAAS held meetings had the largest membership, with the date of such meetings associated with a distinct surge in new registrations from the region. For examples, note that the 1848 meeting was in Philadelphia, the 1849 meeting in Boston, the 1850 meetings in Charleston and New Haven, and the 1851 meetings in Cincinnati and Albany. Table 7 summarizes meeting dates, places, and

TABLE 7

Meeting Dates, Places, and Presidents of the AAAS

This table gives the opening date of each meeting, the city in which the meeting was held, and the presiding officer. The number of members present is only approximate, based on newspaper accounts, signature books at the AAAS Library, or mentioned in the *Proceedings*; it should be used only for general comparisons. Permanent officers were treasurer Alfred Elwyn (1849–60) and secretaries Spencer F. Baird (1851–53) and Joseph Lovering (1854–60). The vice-presidency was initiated in 1857 and held successively by Alexis Caswell, John E. Holbrook, Edward Hitchcock, and Benjamin A. Gould.

Date	Location	President	Number Attending
20 Sept. 1848	Philadelphia	William Redfield	87
14 Aug. 1849	Cambridge	Joseph Henry	128
12 Mar. 1850	Charleston	A. D. Bache	94
19 Aug. 1850	New Haven	A. D. Bache	134
5 May 1851	Cincinnati	A. D. Bache	87
19 Aug. 1851	Albany	Louis Agassiz	194
28 July 1853	Cleveland	Benjamin Peirce	135
26 Apr. 1854	Washington	James D. Dana	168
15 Aug. 1855	Providence	John Torrey	165
20 Aug. 1856	Albany	James Hall	381
12 Aug. 1857	Montreal	Alexis Caswell[a]	351
28 Apr. 1858	Baltimore	Alexis Caswell[a]	190
3 Aug. 1859	Springfield	Stephen Alexander	190
1 Aug. 1860	Newport	Isaac Lea	135
15 Aug. 1866	Buffalo	F. A. P. Barnard	79

[a] Alexis Caswell presided in 1857 because J. W. Bailey died before the meeting; he presided in 1858 when Jeffries Wyman declined to hold the office.

18. This discrepancy was not unusual because the mobility of the scientists was relatively high, and records indicate that AAAS publications frequently had a difficult time reaching members.

presidents. The concept of peripatetic meetings proved valid for broadening the membership base. Local promotional activity and widely distributed broadsides announcing the meeting to local scientists and amateurs contributed to this effect. A second look at Table 6 reveals that the central and northern sections of the eastern seaboard states dominated the AAAS throughout the 1850s, partially because most meetings were held in this region.[19] Of the 2,068 known residences of members, a plurality (822) were from the mid-Atlantic states. The participants from those states north of the Mason-Dixon line and east of the Appalachian Mountains total 1,280. Southerners are most strikingly absent, although sizable numbers joined at the Charleston meeting, and many from the Mississippi valley region joined at the Cincinnati and Cleveland meetings. Lack of adequate data on birthplace makes comparison to adult residence difficult. Data on birthplaces of scientists suggest that New England produced a higher percentage of scientists than she retained, with the mid-Atlantic states and the most settled western regions making dramatic gains in active, practicing scientists by mid-century. Europe was also an important contributor to scientific leadership.

As Table 8 indicates, the concentration of members in particular geographical locations was not simply a reflection of the nation's population distribution in 1850. A chi-square test shows that the distribution is significantly different from that of total population. The Northwest was fourth in regional population yet had proportionately the largest number of members, supporting 17.7 scientists per 100,000 population. New England's longer tradition and advanced outlook on education contributed to her continuing importance as a cultural leader. The South lagged as a residence for both members and leaders of the AAAS. Only one southern meeting was held, the spring session at Charleston in 1850. The lack of meetings might explain the relatively low number of southerners, yet the limited attendance at the Charleston meeting proved that there were not sufficient numbers of southern scientists to encourage further meetings in that section. The still pioneering North-Central region was relatively low in membership and had proportionately fewer leaders than its neighboring eastern states. This finding corroborates those of Beaver, Bruce, and Elliott, who also found the preponderance of their samples living in the mid-Atlantic and New England regions, with the mid-Atlantic area having the largest real number of active

19. Beaver, "American Scientific Community," pp. 177–180; Elliott, "American Scientist," pp. 141–154.

TABLE 8

Regional Membership and Leadership Compared to Total United States Population in 1850

This table indicates the regional participation of the leadership and membership with the total population for that area in 1850 and thus allows for regional comparisons. The general population statistics are derived from *The Statistical History of the United States from the Colonial Times to the Present* (Stamford, Conn., 1957).

Region	Membership			Leadership			United States	
	Number	Number per 100,000	Percentage U.S.	Number	Number per 100,000	Percentage U.S.	Number	Percentage total Population
Northeast	492	17.7	27	97	3.6	31	2,728,000	12
Mid-Atantic	822	12.4	44	137	2.0	43	6,624,000	28
South	283	3.4	15	44	0.5	14	8,266,000ᵃ	36
North-Central	250	5.5	14	37	0.8	12	4,529,000	19
West	6	0.5	<1	1	0.1	<1	1,113,000	5
Foreign	215			21				
Total	2,068			337			23,261,000	100%

ᵃ This figure includes the nearly 2 million slaves in the South in 1850.

scientists at mid-century.[20] The AAAS membership was basically urban; more than 60% lived in population centers of over 10,000. The highest concentration of leaders was in Philadelphia, New York, the Boston-Cambridge nexus, and Washington, D.C., with smaller clusters in Albany, Baltimore, Charleston, and Cincinnati.

Of the 215 foreign members, 81% were Canadian. This group was particularly nonpersistent. Although 154 joined in 1857, 15 did not pay initial dues and were immediately dropped. Then in 1859, after a close scrutiny of the roster for nonpayment of dues, 111 were omitted, with nine more leaving the following year. Thus nearly 90% were inactive after only three years of membership. Geography was a key factor, since most Canadians could not afford the time or money to travel to meetings in the United States. More than 50% of the British North Americans were from the city of Montreal, where the 1857 meeting was held, and another 10% were from Toronto, making these members also principally urban. The AAAS gradually became less enthusiastic about recruiting foreign members and unwilling to carry names of even prominent persons who did not pay dues. By 1860 only 26 foreign members remained on the membership list.

From the long, cumulative roster, a distinct core of men emerged as active leaders. This portion of the AAAS provides a selective index of the persons who thought themselves or were considered by peers to be "men of science." This group designated as leadership is not especially typical of the general AAAS membership. It was a special component of the AAAS made up of persons elected or appointed to hold office by fellow members or members who presented the results of research to an AAAS session. For the purposes of this study any research presented in person or by letter and so indicated in the *Proceedings* was included; some papers mentioned in the public press were not listed in the *Proceedings*, but the press coverage was too uneven to provide usable data for this analysis. The committeemen do not include those giving individual reports on the progress in some branch of science; these were considered publications. When an appointed officer was replaced by another member, only the latter's name is included. While the decision to include all men holding office or presenting papers as leaders meant the inclusion of some individuals not fully accepted as peers by other leaders, any attempt to exclude questionable persons

20. This pattern persisted. See Burton E. Livingston, "Members of the American Association for the Advancement of Science per Million of Population in the United States," *Science*, LX (July–Dec. 1924), 467–469.

would have been equally arbitrary. A contingent of 337 leaders is broad enough to mask anomalies. To the general public this active leadership became the most visible portion of the AAAS membership and appeared as a profile of the organization itself.

As already indicated, the leadership was more persevering in participation than the general membership. Computation of the average length of membership included the year joining and the last date recorded on a roster to 1866. The average time of involvement for the full membership was 4.8 years, while that for the leadership group was 8.2 years.[21] This is consistent with the notion that leaders were more committed to continuation of the association and justifies the assumption that those presenting papers or holding offices were the leading members. This longevity factor also suggests that leadership status for newer members came only after some years so that persons who joined late in the decade would probably not become leaders until the 1860s. The 13 years under consideration here were, however, enough to establish one generation of men as leadership.

In their studies Beaver and Elliott used publication as the criterion for the selection of scientists. Beaver, in his extensive survey of selected pre–Civil War scientific journals, calculated that 10% of the scientists who published were responsible for 50% of all the articles printed in the United States.[22] In the AAAS about 12% of the total membership presented papers; of that group 11% (31 men) were responsible for 51% of the papers presented. Agassiz, Bache, Henry, Rogers, and other leading scientists in this category regularly published in other local journals as well.

The number of persons publishing was greater than the number holding offices in the AAAS. Of the 337 leaders, 283 published or presented papers while only 154 held office; 100 men were in both categories. In order to demonstrate the level of activity in both categories, Table 9 shows the distribution of leaders by the number of papers they presented from 1848 to 1860; the second row gives the percentage of officers in each similar category. The highest frequency agreement between the number of publications and the number of offices held was at one publication and no offices (105 persons); conversely, 35 of the 66 persons holding only one office did not publish. Men active in both groups tended to adhere to either the minimum or maximum number

21. For the membership the standard deviation was 2.8, and for the leadership it was 4.2.
22. Beaver, "American Scientific Community," p. 15.

TABLE 9

Pattern of Leadership Activity

This table presents the activity level of leaders by considering the number of papers presented or offices held by each individual in the leadership category. All figures are rounded to the nearest whole percent.

	Percentage of Leaders at Each Level of Activity										
Activity	None	1	2	3	4	5	6	7	8	9 or More	Total
Percentage Presenting Papers	16	38	13	7	6	5	3	1	2	10	101%
Percentage Holding Offices	53	20	7	4	4	2	3	1	1	5	100%

of both offices and papers. Of those 30 men with eight or nine publications, only 11 held fewer than 10 offices and 19 held more. Of those with eight or nine offices, only one presented less than five papers. The high concentration of men at the extreme ends of both categories suggests that while many men produced at a minimum level, at least within the AAAS, another sizable group was highly prolific. Among the latter were 12 men who were active nine or more times in both categories: Louis Agassiz, Alexander Dallas Bache, James Dwight Dana, Oliver Wolcott Gibbs, Benjamin A. Gould, Jr., Joseph Henry, Benjamin Peirce, William Redfield, Henry D. Rogers, William B. Rogers, and Benjamin Silliman, Jr. Aside from Dana, Redfield, and Silliman, who remained basically aloof from the constitutional debates, the quantitative result verifies the pre-eminence of not only the self-styled Lazzaroni but also their opponents. The clique of Washington and Cambridge scientists led by Bache was very powerful and productive, substantiating the qualitative assertions of Warner and Winslow, but clearly not powerful enough to curb the activity of their leading scientific opponents. The Rogers brothers, especially William, served as an active minority.

Leadership tended toward those men who joined the association in its early years. The reservoir of active scientists wanting a national scientific organization by the end of the 1840s ensured substantial support and a high activity level for the most interested early members. Generally less well established younger scientists joined gradually throughout the decade. Table 10, which shows the percentage of the 337 leaders who joined in each year from 1848 to 1860, substantiates this fact. While the

TABLE 10

Distribution of Leadership Activity by Year Joining

This table cross-tabulates the leaders joining with those presenting papers and those holding offices. The items are presented as a percentage of the total group of 337 leaders.

Activity	Joined												
	1848	1849	1850	1851	1853	1854	1855	1856	1857	1858	1859	1860	Total
Percentage Presenting Papers	31.0	8.6	13.2	6.8	5.0	3.5	2.4	5.6	4.2	1.3	1.8	0.9	84.2
Percentage Holding Office	24.0	5.9	5.0	3.6	1.8	0.9	0.6	1.8	1.4	0.6	0.3	0.0	45.9
Percentage Total Joining	38.0	11.3	13.4	9.2	5.9	3.9	3.0	5.6	5.0	1.8	2.1	0.9	100.1

percentage of leaders in both categories decreased in the latter years, presentation of new research seemed to remain more open to new member participation than was office holding. Similarly, Beaver found that for the older scientists in his sample, their first paper was published at an average age of about 30 years; the last individuals to enter his survey published their first papers at about 26 years, suggesting that productivity was occurring at a younger age for the succeeding generation.[23] In the AAAS the average age of the officers was somewhat older than that of the men publishing, as shown in Table 11. Using 280 persons

TABLE 11

Leadership Activity by Decade of Birth

The number and percentage of scientists whose birth dates are known are tabulated by decade of birth. The respective percentages of men presenting papers (researchers) and of those holding offices are also listed by decade of birth.

Birth Decade	Total Number	Percentage of 280	Researchers	Officers
1770–79	5	1.8%	2.1%	1.3%
1780–89	12	4.3	6.4	3.0
1790–99	38	13.6	20.6	8.6
1800–1809	61	21.8	22.0	22.4
1810–19	83	29.6	32.0	31.5
1820–29	64	22.8	14.9	26.7
1830–39	17	6.1	2.1	6.5
Total	280	100.0	100.1	100.0

23. *Ibid.*, p. 364.

whose birth dates were available, it appears that men in their thirties or early forties at the middle of the century (born between 1810 and 1819) constitute the highest concentration for both groups. While offices tended to be held by the older group, they published less. This pattern is consistent with Elliott's findings that individuals born after 1801 tended to publish proportionately more throughout their lifetimes than those born before that date.[24] The increasing number of journals and the possibility of an occupation in science were critical factors in this change.

Comparing the birth decades for the leadership and membership, the decade from 1810 to 1819 had a plurality for both groups. Table 12 gives the number born in each decade for those members whose birth

TABLE 12

Membership by Decade of Birth

In this table the number and relative percentage of members born in each decade are tabulated for the 522 persons whose date of birth was ascertained.

Birth Decade	Number	Percentage of 522
1760–69	1	0.2%
1770–79	16	3.6
1780–89	29	5.5
1790–99	85	16.2
1800–1809	125	23.9
1810–19	129	24.6
1820–29	84	16.0
1830–39	52	10.0
1840–49	1	0.2
Total	522	100.2

date is known. Nearly half (49%) of those individuals were born before 1810; the proportion of members born in each of those early decades was higher than that for the leadership (compare with Table 11). Leaders were the middle-aged, active group born in the 1810s and 1820s.[25] The membership at large tended to be older. An influx of youn-

24. Elliott, "American Scientist," pp. 55–58.
25. Harris, "Demographic Foundations," pp. 204–213. Harris's fascinating study suggests that a pattern which came full cycle every 22½ years offered increased opportunity to persons from "fairly ordinary homes." In particular, the opportunities for educators and scholars were rising in the 1810s, while those in medicine and the ministry rose from about 1814 to 1824. AAAS leaders had fortuitously been born when opportunities were open, which perhaps accounts for their optimism and enthusiasm about scientific and social progress.

ger men into membership, however, did cause the proportion for members born after 1829 (10.2%) to be larger than for the leaders (6.1%). Many of these persons were potential contributors not yet well enough established to assume leadership positions.

The regional pattern established by considering place of birth for AAAS leaders is more striking than that for adult residence, even allowing for the general population migration within the United States for the intervening years. Because the median birth date of the 224 men was in the 1820s, census figures for that year were used for comparison with the scientists. Table 13 indicates the pre-eminence of the Northeast and

TABLE 13

Regional Birthplace of Leadership by Decade of Birth

This table is based on the biographical information for the 263 leaders whose birthplace and date are available. Regional categories are the same as those in Table 6.

Region	Birth Decade							Total
	1770s	1780s	1790s	1800s	1810s	1820s	1830s	
Northeast	2	6	13	30	27	25	7	110
Massachusetts	1	1	10	17	13	12	4	58
Connecticut	1	4	2	5	8	7	2	29
Mid-Atlantic	2	3	18	14	27	21	6	91
Maryland			1	1	5	1		8
Pennsylvania		3	6	8	9	5	2	33
New York	2		9	5	9	15	4	44
South		1	3	7	7	4		22
South Carolina			2	2	5	1		10
Virginia		1	1	2	1			5
North-Central				1		1		2
Ohio				1		1		2
Foreign	1	1	4	7	11	10	4	38
Total	5	11	38	59	72	61	17	263

the lag of the South, despite nearly equal free populations. The Midwest, still basically territorial, produced only two leaders. Urbanization was an important factor. The combined total of the 77 urban (settlements with more than 10,000 persons) and 49 semi-urban (2,500–10,000) leaders was greater than the rural-born leadership total of 98. Beaver demonstrated that the urban factor contributed to but did not decisively explain the lag in the South by showing that while the Northeast was three times more urban than the South, the disparity in production of leaders was even greater.[26] Philadelphia led with 18 scientists; Boston and Charleston, South Carolina, followed with eight each. Boston by the 1820s was regaining its position over Philadelphia as a cultural and scientific center, despite the latter's spurt at the end of the eighteenth century. After 1800 the Northeast took the lead. The resurgence of the mid-Atlantic region after 1810 was due to increasing activity in New York rather than Pennsylvania. Massachusetts, however, from the 1790s led all other states, including Pennsylvania, in the total number of scientists born. The percentage of foreign-born scientists was greater than the total percentage of foreign-born in the United States census for 1850. The 14% of AAAS leaders in this category corresponds precisely with Elliott's findings.[27]

There are no adequate data on educational levels of the general population for the mid-nineteenth century. The AAAS leadership appears, however, strikingly well educated for the period. Of the 269 men for whom educational experience is known, only three leaders had no apparent schooling and 30 stopped with public or private secondary programs. The self-educated scientist was already virtually a myth. At least 221 of the 337 leaders attended colleges granting bachelor of arts degrees; 85% took degrees. Training for science concentrated at particular institutions. The specific institutions are known for 205 men, and 60% of those attended the ten colleges listed in order of AAAS leadership attendance (Table 14). Although only half of the pre-Civil War liberal arts colleges were in the New England and mid-Atlantic region, more than 75% of the AAAS leaders were educated at colleges in those areas. Most college students took science courses to equip them for professional careers in medicine and education. Moreover, because science was not a prominent part of the liberal arts curriculum for the first half of the century, interested professors teaching natural philosophy would seem a critical factor in encouraging young men into science. Yet a close scrutiny of commencement dates shows only a few points of concentration, such as West Point, where more than half of the AAAS leaders attended from 1822 to 1833. Even allowing for random fluctuations, this lack of concentration differs from Beaver's inference of strong student clusters around particular teachers.[28] Perhaps more important was institutional commitment to scientific instruction. The earliest A.B. held

26. Beaver, "American Scientific Community," pp. 155–159.
27. Elliott, "American Scientist," p. 67.
28. Beaver, "American Scientific Community," pp. 170–171.

TABLE 14

Colleges Attended by Leadership

Ranked in order are the ten degree-granting colleges attended by the AAAS leadership.

Colleges	Number	Percentage of 205
Harvard	36	16.2%
Yale	35	15.8
West Point	17	7.7
Union	11	4.9
Amherst	9	4.1
Williams	9	4.1
Brown	7	3.3
Wesleyan	6	2.7
Rensselaer Polytechnic Institute	5	2.3
University of South Carolina	5	2.3
Total for Top Ten Colleges	140	63.4

by a AAAS member was that taken by Jacob Burnett at Nassau Hall, Princeton, in 1791. Degrees among the Harvard graduates who were AAAS leaders were evenly distributed among the years from 1797 to 1859, and Yale degrees similarly ranged over the period from 1792 to 1852. In 1852, however, five men graduated with bachelor in philosophy degrees from the new program at the Sheffield Scientific School. The educational levels for men presenting papers and for officers are similar, with a somewhat higher proportion of officers having an M.D. In all, 18 men had medical training exclusive of their other college work, and another 66 had both an A.B. and an M.D. Advanced education was of growing professional significance, and 13 of the AAAS leaders held Ph.D.s. This format contrasts sharply with the pattern of a group of prominent pre–Civil War inventors, of whom only 28.8% had A.B. degrees.[29]

Historians of the mid-nineteenth-century scientific community are especially interested in relative changes in fields and subfields of activity because this was a time of transition. Table 3 showed the pattern of publication in the annual AAAS *Proceedings*; Table 15 indicates the general areas of scientific interest or activity for the AAAS leadership during the years from 1848 to 1860. The science categories listed in Table 15 are based on a schematic note, "Class of Papers Presented to

29. Marjorie Ciarlante, "Social Origins of American Inventors" (Paper read to the Society for the History of Technology, New York, 29 Dec. 1971).

Table 15

Major Fields and Subfields of Leadership

The following table presents the major fields and subfields of the AAAS leadership, based on scientific activity of individuals during their period of membership. First the actual number of men in each category is presented. The major fields are then presented as a percentage of the total leadership, while the subfields are calculated as a percentage of the men in the major field. Note that subfields are not additive because some leaders produced generally in a major field wtihout specializing. The last column enumerates the number of individuals with a second major field of interest in that field.

	First Field	Subfield	Percentage of 337	Percentage in Field	Second Field
Mathematics and Physics	85		25.2%		53
Mathematics		22		26.3%	
Mechanics		2		2.4	
Physics		15		17.9	
Astronomy		34		40.7	
Meteorology		12		14.3	
Chemistry and Mineralogy	68		20.4		32
Chemistry		27		39.8	
Mineralogy		29		42.7	
Agriculture		9		13.2	
Natural History	66		19.8		39
Botany		13		19.8	
Zoology		25		38.0	
Physiology		22		33.3	
Geology and Physical Geography	58		17.5		50
Physical Geography		8		13.8	
Geology		43		74.1	
Paleontology		4		6.9	
Mechanical Science	16		4.8		15
Engineering		9		56.2	
Machinery		2		12.5	
Manufactures		1		6.2	
Ethnology and Geography	9		2.7		6
Ethnology		4		43.4	
Geography					
Linguistics		2		22.3	
Statistics	5		1.4		1
General	5		1.4		1
None	4		1.2		
Unknown	21		6.2		

the Am. Assn.," found in Spencer F. Baird's papers.[30] Modern definitions proved inadequate, and this note provided a contemporary organizational pattern for a time when older definitions were being discarded and new distinctions between scientific fields were not yet settled.[31] Identification of the older groupings with modern definitions is made easier by the use of subcategories, most of which are also from Baird. Contemporary usage seemed most pertinent, and so mineralogy is retained under chemistry rather than being linked to geology, for example. While the physical sciences of mathematics and physics (85 individuals) head the list, the biological and earth sciences are so divided that their strength is not immediately evident. Joining the logically linked subgroups of mineralogy and meteorology with geology and physical geography establishes this popular complex of earth studies as a dominant interest of the AAAS leadership (92 persons). Other such groupings readily demonstrate that the natural sciences remained the major interest of the leaders of the organization, whose roots were in the Association of American Geologists and Naturalists. Secondary fields of interest, shown in the last column, were in approximately the same order as the primary fields, with geology moving from fourth to second place in the hierarchy.

Specialization, which accompanied professionalization, was studied as a function of various fields of interest (Table 15). Of those AAAS leaders with more than one field of interest, the proportion whose secondary field lay within their major field of interest was highest in mathematics and physics (58% of the 62 men evidently interested in a secondary field or subfield). An additional nine physical scientists (14%) were interested in applied science or mechanics. In contrast are the 39 persons in geological science with a second field of interest; only four men (8%) had a secondary interest in the same area, while 19 (38%) were interested in chemistry and 16 (32%) did work in natural history. The 39 natural historians who chose secondary fields concentrated in the distant, although related, chemical sciences (14 men; 36%) and the geological sciences (10 men; 26%). Robert Bruce found that 34% of the scientists in his study had secondary fields; AAAS membership shows only 29%, despite a slightly larger number of major field categories.

Recruitment of men into the several dominant sciences varied over

30. Baird MSS, SIA. The undated note is in vol. XI of the Official-Incoming correspondence.

31. Elliott, "The American Scientist," pp. 157–159. Elliott discusses the problem of classifying scientific fields and devises an impressive scheme. His careful delineation has a modern tone and seemed less appropriate for this analysis than that by AAAS secretary Baird.

time. Beaver suggested that the 1850s marked the beginning of a significant change in the distribution of scientific fields, from dominance by
the natural and life sciences to emphasis on the physical and exact
sciences.[32] If true, this trend was not yet evident among the AAAS
leadership in terms of their active interests. Table 16 compares the

TABLE 16

Scientific Interest Compared to Decade of Birth

In this frequency table leaders' research interests are cross-tabulated with their decade
of birth.

Science	Birth Decade							
	1770s	1780s	1790s	1800s	1810s	1820s	1830s	Total
Mathematics and Physics	1	3	8	16	27	14	6	75
Chemistry and Mineralogy	1	2	5	13	14	15	6	56
Natural History	2	1	13	11	16	15	0	58
Geology and Physical Geography	1	0	7	12	16	10	4	50
Mechanical Science	0	2	2	1	3	3	1	12
Ethnology and Geography	0	1	0	0	4	3	0	8
Statistics	0	0	0	1	2	2	0	5
General	0	0	3	1	1	0	0	5
None	0	2	0	2	0	0	0	4
Total	5	11	38	57	83	62	17	273

relative percentages of individuals in all areas of scientific interest by
decade of birth. Every specific area grew in the 1810s, with the physical
sciences of mathematics and astronomy making the most dramatic advance. Natural history, which had led in the 1770s through the 1790s,
dropped dramatically in the 1830s. Geologists maintained a fairly con-
stant proportion of entrants in each decade. Allowing 20 to 30 years for
maturation into full-time scientific productivity, they were still active
into the 1860s. Regular employment on state and federal surveys and the
geologists' ability to specialize while remaining fairly broad-based were
certainly factors in this persistence (see Chapter III). Bruce found that
recruitment into the earth sciences tapered off after the Civil War, which
seems compatible with the AAAS findings. While the change was imminent in the 1850s, the transition seems to have occurred somewhat
later than Beaver postulated.

Sociologists of science posit a theoretical structure for investigating

32. Beaver, "American Scientific Community," p. 145.

the organization of scientists. They recommend that study include the recruitment of scientists, the relationships that exist between scientists, and the patterns of behavior which characterize scientists at a given point in time. The goal of scientific professionals is, according to Robert K. Merton, the "extension of certified knowledge."[33] Data are insufficient to assess the motivation of most men entering specific fields of science. Choice of college or experience in college was undoubtedly a factor. As already indicated in Table 14, a very small percentage of the colleges active in the first half of the century accounted for most of the leading scientific practitioners. Table 17 demonstrates the scientific strengths of these institutions. While Harvard produced the most physical scientists (36% of Harvard's graduates who became AAAS leaders), Yale produced an even greater proportion of chemists, accounting for more than a third of all active AAAS chemists (50% of all leaders who had attended Yale). Graduates of West Point, whose engineering program was modeled on the École Polytechnique in Paris, went predominantly into the physical sciences (11%), with three in mathematics and five in astronomy. Like Harvard, West Point graduated men interested in mechanics and engineering, the applied sciences. Smaller and rural colleges accounted for a larger proportion of men in natural history and the geological sciences, although these schools also graduated men in physical and chemical sciences. Eighty-four AAAS leaders took an M.D. or attended medical school, and this group tended toward the biological and earth sciences; only one M.D. was interested in the exact sciences. Other advanced training was important; only 4% of the AAAS leaders acquired a Ph.D., but at least 79 of the leading scientists traveled abroad and while in Europe visited foreign universities, laboratories, or observatories. Relationships between faculty and students and among students were certainly reinforcing factors for the emerging professionals.

Occupations of the leaders indicate that, not including the 11% in medicine, at least 60% were in positions allowing them to work on their areas of scientific interest. Table 18 gives the total number and percentage of individuals in each occupational category. The order basically agrees with the findings of Bruce, whose study of income sources indicates increasing opportunities in education, entrepreneurship, and consulting by mid-century. Many of the professors were in established northeastern (38 persons) and mid-Atlantic (32) colleges, but a few were also scattered in institutions not so productive of science or sci-

33. Quoted in Storer, "Sociology of Science," p. 201.

TABLE 17

Scientific Interest Compared to College Attended

The distribution of leadership research interest is examined on the basis of the colleges training these men. Only one undergraduate college was recorded for each AAAS leader; medical and other advanced degrees are not included in this tabulation of colleges attended.

College	Mathematics and Physics	Chemistry and Mineralogy	Natural History	Geology and Physical Geography	Ethnology and Geography	Statistics	Mechanical Science	General	None	Total
Harvard	13	3	11	3	1	0	2	2	1	36
Yale	12	17	2	2	0	1	0	0	0	34
West Point	11	0	1	1	0	0	2	2	0	17
Union	2	1	3	0	1	1	0	0	1	10
Amherst	2	2	2	3	0	0	0	0	0	9
Williams	1	2	4	1	1	0	0	0	0	9
Brown	2	1	2	0	0	1	1	0	0	7
Wesleyan	1	2	0	2	0	0	0	0	0	5
Rensselaer	0	1	1	3	0	0	0	0	0	5
University of South Carolina	1	0	4	0	0	0	0	0	0	5
Other	14	17	10	11	4	2	2	0	0	57
Total	60	46	40	26	4	5	7	4	2	194

TABLE 18

Leadership Scientific Interest Compared to Occupation

This table permits quantitative statements concerning the correlation between the occupations of the AAAS leadership and their areas of scientific interest.

Occupation	Mathematics and Physics	Chemistry and Mineralogy	Natural History	Geology and Physical Geography	Ethnology and Geography	Statistics	Mechanical Science	Other	Total	Percentage Occupation
Higher Education[a]	36	29	20	15	3	0	1	2	108	32.1%
Government[b]	19	4	4	17	0	1	1	0	47	14.0
Medicine[c]	1	7	17	5	2	2	1	0	38	11.3
Business[d]	8	8	3	4	1	1	0	1	33	9.8
Consulting Scientist	2	7	2	3	1	0	0	0	17	5.1
Military	5	1	0	4	0	0	1	0	15	4.5
Secondary Education	4	2	4	2	0	0	0	1	14	4.2
Law	1	0	1	1	1	1	0	3	11	3.3
None (independent income)	1	3	5	1	0	0	0	0	10	3.0
Theology	1	1	3	1	0	0	1	0	8	2.4
Agriculture	1	1	2	0	0	0	0	0	5	1.5
Journalism	0	1	0	1	1	0	0	2	5	1.5
Student	0	1	1	1	0	0	1	0	4	1.2

a Of the educators, 77 taught science, 16 taught medicine, and 11 were administrators.

b State governments hired 14 of these scientists, the Coast Survey followed with 11, then the Smithsonian Institution and the Nautical Almanac with 7 each.

c Of those practicing medicine, 29 were physicians and 4 were surgeons.

d These businessmen included 13 engineers, 3 inventors, and 4 instrument makers.

entists in the South (11) and the newly settled Midwest (13). Often these were young men unable to obtain appointments at the older, more prestigious schools, which were highly competitive and had less turnover. Silliman Jr., for example, spent nearly six years at the University of Louisville in Kentucky before returning to a position at Yale. Government, taking local, state, and national levels together, ranked second as an employer of scientists, using such personnel for research not only on the Coast Survey but also in the Patent Office, the U.S. Mint, and as civilians on military surveys. A surprisingly large number of these government employees (42%) were born in the 1820s. A small sample of men born in the 1830s who went into government posts suggests that the younger men were entering business and independent consulting in greater numbers and perhaps that government hiring had reached a plateau. Other scientists in this study, especially professors and private researchers, worked as government consultants, but because this was not their major occupation, they were not included in this tabulation. State and federal governments as well as the military hired primarily physical and geological scientists to do topographical work on the coasts and in the West. In fact, while these two areas account for only 18.5% of all scientific hiring, they employed 17 of the 58 geologists (29.2%) and 19 of the 85 physical scientists (22.3%). Beaver found that academic and government employment were the most congenial or desirable because they allowed work in a person's area of specialty. Medicine was closely correlated to natural history (45% of the medical practitioners were interested in this area) because zoology included human anatomy and botany aided pharmaceutical study. There was a slight decline in medicine and secondary education as sources of income for active scientists. This might be because these two occupations were becoming increasingly professional and because their members began to concentrate in organizations more directly related to their occupation. Natural history, which drew retired and independently wealthy individuals, ministers, and farmers, lost its following as these groups faded from a more professionally oriented organization. Because chemistry was applicable in several occupations, interest in the subject was widely distributed by occupation, and nearly half of the independent, consulting scientists were in that field.

For comparison, Table 19 shows the occupational distribution of the general membership where known. National census data are inadequate for systematic comparison. Although the membership sample represents a smaller proportion of the whole group surveyed than does the leader-

TABLE 19

Occupations of AAAS Members

The occupations of the American membership, of the 1,109 for whom such information was available, are listed below in rank order. Engineering is dispersed into particular categories of employment.

Occupation	Number of American Members	Percentage of 1,109
Medicine	258	23.2%
Higher Education	229	20.6
Theology	116	10.4
Business	107	9.6
Preparatory Education	100	9.0
Law	67	6.0
Military	54	4.8
Journalism	37	3.3
Consulting Scientist	34	3.1
Agriculture	31	2.8
Government	30	2.7
None (independent wealth)	27	2.4
Student	12	1.0
Other	7	0.6
Total	1,109	99.5

ship, the results offer an interesting contrast (see occupational totals in Table 18). Government employees, professors, and consulting scientists were especially apt to rise to leadership positions, while business and military men had proportionate representation among the leadership. The theologians, secondary school educators, lawyers, and journalists were less likely to participate so actively. The high number of individuals in law (6%) and theology (10.4%) suggests that such members perceived their AAAS membership and an implied interest in science as an intellectual rather than an exclusively scientific enterprise and also that for many persons science was by necessity still an avocational pursuit.

Historians of modern science have investigated the relationship between science and religion with varying conclusions. Comparative national data are not available in the 1850 census, nor is there any way to test regional backgrounds. Some observations are pertinent. Two hundred fifty-three leaders (75% of the total group) were identified by biographers as religious or as church members, while only 31 were avowedly nonreligious. None was Jewish and only 16 were Roman Catholic; the rest were Protestant, dominated by the Episcopalians

(50.9%), and with a substantial grouping of Congregationalists (17.7%), Presbyterians (9.4%), and Lutherans (7.9%). The United States was predominantly Protestant and the AAAS scientists reflected that fact.

At this point it should be added that separate analyses were completed for two groups whose members did not fit comfortably into either the leadership or membership categories, but who nonetheless contributed to the AAAS and whose characteristics suggest something further about the nature of the AAAS itself. The first is the local committee appointees who either nominally or actually helped with arrangements for the annual meetings. The second is the surprisingly large number of persons who held office, presented a paper, or sent data by letter to the AAAS yet never became formal members.

The 257 men serving on local committees, some more than once, fell into all three categories of AAAS activity: 41% were regular members, 27% had other leadership responsibilities, and 32% were nonmembers. The total of 73% outside the normal leadership group demonstrates quite conclusively that such appointments were a part of the popularizing function of the AAAS. Because these men were to arrange local meeting places and receptions, their residence naturally coincided with the meeting place for their year of activity. The honorary status of local committee selection is clear when the known occupations for 205 committeemen are tabulated in Table 20. The selection of nonmember busi-

TABLE 20

Local Committee Occupation Compared to AAAS Role

Occupations for 257 local committee members are tabulated by the individual participant's role as nonmember, member, or leader in the AAAS.

	Non-member	Member	Leader	Total	Percentage of Total
Business	19	19	5	43	16.6%
Higher Education	0	15	24	39	15.1
Law	11	11	2	24	9.5
Government	1	7	14	22	8.6
Military	3	12	7	22	8.6
Medicine	4	12	5	21	8.2
Theology	1	3	6	10	3.9
Editor	4	2	1	7	2.7
Secondary Education	2	4	1	7	2.7
Consulting Scientist	0	2	4	6	2.3
Other	1	2	1	4	1.6
Unknown	36	16	0	52	20.3
Total	141	98	70	257	99.9

nessmen and lawyers, many of whom were local politicians, skewed the distribution away from the pattern of general membership or leadership profiles. Requesting prominent citizens to help with local arrangements ensured wider promotion for the meeting and better hospitality. Those persons showed markedly less persistence in membership in the association than did the doctors, professors, and government employees who were similarly appointed.

Participating nonmembers were in large part, but not exclusively, local committee members. Most of these nonmembers participated in the middle to latter part of the decade (Table 21). The high proportion of

TABLE 21

Participation of Nonmembers by Year of Initial Activity

This table indicates the number of nonmembers by the year they first participated in a AAAS meeting by presenting a paper or holding an office. Some persons were involved in more than one activity, but only the first event is recorded below.

Year of First Activity	Presentation	Office	Local Committee	Total
1848	1	0	1	2
1849	1	4	3	8
1850	6	0	1	7
1851	4	5	1	10
1853	1	0	3	4
1854	3	2	16	21
1855	1	0	23	24
1856	0	0	9	9
1857	5	1	4	10
1858	5	0	16	21
1859	2	0	12	14
1860	2	0	0	2
Total	31	12	89	132

nonmembers appointed to the local committees accounts for the concentrations of active nonmembers in 1854, 1855, and 1858; when there was a larger committee, there tended to be a greater proportion of nonmembers. Beginning with a nine-person local committee in 1848, the AAAS expanded to 56 appointees in 1854. Of the total group, 89 nonmembers were on the local committees in their respective home towns. As a result, of the 116 persons with known residences, 94% (109 persons) were urban dwellers. Only 12 nonmembers held an office other than local committee, usually on a committee charged with memorializing a state legislature or Congress. Several of the 33 persons presenting papers did so by letter. Papers were given in the following fields, listed in order of frequency: physical science, chemistry, geology, and natural history. Only two persons presented more than three papers. As indicated in Chapter V, the single female participant was nonmember Eunice Foote, who presented a paper "On a New Source of Electrical Excitation." Given the vague skepticism and outright hostility of some leaders, the women's failure to participate is not surprising. Table 21 indicates the tendency to increase the number of prominent nonmembers on local committees after 1854, but at the same time to avoid such appointments to other offices. Publication activity remained equal before and after that median year.

No simple or single conclusion can be made about AAAS leadership and membership. Yet the findings of this chapter suggest a series of fundamental characteristics about the men who established the first national scientific organization.

The cumulative list of members and participants at AAAS meetings indicates that the number of persons interested in science was sizable by the 1850s. Turnover among the 2,073 affiliated members, however, was rapid. Many of the nonparticipating members were disenchanted when they realized that they were basically observers of an emerging professional discipline, while others could not or would not make the long journeys to meetings distant from their homes. Southerners were rare in the membership, and the South in the pre–Civil War decades displayed an essentially negative attitude toward scientific ideas; this posture was reinforced, according to Clement Eaton, by the romanticism of the educated classes and the inhibiting influence of slavery.[34] In the rest of the nation, however, the opportunity to participate with or observe and meet the eminent scientists of the day did draw local crowds, as the pattern of residence and year of joining demonstrates in Table 6. Such interest rapidly dissipated after the week-long meeting ended, and even a regular journal subscription could not keep the majority of onlookers involved. Insufficient data precluded further analysis on this broad population. The limited information on occupations, however, suggests that the general membership had an avocational rather than a vocational interest in science which was fueled by the peripatetic scientific meetings.

A quick review of the major findings for the AAAS leadership reveals

34. Clement Eaton, *The Mind of the Old South*, rev. ed. (Baton Rouge, La., 1967), pp. 224–244. Also see W. J. Cash, *The Mind of the South* (New York, 1941).

that it was comprised primarily of men who could earn their living as scientists. Born in the Era of Good Feeling, when occupational opportunity in academe and medicine was relatively open, these scientists came of age during the decade of the 1840s, when science gained an institutional foothold in the United States in the colleges, the Smithsonian Institution and government agencies, and the American Association for the Advancement of Science. Better educated than their fellow Americans, the majority of the leaders graduated from colleges, usually on the eastern seaboard, and many had medical training. Typically, they concentrated on fields of science that were directly related to their occupations. Only a few men actively participated year after year by holding office and presenting papers, but those few were critical in directing the young AAAS along professional lines. The other leaders were important, too, as they helped sustain the broader base of the AAAS and provided a variety of reports and local sponsorship.

The profile of the AAAS is a blend of the leaders, members, and participating nonmembers who totaled more than 2,200 persons. The AAAS leadership is similar in general outline to the pattern of characteristics found by Beaver's and Elliott's quantitative analyses. Because these studies both described the most productive scientists of the pre–Civil War period, it is evident that the AAAS did indeed represent the leading scientists of the nation while at the same time providing an opportunity for learning to a very sizable number of interested amateurs and onlookers who could attend meetings and even become members if they so desired. The data in this chapter complement the earlier studies of the American scientific community, verifying their findings, and in addition provide a profile of the scientific community for the period from 1848 to 1860.

IX

A Question of Survival

By the late 1850s the AAAS had lost the vigor of its early years and failed to formulate satisfactory resolutions to many of the issues raised under its auspices. Local public enthusiasm was tempered by negative editorials in the press and outright criticism by some scientists. Among themselves the leading members of the AAAS expressed skepticism about its future. Attendance in 1860 dropped to about the 1854 level. But the clouds which diffused the earliest hopes of the scientific community cannot completely overshadow the association's important contributions from 1848 to 1860 as the representative of American science. At the same time the successes of its early years must be critically measured against the dissatisfactions of members later in the decade and its inability to draw together the full membership into working harmony. Although few in 1860 denied that the AAAS had filled an institutional vacuum and created a self-aware scientific community, visions of its future were less optimistic than they had been a decade earlier. Subsequent history of the AAAS matched neither the worst fears nor the more naive hopes of members in 1860; the association survived internal trouble and outside competition, but it never regained the dominant position among American scientists that it had enjoyed in the 1850s.

By the end of the decade it was evident to even casual observers that the association had not met the highest expectations of any of its founders. Diligent efforts by the standing committees had not kept inept research results from presentation or guaranteed an absolutely professional publication. The association had not become an effective or totally accepted arbiter for scientific disputes. In fact, it seemed rather to avoid controversial problems, as it did in 1860 by ignoring the publication of Charles Darwin's *On the Origin of Species*. This omission led the *New York Times* to report sarcastically,

> The American Association now sitting at Newport, seems equally disposed to give the propositions of Mr. Darwin a wide berth. . . . Yet Agassiz, to whose doctrines that of Darwin is in open antagonism, is present at Newport, and Prof. Gray, who is supposed from his contributions to Silliman's *Journal* and other indications, to sympathize with Darwin, is there also; and a host of others fully and peculiarly competent to treat the subject with the enlightment and candor which it merits, and which the advancement of science requires.[1]

Gray and Agassiz took their controversy elsewhere, but the reporter's frustration over their failure to speak out on the scientific issues of the day marked the other apparent problem of the AAAS. While the meetings were social and drew a "turbulent troop" of followers,[2] the size of the crowds and the tone of the American Association's general sessions never matched those of the BAAS. Compounding internal disillusion and a public dissatisfaction with the AAAS was the threat of Civil War and a sluggish national economy.

The 1860 Newport meeting symbolized the association's lassitude. Less contentious and without the constitutional wrangling of earlier meetings,[3] the general sessions were "smaller in number" and "smaller in talk" than most of the previous meetings.[4] Newport provided a congenial, fashionable setting, and attending members were in a conciliatory mood. When the geologists proposed J. S. Newberry as their vice-presidential nominee, the Bache contingent recommended F. A. P. Barnard as president, for a balanced ticket.[5] Because William Rogers and his cohort John W. Foster were on the standing committee, controversial matters were quietly resolved there rather than in open meetings. The almost unnatural quiescence was due as much to the impending national critics as to simple fatigue or disinterest on the part of the attend-

---

1. 6 Aug. 1860.
2. J. H. Alexander to [Bache], Baltimore, 5 Aug. 1860, Bache MSS, LC. Alexander deliberately avoided the meeting because of the clamor and wondered whether "Uriel ever attends an association of cherubim who delight in knowledge."
3. William Whitney to Elizabeth Whitney, Newport, 2 Aug. 1860, Whitney MSS, YUA. In evident surprise Whitney wrote, "I have been at no meeting of the Association at which a more quietly & reasonably pleasant time was enjoyed & promised."
4. Bache to Peirce, Boston, 9 Aug. 1860, Peirce MSS, HLH.
5. *Ibid.*

ing members. John LeConte, then a South Carolina resident, remembered that the Newport meeting's atmosphere was "like a stifling air before a storm"; everyone felt a "deep suppressed uneasiness" about the political condition of the nation.[6] This effect tempered the meeting. As David Wells reported, "The conviction prevailed among many who were present at Newport of a decadence in the scientific character of the Association, of a loss of tone, which, if not already a demoralization, threatened soon to become such."[7]

The decision to hold the 1861 meeting in Nashville, Tennessee, was a reassertion of the national aspirations of the AAAS and expressed a hope for greater unity of the eastern scientists with those in the West and South. The selection of Nashville was not unanimous, however. The "young men" who wanted the AAAS to be a conference for leading research scientists resented the decision to meet in a scientific backwater far from home and approached Bache about challenging both the vice-presidential nomination of Newberry and the Nashville proposal.[8] These young scientists, disillusioned by the Cleveland meeting nearly a decade before, contended that the "Gnashville" meeting would be a "horror, scientifically speaking," and that few prominent northern scientists would bother to attend.[9] They went on to raise serious questions about the value of the AAAS itself. Gould confessed that he was "sorry the Association exists" while admitting that "its present exit from this world would do more harm than its continuation."[10] Previously in close agreement with the pessimistic Wolcott Gibbs, Gould now feared that a "blatant successor" might rise "phoenix-like from its ashes."[11] Some of the dissidents like Gibbs informed Bache that they intended to resign from the association in any case and thought a resignation movement would increase.[12] Bache, however, recognized that overturning the nominating committee would be a dangerous precedent and suppressed the opposition to Newberry and Nashville.[13] Despite his own dissatisfaction with the organization, Bache seemed determined to keep it afloat.

Gibbs's negativism reflected disillusion with the association, not simply

6. Armes, *Autobiography of Joseph LeConte*, p. 178.
7. *Annual of Scientific Discovery . . . for 1860* (1861).
8. R. W. Gibbes of South Carolina was elected vice-president.
9. Gould to Silliman Jr., Cambridge, 21 Sept. 1860, Gould MSS, BLY.
10. *Ibid.*
11. *Ibid.*
12. Gibbs to [Bache], New York, 3 Feb. 1861, Rhees MSS, HHL.
13. Bache to Peirce, n.p., n.d., Peirce MSS, HLH.

disapproval about the inconvenient meeting place for 1861. After the Newport meeting he rhetorically asked Bache, "Why not let the monster die a natural death?"[14] Peirce also wrote to his "Chief" from Cambridge asking whether the Nashville meeting should be avoided.[15] Bache retorted with the assertion, "Is there an American Association? Yes."[16] The question was moot, however. Outbreak of the Civil War during the winter of 1860–61 forced a reconsideration of the plans for the 1861 meeting. Joseph Henry, after discussion with Bache in Washington, recommended to incoming president Barnard that the scheduled Nashville meeting be postponed.[17] That postponement remained in effect throughout the Civil War.[18]

All scientific aspiration and institutions were not quiescent during the war, however; Washington scientists were unusually active. A. Hunter Dupree has demonstrated that the federal government used applied science to an unprecedented degree for special problems during the war, although support for science was not available in all fields of research.[19] In the midst of the war effort Joseph Henry persuaded political leaders and federal administrators that "the art of destroying life, as well as that of preserving it, calls for the application of scientific principles, and the institution of scientific experiments on a scale of magnitude which would never be attempted in time of peace. New investigations as to the strength of materials, the laws of projectiles, the resistance of fluids, the applications of electricity, light, heat, and chemical actions as well as of aerostation, are all required."[20] Henry hoped that "independent of the political results which may flow from it [war] scientific truths are frequently developed during its existence of much theoretical as well as of practical importance."[21] Henry was too optimistic, because many areas of basic research suffered from lack of funding, and the normal pattern of scientific activity was interrupted

14. Gibbs to [Bache], New York, 3 Feb. 1861, Rhees MSS, HHL.
15. Peirce to Bache, 14 Feb. 1861, Rhees MSS, HHL.
16. Bache to Peirce, Washington, 22 Dec. 1860, Peirce MSS, HLH.
17. Henry to F. Barnard, undated but early 1861, Barnard MSS, Columbia University Library. Also see Bache to Peirce, Washington, 25 Feb. 1861, Peirce MSS, HLH. Bache felt that changing the meeting place to Washington was inexpedient.
18. A prefatory note in the *Proceedings* for 1860, published the following spring, informed members, "By order of the Standing Committee, the Nashville meeting of the Association, appointed for April 1861, has been postponed for one year."
19. Dupree, *Science in the Federal Government*, pp. 115–119, 135–148.
20. Smithsonian Institution, *Annual Report*, Washington, D.C. (1863), p. 13.
21. *Ibid.*

as scientists participated in military operations.[22] The success of certain applied programs and the need for more support for basic research became spurs to activity in Washington in the winter of 1862–63. There Alexander Bache regrouped a select few of his Lazzaroni in an effort to create a national institute for science. Together with Benjamin Peirce, Louis Agassiz, and Senator Henry Wilson of Massachusetts, he drafted a charter for the National Academy of Sciences, which was passed by Congress that spring in its last session.[23] Rapid passage of the bill and the private selection of the academy's fifty incorporators—purportedly the most eminent American scientists—aroused instant furor.[24] The exclusion of Harvard astronomer George Bond, Smithsonian zoologist Spencer F. Baird, and New York's John W. Draper revived anger about the elitist tactics of the Washington-Cambridge clique and reinforced the old opposition that had challenged the Lazzaroni in the AAAS. William B. Rogers, for example, refused to attend after the organizational meeting in New York and was finally excluded for "nonattendance." While Bache remained at the helm, the academy worked as an advisory body to the government on several Bache-inspired federal projects. After his debilitating illness in 1864, however, activity slackened, and internal dissension threatened the young organization with dissolution.[25] Probably only Joseph Henry's initiative as leader and mediator sustained it as membership participation and support fell steadily from 1864 to 1867.

From its origin in 1863 many scientists assumed that the new academy was established as a replacement for the AAAS. Peter Lesley of Philadelphia observed, "It is, I suppose, intended to suppress the American Association." Shrewdly he continued, "I think it cannot be done. This new organization is too much of a close [sic] corporation or Oligarchy to be tolerated by the men of science in America."[26] James Dwight Dana wrote similarly to the excluded Baird: "What a sly birth for the

22. In a critique following A. Hunter Dupree's "Science and Technology" essay, Robert V. Bruce pointed out that the Civil War was a distinct detriment to basic scientific research. See Gilchrist and Lewis, eds., *Economic Change in the Civil War Era*, pp. 123–128.

23. A. Hunter Dupree, "The Founding of the National Academy of Sciences: A Reinterpretation," American Philosophical Society, *Proceedings*, CI (Oct. 1957), 434–440.

24. Only three incorporators had not been members of the AAAS; forty were in the leadership category and seven were simply members. The physical sciences were the dominant area of interest of the NAS's founding members.

25. See Reingold, "Alexander Dallas Bache," pp. 163–164.

26. J. Peter Lesley to Benjamin S. Lyman, n.p., 7 May 1863, Lesley MSS, APS.

N. A. S. . . . I have been glad that the old Amer. Assoc. had died out. . . .
But I should not now wonder if it were revived with new zeal. Those
who were active in it will hardly feel inclined to tail on to the N. A. S."[27]

More than discontent was needed to revive the AAAS, however, at the
end of the war. As suggested earlier, criticism of the association itself had
been particularly strong among European-educated and younger
members. The general membership expressed more dissatisfaction with
its autocratic leadership and continuous contention than with the AAAS
as such.[28] When in the winter of 1866–67 F. A. P. Barnard, by then
president of Columbia University, suggested reconvening the association,
response was positive. B. A. Gould replied enthusiastically, "Your circular letter seems to me just the thing, and as you & I (who will give
you my proxy) make with Lea[,] Trowbridge & Elwyn an absolute
majority we can authorize you to call a meeting.—Indeed by summon-
ing a meeting of the Standing Committee, from which LeConte and
R. W. Gibbes would probably be absent, Trowbridge with either Dr.
Lea *or* Dr. Elwyn, who is a hightoned man & not likely to be in any such
plot *understandingly*, would make up a working majority."[29] Gould's
letter conjured up images of the conspiratorial attitude of the 1850s and
probably indicates the fact that he was no longer an intimate of the
old Lazzaroni fellows. For most of the men interested in reviving the
AAAS, however, the old hostilities were no longer particularly relevant.
In fact, the revival by Barnard seemed more in spite of rather than to
spite the old AAAS leadership; Barnard was an incorporating member
of the academy and had been the old leadership's preference for AAAS
president in 1860. At about the same time James Hall urged Joseph
Henry and William Rogers to help in reviving the AAAS, arguing that
the "Am. Association [is] one of the best means of advancing science
and promoting communication."[30] In an apparent effort to stem criticism
and perhaps eliminate the need for the AAAS, the National Academy
considered a plan to expand its membership beyond the charter's maximum of fifty persons. Hall felt, however, that this would not "prevent
or obviate the necessity for a reorganization" of the AAAS.[31] Rogers, in
reply, agreed: "Whatever may have been its faults, its shortcomings,
it was of great service, and its free & unexclusive spirit ought I think

27. Dana to Baird, New Haven, 10 Mar. 1863, Baird OFF, SIA.
28. A. Caswell to Bache, Berlin, Germany, 17 Sept. 1860, Rhees MSS, HHL.
29. Gould to Barnard, 7 Feb. 1866, Barnard MSS, Columbia University Library.
30. Hall to Rogers, Albany, 18 Jan. 1866, Rogers MSS, MITA.
31. *Ibid.*

to be perpetuated."[32] Buffalo replaced Nashville as meeting place,[33] and the New York–based scientists organized to promote the meeting. The Buffalo Society of Natural Science as hosts were determined that "nothing but the cholera or some other pestilence" would prevent the AAAS meeting.[34]

The meeting was strategically arranged for the week following the National Academy of Science's Northampton meeting in order to allow interested scientists to attend both. Many of the academy's members could agree with William Whitney when he wrote to Wolcott Gibbs that the organization had not fulfilled the function of European academies:

> For our Academy . . . though called National, no official activity in connection with the national government is to be anticipated; the lamented failure in health of our President cuts off any such hope, if it was before to have been entertained; neither Congress nor the Department are wont or inclined to seek external counsel. A national endowment, or a continuous and satisfactory provision for publication is still less to be looked for. Our meetings, in the third place are too rare and too thinly attended to possess the consequences which alone ought to satisfy us: of hardly any department is there a quorum in attendance, who could criticize a communication in that department or constitute a fit auditory for it. With neither an auditory or means of full and prompt publication, the Academy is not a power in the scientific activity of the country: its membership is rather of the nature of a *decoration*, it is an honorable and gratifying recognition of scientific merit but little more. The revived Association will be likely to overshadow it, both in popular estimation and in real activity.[35]

As Whitney implied, revival of the association was more than a reaction to the academy; it was recognition that the AAAS had exercised a very real function in the 1850s, a function that might be reasserted. The Northampton meeting was small and ineffectual.[36] Anticipating that the revived association would overshadow the National Academy, Whitney

32. [Rogers to Hall], Boston, 30 Jan. 1866, Rogers MSS, MITA. Rogers, preoccupied as president of the newly founded Massachusetts Institute of Technology and with a proposal to link the new college with Harvard, does not seem to have been very active in the reorganization of the AAAS.

33. Perhaps as a compensation, Nashville became host for the 1877 meeting.

34. William W. Stewart to Henry A. Ward, Buffalo, (?) July 1866, Ward MSS, University of Rochester Archives, Rochester, N.Y.

35. William D. Whitney to W. Gibbs, Northampton, 4 Aug. 1866, Whitney MSS, YUA.

36. W. D. Whitney to J. D. Whitney, Northampton, 10 Aug. 1866, Whitney MSS, YUA.

went directly from the Northampton meeting to Buffalo.[37] En route he was joined by Benjamin Gould, Barnard, and the University of Vermont's new chemistry professor, Eli Blake. Although Agassiz, Peirce, Rogers, Dana, and Henry were all absent, Whitney reported, "Of clashing with the Academy there is no sign whatever; nor have I heard any or heard of any ill-toned remarks or discussions, public or private. That is partly because the Ac. has the president . . . and enough of its members on hand to exercise a powerful influence, and to give most acceptable help."[38] Attending academy members were not hostile to but, rather, supportive of the association's revival.

If not an overwhelming success, the 1866 meeting firmly re-established the association. Local reception was excellent. A fair and horse show had crowded Buffalo hotels so that most AAAS members were entertained in private homes.[39] Attendance was limited, perhaps in part because Lovering was late in sending out circulars.[40] Geologists moved once again to center stage in the organization they had been so instrumental in founding. J. S. Newberry, A. H. Worthen, Charles Whittlesey, James Hall, and T. S. Hunt attended, with the latter two engaging in an extensive debate about glaciation. Only one-fourth of the sixty-three papers presented were published, and most were by minor scientists. Undoubtedly many skeptics decided to wait and see if the revived association was successful. Eugene Hilgard wrote from Mississippi that many southerners would not attend because the distance was too great and the railroads had rescinded their earlier policy of allowing free passage one way; but he insisted, "That is the way you must interpret our absence; we don't mean to secede."[41] Most attending members were fully satisfied with the meeting as a fresh beginning. Whitney supported this view but observed that the dynamic of the 1850s was absent: "All had been quiet and harmony . . . no cliques; no opposition to the administering party; no party; not much of anything save insignificance."[42]

37. W. D. Whitney to J. D. Whitney, Buffalo, 19 Aug. 1866, and George L. Squier to W. D. Whitney, Buffalo, 7 Aug. 1866, both in Whitney MSS, YUA.
38. W. D. Whitney to J. D. Whitney, Buffalo, 19 Aug. 1866, Whitney MSS, YUA.
39. Perhaps the most prestigious invitation was that from former President Millard Fillmore to Joseph Henry. See Fillmore to Henry, Buffalo, 4 Aug. 1866, Indiana Historical Society, Indianapolis.
40. F. A. P. Barnard to Chester Dewey, New York, 19 July 1866, Dewey MSS, University of Rochester Archives.
41. Hilgard to George J. Brush, Oxford, Miss., 22 Aug. 1866, Brush Family MSS, YUA.
42. William D. Whitney to W. Gibbs, Northampton, 4 Aug. 1866, Whitney MSS, YUA.

The next few years were similar; the revival was not complete and the AAAS would never regain the stature it had enjoyed in the 1850s.

In the post–Civil War years, the annual meetings of the AAAS continued to provide fellowship across disciplinary boundaries. In order to extend its visibility and accessibility, the association deliberately scheduled provincial meetings. By 1875 sessions had been held in Chicago, Indianapolis, Dubuque, and Detroit, with no revisits to the major cities of the pre–Civil War eastern hegemony. However, local societies were dwindling and could offer little support. In their place was a multiplicity of specialized societies, several of which developed out of the AAAS sectional meetings. These new societies handled specific research problems while the AAAS provided the place at which general problems were discussed.[43] It was also increasingly a showcase for comprehensive review papers and remained an active political force, particularly during the Progressive period.

The AAAS had been a crucial leader of the national scientific community in the 1850s. It reflected the scientific community because major scientists and amateurs alike managed to attend its meetings. Most significantly, it had attempted to join the components for scientific progress, designated "diffusion" and "advancement" of science, into a single organization while clarifying the goals and activities for each more precisely. Despite some internal dissension, the association had increased the broadly based *esprit de corps* among leading scientists. By linking scientists and civic leaders on lobbying committees, the AAAS had helped consolidate support for government surveys. Its prestige was evident in the attention paid its recommendations by the state and federal governments, the honor associated with the presidency, and the local receptions given visiting savants. The confidence level of the scientific community was considerably higher than it had been in the previous decade. These legacies were not lost, but the optimism which had permeated the antebellum nation was not recaptured in American science or society after the Civil War.

The relative number of men in science rapidly increased at the end of the nineteenth century, and women were also in greater evidence.[44]

43. Marcus Benjamin, "Organization and Development of the Chemical Section of the American Association for the Advancement of Science," American Chemical Society, *Twenty-fifth Anniversary Volume* (1902), pp. 86–98.

44. Using *Who's Who* as an index, Beaver points out that native-born persons interested in science were 6 percent of the population before 1870 and rose to 12 percent after that date. See Beaver, "American Scientific Community," p.

The transformation of the scientists from a few isolated individuals reading and researching because of an avocational interest in science to a
well-educated, active, and employable community had occurred by
1860. Various institutions—journals, local societies, the growing federal
and state agencies, and a changing college curriculum—were all causal
factors in the change. The American Association for the Advancement
of Science played a unique role, complementing and extending these
developments by offering a forum in which to discuss and to formulate
professional aspirations and a podium from which to announce new
policy to the general public. As statistical data have indicated, the AAAS
was led in the 1850s by the major research scientists in the nation. These
men were characteristically urban, resided near the Atlantic seaboard,
had college degrees, and were employed in occupations related to their
areas of scientific interest. Society membership offered direct and regular communication among AAAS members. Quantitative data suggest
that the similar origins and productivity patterns of the leadership were
those of the leading American scientists. Their new cohesiveness as a
group encouraged adoption of their ideas by the general membership
and, finally, an observing public.

Although the AAAS was not judged an unqualified success by many
contemporaries, history is a less harsh critic. The intensity of discussion
about the merits of the AAAS indicates the importance associated with
its functions and goals. Moreover, the AAAS was obviously the central
arena for scientific debate in the middle of the nineteenth century.
While the overall membership pattern was irregular, a visible group of
leading scientists sustained the association by regular participation and
enhanced the reputation of American science. The quality of the per-
sistently participating members, the establishment of a reputable series
of *Proceedings*, public approval, and expanding international recognition are all indices of the success of the group. The American Association for the Advancement of Science, alive with enthusiasm and
controversy, was, in effect, the pre-eminent scientific communiy of mid-
nineteenth-century America.

---

19. On women and science, see Margaret W. Rossiter, "Women Scientists in
America before 1920," *American Scientist*, LXII (May–June 1974), 312–323.

Bibliography

PRIMARY SOURCES

Manuscripts

Charles Baker Adams MSS. Amherst College Archives, Amherst, Mass. A small collection.

Jean Louis Agassiz MSS. Houghton Library, Harvard University, Cambridge, Mass.
Largest single collection of Agassiz material but disappointingly unrevealing about the AAAS's early years.

———. Museum of Comparative Zoology, Harvard University, Cambridge, Mass.
Chiefly miscellaneous papers and some letterbooks of Alexander Agassiz.

American Association for the Advancement of Science MSS. American Association for the Advancement of Science Library, Washington, D.C.
Two bound volumes: one contains handwritten minutes of some meetings, a few copies of circulars, and several letters for the period 1848 to 1856; the other is a signature book of new members to 1874 which is inaccurate and incomplete.

Appleton Family MSS. Massachusetts Historical Society, Boston.

Association of American Geologists and Naturalists MSS. Academy of Natural Sciences of Philadelphia, Philadelphia.
Handwritten minutes for some annual meetings and numerous letters for the years from 1844 to 1847.

Charles Babbage MSS. Microfilm of correspondence with Americans. American Philosophical Society, Philadelphia.

Alexander Dallas Bache MSS. American Philosophical Society, Philadelphia.
Letters from 1853 to 1861, several on the Albany university plans.

———. Library of Congress, Washington, D.C.
Much of this correspondence to Bache deals with the Coast Survey; also a scrapbook and two-year diary.

———. Smithsonian Institution Archives, Washington, D.C.
This mixed collection of letters, letterpress books, notebooks, and mem-

bership certificates complements the holdings in the Library of Congress and the Rhees MSS at the Henry L. Huntington Library.

Jacob W. Bailey. Typescript of "Selected Letters of Jacob W. Bailey." New York Public Library, New York.

Spencer F. Baird MSS. Smithsonian Institution Archives, Washington, D.C. An outstanding collection of both personal and official correspondence for the years during which Baird was affiliated with the Smithsonian.

Breckenridge Family MSS. Library of Congress, Washington, D.C.

Brush Family MSS. Yale University Library, New Haven, Conn. This substantial collection of family materials contains a series of useful letters from George J. Brush concerning his scientific study in Europe and his subsequent enterprises.

Coast and Geodetic Survey MSS. U.S. National Archives, Washington, D.C.

James Dwight Dana MSS. Yale University Library, New Haven, Conn. The Sterling Library has family correspondence, and the Beinecke Rare Book and Manuscripts Library holds several volumes of Dana's scientific correspondence.

Chester Dewey MSS. University of Rochester Archives, Rochester, N.Y.

Eaton Family MSS. Yale University Library, New Haven, Conn. Only letters of Amos Beebe Eaton and Daniel Cady Eaton were used.

Charles Ellet MSS. Transportation Library, University of Michigan, Ann Arbor.

Edward Everett MSS. Massachusetts Historical Society, Boston.

Cornelius C. Felton MSS. Houghton Library, Harvard University, Cambridge, Mass. Helpful only in revealing Felton's interests and personality; there is no mention of the AAAS or the Lazzaroni.

John Fries Frazer MSS. American Philosophical Society, Philadelphia.

Galloway-Maxcy-Markoe Family MSS. Library of Congress, Washington, D.C.

Lewis R. Gibbes MSS. Library of Congress, Washington, D.C. Large collection, especially helpful for the years in which Gibbes served as an officer in the AAAS.

Oliver Wolcott Gibbs MSS. Franklin Institute, Philadelphia. Most letters are for the period after 1860.

Gibbs Family MSS. Wisconsin Historical Society, Madison. Letters from the 1840s reveal Oliver Wolcott Gibbs as a student.

Daniel Coit Gilman MSS. Johns Hopkins University Archives, Baltimore. A series of Dana and Silliman letters were apparently collected by Gilman for use in his biography of Dana.

Augustus A. Gould MSS. Houghton Library, Harvard University, Cambridge, Mass.

Benjamin A. Gould MSS. Yale University Library, New Haven, Conn.
No collection of Gould materials has been found, and these four letters
are symbolic of the scarcity of sources on this intriguing person.

Simon Gratz Autograph Collection. Historical Society of Pennsylvania,
Philadelphia.
One major section of the collection relates to men of science in America.

Asa Gray MSS. Gray Herbarium, Harvard University, Cambridge, Mass.
Extensive letter collection has correspondence from nearly every important naturalist of the period.

Samuel S. Haldeman MSS. Academy of Natural Sciences of Philadelphia,
Philadelpia.

James Hall MSS. New York State Library, Albany.
Incoming Hall correspondence reveals much about both the AAGN and
the AAAS, and the collection contains an interesting account of the founding of the former organization.

Robert Hare MSS. American Philosophical Society, Philadelphia.

Joseph Henry MSS. Smithsonian Institution Archives, Washington, D.C.
A fire at the Smithsonian in 1865 unfortunately destroyed much of Henry's
personal and official correspondence; what remains is suggestive. A project
to collect and publish scattered Henry material is presently headed by
Nathan Reingold at the Smithsonian.

Edward C. Herrick MSS. Yale University Library, New Haven, Conn.
Several collections, individually catalogued and indexed, contain Herrick's
correspondence with William Redfield, James D. Dana, Elias Loomis, and
others.

Edward Hitchcock MSS. Amherst College Archives, Amherst, Mass.

Jebediah Hotchkiss MSS. Library of Congress, Washington, D.C.

Douglass Houghton MSS. Michigan Historical Collections, Ann Arbor.

Autograph Collection in Houghton Library, Harvard University, Cambridge,
Mass.

Bela Hubbard MSS. Michigan Historical Collections, Ann Arbor.

Charles T. Jackson MSS. Library of Congress, Washington, D.C.
Most letters are from the early 1830s, when Jackson studied in Paris.

Increase A. Lapham MSS. Wisconsin Historical Society, Madison.

Isaac Lea MSS. Academy of Natural Sciences of Philadelphia, Philadelphia.

John Lawrence LeConte and John Eaton LeConte MSS. American Philosophical Society, Philadelphia.

Joseph Leidy MSS. Academy of Natural Sciences of Philadelphia, Philadelphia.

[John] Peter Lesley, Jr., MSS. American Philosophical Society, Philadelphia.

Elias Loomis MSS. Yale University Library, New Haven, Conn.

Matthew Fontaine Maury MSS. Library of Congress, Washington, D.C.

George P. Merrill Collection. Library of Congress, Washington, D.C.
Merrill collected about 1,800 miscellaneous letters of American and foreign scientists, chiefly geologists.

Ormsby MacKnight Mitchel MSS. Cincinnati Historical and Philosophical Society Library, Cincinnati, Ohio.

Scientific Letters in Museum of Comparative Zoology, Harvard University, Cambridge, Mass.

National Institution for the Promotion of Science MSS. Smithsonian Institution Archives, Washington, D.C.

Miscellaneous MSS in New York Historical Society, New York.

John Pitkin Norton MSS. Yale University Library, New Haven, Conn.
This newly received collection is still uncatalogued and is especially helpful in suggesting Norton's role in promoting the Albany university scheme.

James A. Pearce MSS. Maryland Historical Society, Baltimore.

Benjamin Peirce MSS. Houghton Library, Harvard University, Cambridge, Mass.
An important collection, recently combined and organized by Ms. Fisch, which reveals much about the Cambridge scientists and the Lazzaroni.

John V. L. Pruyn MSS. New York State Library, Albany.

William C. Redfield MSS. Yale University Library, New Haven, Conn.
A collection of correspondence and letterbooks that is especially informative on AAGN organizational matters.

William J. Rhees Collection. Smithsonian Institution Archives, Washington, D.C.
A folder on the AAAS contains some circulars and miscellaneous letters.

William J. Rhees MSS. Henry L. Huntington Library, San Marino, Calif.
With numerous Bache letters, some Henry correspondence, and records of the National Institution for the Promotion of Science, the collection supplements holdings of the Smithsonian Institution Archives.

William Barton Rogers MSS. Massachusetts Institute of Technology Archives, Cambridge.
The bulk of Rogers's papers are here, and some have been reproduced in Emma Rogers's biography of her husband.

———. University of Virginia Archives, Charlottesville.
A small, miscellaneous collection.

———. Virginia State Library, Richmond.
The collection is primarily from the years 1836 to 1842, when Rogers served as state geologist.

Henry R. Schoolcraft MSS. Library of Congress, Washington, D.C.

————. "Scientific Associations Abroad." New York Historical Society Archives, New York.

Benjamin Silliman Family MSS. Historical Society of Pennsylvania, Philadelphia.
Chiefly notebooks and letters of the senior Silliman's early years.

————. Yale University Library, New Haven, Conn.
Extensive collection more useful for the AAGN than the AAAS.

Eban S. Snell MSS. Amherst College Archives, Amherst, Mass.

D. Humphrey Storer MSS. Museum of Science, Boston.

Benjamin Tappan MSS. Library of Congress, Washington, D.C.

John Torrey MSS. Academy of Natural Sciences of Philadelphia, Philadelphia.

————. New York Botanical Garden, New York.
Large collection particularly useful for gossip among naturalists.

Henry A. Ward MSS. University of Rochester Archives, Rochester, N.Y.

John Warner MSS. American Philosophical Society, Philadelphia.
Two boxes of letters outline the controversy between Warner and Peirce and contain information on the Dudley Observatory debates from 1857 to 1859.

John Collins Warren MSS. Massachusetts Historical Society, Boston.

————. Yale University Library, New Haven, Conn.
Small collection of letters to Benjamin Silliman, Sr.

David A. Wells MSS. Library of Congress, Washington, D.C.
Almost the entire collection deals with the period after 1860.

William Whewell MSS. Trinity College, Cambridge, England.
A series of twenty-eight letters from Edward Everett discusses science in the Boston-Cambridge community and occasionally mentions the AAGN.

William Dwight Whitney MSS. Yale University Library, New Haven, Conn.
A large and useful collection also containing some letters of his brother Josiah Dwight Whitney.

Charles D. Wilkes Family MSS. Library of Congress, Washington, D.C.

Alexander Winchell MSS. Michigan Historical Collections, Ann Arbor.

Scientific Periodicals and Related Reviews

Academy of Natural Sciences of Philadelphia, *Journal.*
American Academy of Arts and Sciences, *Proceedings.*
American Association for the Advancement of Education, *Proceedings.*
American Association for the Advancement of Science, *Proceedings.*
American Journal of Science.
Known as *American Journal of Science and Arts* from 1818 to 1879 and popularly referred to as Silliman's *Journal.*

American Philosophical Society, *Proceedings.*
American Quarterly Register.
Annual of Scientific Discovery.
Association of American Geologists and Naturalists, *Reports of the Trans-actions of the Association of American Geologists and Naturalists, at Phil-adelphia in 1840 and 1841, and in Boston, 1842. Embracing Its Proceedings and Transactions.* Boston, 1843.
Boston Society of Natural History, *Journal; Proceedings.*
British Association for the Advancement of Science, *Report.*
Edinburgh Journal of Science.
Geological Society of Pennsylvania, *Transactions.*
National Institution for the Promotion of Science, *Bulletin; Proceedings.*
New York Review.
North American Review.
Pottsville Scientific Association, *Bulletin.*
Smithsonian Institution, *Annual Report; Collections.*

Newspapers

Albany Daily Advertiser.
Albany Evening Journal.
Baltimore Sun.
Boston Daily Courier.
Boston Daily Journal.
Boston Daily Traveller.
Charleston Courier.
The Christian Examiner.
Cleveland Plain Dealer.
Daily Cincinnati Commercial.
New York Daily Tribune
New York Evening Express.
New York Herald.
Philadelphia North American and United States Gazette.
Philadelphia Public Ledger and Daily Transcript.
Providence Journal.
Springfield Daily Republican.
Washington Intelligencer.

Published Books and Articles

Alexander, John H. *International Coinage for Great Britain and the United States: A Note Inscribed to the Hon. James A. Pearce.* Baltimore, 1855.
Allen, William H. *An Address before the Cuvierian Society of the Wesleyan University, Middletown, Connecticut, July 31, 1838.* New York, 1838.
American Lyceum, or Society for the Improvement of Schools and Diffusion of Useful Knowledge. Boston, 1831.
Babbage, Charles. "Account of the Great Congress of Philosophers at Berlin on the 18th of September 1828." *Edinburgh Journal of Science,* XX (1828–29), 31, 225–234.

————. *Reflections on the Decline of Science in England, and on Some of its Causes.* London, 1830.

Bache, Alexander D. *Anniversary Address before the American Institute, of the City of New York at the Tabernacle, October 28th, 1856, during the Twenty-eighth Annual Fair.* New York, 1857.

Barlow, Joel. *Prospectus of a National Institution to Be Established in the United States.* Washington, D.C., 1806.

Bigelow, Jacob. *The Useful Arts, Considered in Connexion with the Applications of Science.* New York, 1863.

Brewer, William H. "The Debt of This Century to Learned Societies." Connecticut Academy of Arts and Sciences, *Transactions*, XI, part 1 (1901–3), xlvi–liii.

Channing, William E. *Works.* 6th ed. Boston, 1848.

Dall, Caroline Wells. *The College, the Market, and the Court; or, Women's Relation to Education, Labor and the Law.* Boston, 1967.

DeKay, James. *Anniversary Address on the Progress of the Natural Sciences of the United States, Delivered before the Lyceum of Natural History of New York, February, 1826.* New York, 1826.

D[wight], S. E. "Notice of Scientific Societies in the United States." *American Journal of Science*, vol. X (1826).

Featherstonehaugh, George W. *The Monthly American Journal of Geology and Natural Science.* Reprint introduced by George W. White. New York, 1969.

Gould, Augustus A. "Notice of the Origin and Progress and Present Condition of the Boston Society of Natural History." *American Quarterly Register*, XIV (1842), 236–241.

Gould, Benjamin A. "An American University." *American Journal of Education*, II (Sept. 1856), 29ff.

Greenwood, F. W. P. "An Address Delivered before the Boston Society of Natural History." Boston Society of Natural History, *Journal*, 1 (1834–37), 7–12.

Hare, Robert. *Experimental Investigation of the Spirit Manifestations, Demonstrating the Existence of Spirits and Their Communication with Mortals.* N.p., 1855?

Hitchcock, Edward. "Address at the Opening of the Geological Hall at Albany, New York, August 27, 1856." State Cabinet of Natural History, *Tenth Annual Report* (Albany, N.Y., 1857), pp. 23–25.

————. *Religious Bearings of Man's Creation, a Discourse Delivered at the Second Presbyterian Church, Albany, on Sabbath Morning,* Albany, N.Y., 1856.

————. *Reminiscences of Amherst College, Historical, Scientific, Biographical and Autobiographical; Also of Other and Wider Life Experiences.* Northampton, Mass., 1863.

Hopkins, J. H. *Relations of Science and Religion, a Discourse Delivered in Albany during the Session of the American Association for the Advancement of Science.* Albany, N.Y., 1856.

Hopkins, Mark. *Science and Religion, a Sermon Delivered in the Second*

Presbyterian Church, Albany, on Sabbath Afternoon, August 24, 1856. Albany, N.Y., 1856.

Hunt, Washington. *Communication from the Governor, Transmitting a Memorial of the Committee of the American Association for the Advancement of Science, on the Subject of a Geographical Survey of the State.* Albany, N.Y., 1852.

Johnston, James F. W. *Notes on North America: Agricultural, Economical and Social.* 2 vols. Boston, 1851.

Lachlan, Robert. *A Paper and Resolutions in Advocacy of the Establishment of a Uniform System of Meteorological Observations, throughout the Whole American Continent,* Cincinnati, 1859.

Lyell, Charles. *Travels in North America; with Geological Observations on the United States, Canada, and Nova Scotia.* London, 1845.

Morton, Samuel George. "History of the Academy of Natural Sciences of Philadelphia." *American Quarterly Register,* XIII (1841), 433–438.

The New American Encyclopedia: A Popular Dictionary of General Knowledge. Edited by George Ripley and Charles A. Dana. New York, 1858–63.

Olmsted, Denison. *Address on the Scientific Life and Labors of William C. Redfield, A.M., First President of the American Association for the Advancement of Science.* New Haven, Conn., 1857.

Patterson, Robert M. *Early History of the American Philosophical Society; a Discourse Pronounced by Appointment of the Society, at the Celebration of Its Hundredth Anniversary.* Philadelphia, 1843.

Peirce, Charles. "The Century's Great Men in Science." Smithsonian Institution, *Annual Report . . . for 1900* (Washington, D.C., 1901), pp. 693–699.

Philadelphia: or, Glances at Lawyers, Physicians, First-Circle, Wistar Parties, &c. &c. Philadelphia, 1826.

Report on the History and Progress of the American Coast Survey up to the Year 1858. N.p., 1858.

Rogers, Henry D. *Address Delivered at the Meeting of the Association of American Geologists and Naturalists, Held in Washington, May, 1844, with an Abstract of the Proceedings of Their Meeting.* New York, 1844.

Ruschenberger, Walter. *A Notice of the Origin, Progress, and Present Condition of the Academy of Natural Sciences of Philadelphia.* Philadelphia, 1852.

Stansbury, Charles F. *Report of the Recording Secretary of the National Institute for the Year 1850.* Washington, D.C., 1850.

Tocqueville, Alexis de. *Democracy in America.* Edited by Phillips Bradley. 2 vols. New York, 1945.

———. *Journey to America.* Edited by J. P. Mayer. New Haven, Conn., 1960.

Van Rensselaer, Jeremiah. *Lecture on Geology: Being an Outline of the Science Delivered in the New York Athenum.* New York, 1825.

Warner, John. *New Theorems, Tables, and Diagrams for the Computation of Earthworks.* Philadelphia, 1863.

———. *Studies in Organic Morphology: An Abstract of Lectures Delivered before the Pottsville Scientific Association in 1855 and 1856.* Philadelphia, 1857.

Warren, Edward. *The Life of John Collins Warren, M.D., Compiled Chiefly from His Autobiography and Journals.* 2 vols. Boston, 1860.

Winslow, Charles. *Cosmography; or, Philosophical Views of the Universe.* Boston, 1854.

SECONDARY SOURCES

Unpublished Monographs and Dissertations

Abramos, Albertina. "The Policy of the National Education Association toward Federal Aid to Education (1857–1953)." Ph.D. dissertation, University of Michigan, 1954.

Beaver, Donald DeB. "The American Scientific Community, 1800–1860: A Statistical-Historical Study." Ph.D. dissertation, Yale University, 1966.

Bruce, Robert V. "Universities and the Rise of the Professions: Nineteenth-Century American Scientists." Paper presented at the American Historical Association meeting, Dec. 1970.

Ciarlante, Marjorie. "Social Origins of American Inventors." Paper read to the Society for the History of Technology, New York, 29 Dec. 1971.

Dahm, John J. "Science and Religion in Eighteenth-Century England: The Early Boyle Lectures and the Bridgewater Treatises." Ph.D. dissertation, Case Western Reserve University, 1969.

DeJong, John A. "American Attitudes toward Evolution before Darwin." Ph.D. dissertation, State University of Iowa, 1962.

Elliott, Clark. "The American Scientist, 1800–1863: His Origins, Career, and Interests." Ph.D. dissertation, Case Western University, 1970.

Flack, James K., Jr. "The Formation of the Washington Intellectual Community, 1870–1898." Ph.D. dissertation, Wayne State University, 1969.

Gerstner, Patsy Ann. "The 'Philadelphia School' of Paleontology: 1820–1845." Ph.D. dissertation, Case Western Reserve University, 1967.

Guralnick, Stanley. "Science and the American Colleges, 1828–1860." Ph.D. dissertation, University of Pennsylvania, 1969.

Kaufman, Martin. "Homeopathy and the American Medical Profession, 1820–1860." Ph.D. dissertation, Tulane University, 1969.

Loveday, Amos J. "A Statistical Study of the American Scientific Community, 1855–1870." M.A. thesis, Ohio State University, 1970.

Meyer, Donald H. "The American Moralists: Academic Moral Philosophy in the United States, 1835–1880." Ph.D. dissertation, University of California at Berkeley, 1967.

Numbers, Ronald L. "The Nebular Hypothesis in American Thought." Ph.D. dissertation, University of California at Berkeley, 1969.

Olson, Richard. "The Scientific Lazzaroni: A Case Study in the Social History of American Science." Mimeographed.

Rossiter, Margaret W. "Louis Agassiz and the Lawrence Scientific School." A.B. thesis, Radcliffe College, 1966.

Shapiro, Henry. "The Western Academy of Natural Sciences of Cincinnati and the Structure of Science in the Ohio Valley, 1810–1850." Paper read at the Conference on the Early History of Societies for the Promotion of Knowledge in the United States, June 1973.

Thomas, Milton H. "The Gibbs Affair at Columbia in 1854." M.A. thesis, Columbia University, 1942.

Towle, Edward L. "Science, Commerce and the Navy on the Seafaring Frontier (1842–1861)—The Role of Lieutenant M. F. Maury and the U.S. Navy Hydrographic Office in Naval Exploration, Commercial Expansion and Oceanography before the Civil War." Ph.D. dissertation, University of Rochester, 1966.

Books

Agassiz, Elizabeth C., ed. *Louis Aggasiz: His Life and Correspondence.* 2 vols. Boston, 1885.

The American Academy of Arts and Sciences, 1780–1940. Boston, 1941.

Anderson, Paul R., and Max H. Fisch, eds. *Philosophy in America from the Puritans to James with Representative Selections.* New York, 1939.

Appleton's Cyclopedia of American Biography. 7 vols. Rev. ed. New York, 1900.

Armes, William D., ed. *The Autobiography of Joseph LeConte.* New York, 1903.

Ashby, Eric. *Technology and the Academics: An Essay on Universities and the Scientific Revolution.* New York, 1963.

Bailey, Edward. *Charles Lyell.* New York, 1962.

Barber, Bernard. *Sociology of Science.* New York, 1962.

Basalla, George, *et al.,* eds. *Victorian Science: A Self-Portrait from the Presidential Addresses of the British Association for the Advancement of Science.* New York, 1970.

Bates, Ralph S. *Scientific Societies in the United States.* 3rd ed. Cambridge, Mass., 1965.

Bell, Whitfield J., Jr. *Early American Science: Needs and Opportunities for Study.* Williamsburg, Va., 1955.

Bemis, Samuel Flagg. *John Q. Adams and the Union.* New York, 1956.

Berkhofer, Robert F., Jr. *A Behavioral Approach to Historical Analysis.* New York, 1969.

Bode, Carl. *The American Lyceum: Town Meeting of the Mind.* New York, 1956.

Bolzan, Emma Lydia. *Almira Hart Lincoln Phelps: Her Life and Work.* New York, 1956.

Bottomore, T. B. *Elites and Society.* New York, 1964.

Brewster, Edwin T. *Life and Letters of Josiah Dwight Whitney.* Boston, 1909.

Brown, Harcourt. *Scientific Organizations in Seventeenth Century France, 1620–1680.* Baltimore, 1934.

Bruchey, Stuart. *The Roots of American Economic Growth, 1607–1861.* New York, 1965.

Calhoun, Daniel H. *The American Civil Engineer: Origins and Conflict.* Cambridge, Mass., 1960.

———. *Professional Lives in America: Structure and Aspiration, 1750–1850.* Cambridge, Mass., 1965.

Calvert, Monte A. *The Mechanical Engineer in America, 1830–1910: Professional Cultures in Conflict.* Baltimore, 1967.

Cardwell, D. S. L. *The Organization of Science in England: A Retrospect.* London, 1957.

Cash, W. J. *The Mind of the South.* New York, 1941.

The Century, 1847–1946. New York, 1947.

Clarke, John M. *James Hall of Albany, Geologist and Paleontologist, 1811–1898.* 2nd ed. Albany, N.Y., 1923.

Dall, William H. *Spencer Fullerton Baird, a Biography, Including Selections from His Correspondence with Audubon, Agassiz, Dana, and Others.* Philadelphia, 1915.

Dana, Edward S., *et al. A Century of Science in America, with Special Reference to the American Journal of Science, 1818–1918.* New Haven, Conn., 1918.

Daniels, George. *American Science in the Age of Jackson.* New York, 1968.

———, ed. *Nineteenth-Century American Science: A Reappraisal.* Evanston, Ill., 1972.

Dictionary of American Biography, New York, 1927—.

Dictionary of Canadian Biography. Toronto, 1966—.

Dictionary of Scientific Biography. New York, 1970—.

Dupree, A. Hunter. *Science in the Federal Government: A History of Policies and Activities to 1940.* Cambridge, Mass., 1957.

Eaton, Clement. *The Mind of the Old South.* Rev. ed. Baton Rouge, La., 1967.

Ekirch, Arthur A. *The Idea of Progress in America, 1815–1860.* New York, 1944.

Emerson, Edward W. *The Early Years of the Saturday Club, 1855–1870.* Boston, 1917.

Fairchild, Herman L. *A History of the New York Academy of Sciences, Formerly the Lyceum of Natural History.* New York, 1887.

Farrington, Benjamin. *Francis Bacon: Philosopher of Industrial Science.* New York, 1949.

Fisher, George P. *The Life of Benjamin Silliman.* 2 vols. New York, 1866.

Fleming, Donald H. *John William Draper and the Religion of Science.* Philadelphia, 1950.

———. *Science and Technology in Providence, 1760–1914: An Essay in the History of Brown University in the Metropolitan Community.* Providence, R.I., 1952.

Fulton, John. *Memoirs of Frederick A. P. Barnard . . . Tenth President of Columbia College in the City of New York.* New York, 1896.

Fulton, John, and Elizabeth Thomson. *Benjamin Silliman, 1779–1864: Pathfinder in American Science.* New York, 1968.

Gabriel, Ralph H. *The Course of American Democratic Thought: An Intellectual History since 1815.* New York, 1940.

Geike, Archibald. *Annals of the Royal Society Club: The Record of a London Dining-Club in the Eighteenth and Nineteenth Centuries.* London, 1917.

Gilchrist, David T., and W. David Lewis, eds. *Economic Change in the Civil War Era: Proceedings of a Conference on American Economic and Institutional Change, 1860–1873, and the Impact of the War.* Greenville, Del., 1965.

Gillispie, Charles. *Genesis and Geology: A Study in the Relations of Scientific Thought, Natural Theology, and Social Opinion in Great Britain, 1790–1850.* Cambridge, Mass., 1951.

Goetzmann, William H. *Army Exploration in the American West, 1803–1863.* New Haven, Conn., 1959.

————. *Exploration and Empire: The Explorer and the Scientist in the Winning of the American West.* New York, 1966.

Goode, George Brown. *The Smithsonian Institution, 1846–1896: The History of Its First Half Century.* Washington, D.C., 1897.

————, ed. *A Memorial of George Brown Goode.* Smithsonian Institution, Annual Report . . . for 1897, part II. Washington, D.C., 1901.

Goodrich, Carter. *Government Promotion of American Canals and Railroads.* New York, 1960.

Hahn, Roger. *Anatomy of a Scientific Institution: The Paris Academy of Sciences, 1666–1803.* Berkeley, Calif., 1971.

Haller, John S. *Outcasts from Evolution: Scientific Attitudes of Racial Inferiority, 1859–1900.* Urbana, Ill., 1971.

Harding, Walter. *Mr. Thoreau Declines an Invitation: Two Unpublished Papers by Thoreau: An Original Letter and an Annotated Questionnaire.* Richmond, Va., 1956.

Haskell, Daniel. *The United States Exploring Expedition, 1838–1842, and Its Publications, 1844–1874.* New York, 1942.

Herringshaw's Encyclopedia of American Biography of the Nineteenth Century. Edited by Thomas S. Herringshaw. Chicago, 1905.

Hindle, Brooke. *The Pursuit of Science in Revolutionary America, 1735–1789.* Chapel Hill, N.C., 1956.

Hofstadter, Richard. *Social Darwinism in American Thought.* New York, 1955.

————, and Walter P. Metzger. *The Development of Academic Freedom in the United States.* New York, 1955.

Hornberger, Theodore. *Scientific Thought in American Colleges, 1638–1800.* Austin, Tex., 1945.

Howarth, O. J. R. *The British Association for the Advancement of Science: A Retrospect, 1831–1931.* London, 1931.

International Encyclopedia of the Social Sciences. Edited by David Sills. New York, 1968.

Jaffe, Bernard. *Men of Science in America.* New York, 1958.

James, Edward T., and Janet James, eds. *Notable American Women, 1607–1950.* 3 vols. Cambridge, Mass., 1971.

Johnson, Thomas C., Jr. *Scientific Interests in the Old South.* New York, 1936.

Jones, Howard M., and I. Bernard Cohen, eds. *Science before Darwin: A Nineteenth-Century Anthology.* London, 1963.

Jordan, David Starr. *Leading American Men of Science.* New York, 1910.

Kelley, Howard, and Walter Burrage, eds. *American Medical Biographies.* Baltimore, 1920.

Kendall, Phebe M. *Maria Mitchell, Life, Letters and Journals.* Boston, 1896.

Kett, Joseph. *The Formation of the American Medical Profession: The Role of Institutions, 1780–1860.* New Haven, Conn., 1968.

Knapp, Robert H., and H. B. Goodrich. *Origins of American Scientists.* Chicago, 1952.

The Lazzaroni: Science and Scientists in the Mid-Nineteenth Century. National Portrait Gallery, Washington, D.C., 1972.

Lenzen, Victor. *Benjamin Peirce and the Coast Survey.* San Francisco, 1969.

Lockwood, George B. *The New Harmony Communities.* Marion, Ind., 1902.

Lurie, Edward. *Louis Agassiz: A Life in Science.* Chicago, 1960.

McAllister, Ethel. *Amos Eaton: Scientist and Educator.* Philadelphia, 1941.

Madsen, David. *The National University: Enduring Dream of the U.S.A.* Detroit, Mich., 1966.

Mason, S. F. *Main Currents of Scientific Thought: A History of the Sciences.* New York, 1953.

Marx, Leo. *Machine in the Garden.* New York, 1964.

Meisel, Max. *A Bibliography of American Natural History: The Pioneer Century, 1769–1865.* 3 vols. New York, 1924–29.

Merrill, George P. *Contributions to a History of American State Geological and National History Surveys.* Smithsonian Institution, *Bulletin 109.* Washington, D.C., 1920.

———. *The First One Hundred Years of American Geology.* New Haven, Conn., 1924.

Merritt, Raymond H. *Engineering in American Society, 1850–1875.* Lexington, Ky., 1969.

Merritt, Richard L. *Systematic Approaches to Comparative Politics.* Chicago, 1969.

Miller, Howard S. *Dollars for Research: Science and Its Patrons in Nineteenth-Century America.* Seattle, Wash., 1970.

Miller, Perry. *The Life of the Mind in America from the Revolution to the Civil War.* New York, 1965.

National Academy of Sciences. *Biographical Memoirs.* Washington, D.C., 1877—.

National Cyclopedia of American Biography. 47 vols. New York, 1898–1965.

Ornstein, Martha. *The Role of Scientific Societies in the Seventeenth Century.* Chicago, 1928.

Parsons, Talcott, ed. *American Sociology: Perspectives, Problems, Methods.* New York, 1968.

Persons, Stow. *American Minds: A History of Ideas.* New York, 1958.

Phillips, Venia T. *Guide to the Microfilm Publication of the Minutes and Correspondence of the Academy of Natural Sciences of Philadelphia, 1812–1942.* Philadelphia, 1967.

Purver, Margery. *The Royal Society: Concept and Creation.* Cambridge, Mass., 1967.

Rathbun, Richard. *The Columbian Institute for the Promotion of Arts and Sciences: A Washington Society of 1816–1838, Which Established a Museum and Botanic Garden under Government Patronage.* Smithsonian Institution, *Bulletin 101.* Washington, D.C., 1917.

Reingold, Nathan. *Science in Nineteenth-Century America: A Documentary History.* New York, 1964.

———, ed. *The Papers of Joseph Henry.* Washington, D.C., 1972—.

Rezneck, Samuel. *Education for a Technological Society: A Sesquicentennial History of Rensselaer Polytechnic Institute.* Troy, N.Y., 1968.

Rhees, William J., ed. *The Smithsonian Institution: Documents Relative to Its Origin and History, 1835–1899.* 2 vols. Washington, D.C., 1901.

Rintala, Edsel K. *Douglass Houghton: Michigan's Pioneer Geologist.* Detroit, Mich., 1954.

Rogers, Emma, ed. *Life and Letters of William Barton Rogers.* 2 vols. Cambridge, Mass., 1896.

Rosenberg, Charles E. *The Cholera Years: The United States in 1832, 1849, and 1866.* Chicago, 1962.

Rossiter, Margaret. *The Emergence of Agricultural Science: Justus Liebig and the Americans, 1840–1880.* New Haven, Conn., 1975.

Rudolph, Frederick. *The American College and University: A History.* New York, 1962.

The Rumford Fund of the American Academy of Arts and Sciences. Boston, 1905.

Sanford, Charles L. *The Quest for Paradise: Europe and the American Moral Imagination.* Urbana, Ill., 1961.

Schneer, Cecil J., ed. *Toward a History of Geology.* Cambridge, Mass., 1969.

Sinclair, Joseph B. *Early Research at the Franklin Institute: The Investigation into the Causes of Steam Boiler Explosions, 1830–1837.* Philadelphia, 1966.

———. *Philadelphia's Philosopher Mechanics: A History of the Franklin Institute, 1824–1865.* Baltimore, 1974.

Smith, Edgar F. *The Life of Robert Hare: An American Chemist, 1781–1858.* Philadelphia, 1917.

Solberg, Winton U. *The University of Illinois, 1867–1894: An Intellectual and Cultural History.* Urbana, Ill., 1968.

Somkin, Fred. *The Unquiet Eagle: Memory and Desire in the Idea of American Freedom, 1815–1860.* Ithaca, N.Y., 1967.

Spence, Clark C. *Mining Engineers and the American West: The Lace-Boot Brigade, 1849–1933.* New Haven, Conn., 1970.

Stanton, William R. *The Leopard's Spots: Scientific Attitudes toward Race in America, 1815–1859.* Chicago, 1960.

Stearns, Raymond P. *Science in the British Colonies of North America.* Urbana, Ill., 1970.

Storr, Richard J. *The Beginnings of Graduate Education in America.* Chicago, 1953.

Struik, Dirk J. *Yankee Science in the Making*. New York, 1962.

Sudhoff, Karl. *Hundert Jahre deutscher Naturforscher-Versammlunger*. Leipzig, 1922.

Taylor, George R. *The Transportation Revolution, 1815–1860*. New York, 1957.

Tewksbury, Donald G. *The Founding of American Colleges and Universities before the Civil War with Particular Reference to the Religious Influences Bearing upon the College Movement*. New York, 1932.

Thistlethwaite, Frank. *The Anglo-American Connection in the Early Nineteenth Century*. Philadelphia, 1959.

Thomas, R. Hinton. *Liberalism, Nationalism, and the German Intellectuals, 1822–1847: An Analysis of the Academic and Scientific Conferences of the Period*. Cambridge, 1951.

Tyler, Alice Felt. *Freedom's Ferment: Phases of American Social History to 1860*. Minneapolis, Minn., 1944.

Tyler, David B. *The Wilkes Expedition: The First United States Exploring Expedition, 1838–1842*. Philadelphia, 1968.

Van Alstyne, Richard W. *Genesis of American Nationalism*. Waltham, Mass., 1970.

Van Tassel, David D., and Michael G. Hall, eds. *Science and Society in the United States*. Homewood, Ill., 1966.

Veysey, Lawrence R. *The Emergence of the American University*. Chicago, 1965.

Ward, Julius H. *The Life and Letters of James Gates Percival*. Boston, 1866.

Washburn, Wilcomb E., ed. *The Great Design: Two Lectures on the Smithsonian Bequest by John Quincy Adams*. Washington, D.C., 1965.

Weeks, Edward. *The Lowells and Their Institute*. Boston, 1966.

Wesley, Edgar B. *The NEA: The First Hundred Years: The Building of the Teaching Profession*. New York, 1957.

Wilson, Leonard G. *Charles Lyell, the Years to 1841: The Revolution in Geology*. New Haven, Conn., 1972.

Woodward, Horace B. *The History of the Geological Society of London*. London, 1907.

Woody, Thomas. *A History of Women's Education in the United States*. 2 vols. New York, 1929.

Wright, Helen. *Sweeper in the Sky: The Life of Maria Mitchell, First Woman Astronomer in America*. New York, 1949.

Youmans, E. L. *Pioneers of Science in America*. New York, 1896.

Articles

Aldrich, Mark. "Earnings of American Civil Engineers, 1820–1859." *Journal of Economic History*, XXXI (June 1971), 407–419.

Back, William. "The Emergence of Geology as a Public Function, 1800–1879." Washington Academy of Sciences, *Journal*, XLIX (July 1959), 205–209.

Bell, Whitfield J., Jr. "The American Philosophical Society as a National

Academy of Sciences, 1780–1846." *Ithaca: Proceedings of the Tenth International Congress of the History of Science...*, I (1962), 165–177.

———. "Astronomical Observatories of the American Philosophical Society, 1769–1843." American Philosophical Society, *Proceedings*, CVIII (1964), 7–14.

Benjamin, Marcus. "The American Association for the Advancement of Science." *Chautauquan: A Monthly Magazine*, XIII (Aug. 1891), 727–731.

———. "The Early Presidents of the American Association." American Association for the Advancement of Science, *Proceedings*, XLVIII (Aug. 1899), 397–499.

———. "Organization and Development of the Chemical Section of the American Association for the Advancement of Science." In American Chemical Society, *Twenty-fifth Anniversary Volume*. 1902. Pp. 86–98.

Bouvé, Charles. "The Manuscripts of C. S. Rafinesque (1783–1840)." American Philosophical Society, *Proceedings*, CII (Oct. 1958), 590–595.

Bouvé, Thomas T. "Historical Sketch of the Boston Society of Natural History; with a Notice of the Linnaean Society, Which Preceded It." *Anniversary Memoirs of the Boston Society of Natural History, 1830–1880*. Boston, 1880.

Cannon, Walter. "History in Depth. The Early Victorian Period." *Journal of the History of Science*, III (1964), 20–38.

Cohen, I. Bernard. "Science in America. The Nineteenth Century." In Arthur M. Schlesinger, Jr., and Morton White, eds., *Paths of American Thought*. Boston, 1963. Pp. 167–189.

Cox, Edward F. "The International Institute. First Organized Opposition to the Metric System." *Ohio Historical Quarterly*, LXIII (Jan. 1959), 54–83.

———. "The Metric System. A Quarter-Century of Acceptance (1851–1876)." *Osiris*, VIII (1958), 358–379.

Daniels, George. "The Process of Professionalization in American Science. The Emergent Period, 1820–1860." *Isis*, LVIII (Summer 1967), 151–166.

Dupree, A. Hunter. "The Founding of the National Academy of Sciences: A Reinterpretation." American Philosophical Society, *Proceedings*, CI (Oct. 1957), 434–440.

———. "Science and Technology." In David T. Gilchrist and W. David Lewis, eds., *Economic Change in the Civil War Era: Proceedings of a Conference on American Economic and Institutional Change, 1850–1873, and the Impact of the War*. Greenville, Del., 1965. Pp. 123–128.

Eyles, Joan M. "William Smith: Some Aspects of His Life and Work." In Cecil J. Schneer, ed., *Toward a History of Geology*. Cambridge, Mass., 1969. Pp. 142–158.

Fairchild, Herman L. "The History of the American Association for the Advancement of Science." *Science*, LIX, nos. 1527–29 (1924), 365–369, 385–390, 410–415.

Fisher, Marvin. "The Iconology of Industrialism, 1830–1860." In Hennig Cohen, ed., *The American Culture: Approaches to the Study of the United States*. Boston, 1968. Pp. 228–245.

Foote, George A. "The Place of Science in the British Reform Movement, 1830–1850." *Isis*, XLII (Oct. 1951), 192–208.

Forman, Sidney. "West Point and the American Association for the Advancement of Science." *Science*, CIV (19 July 1946), 47–48.

Goldfarb, Stephen. "Science and Democracy: A History of the Cincinnati Observatory, 1842–1872." *Ohio History*, LXXVIII (Summer 1969), 172–178.

Goode, George Brown. "The First National Scientific Congress (Washington, April 1844) and Its Connection with the Organization of the American Association." American Association for the Advancement of Science, *Proceedings*, XL (Aug. 1891), 39–47.

Greene, John C. "The Development of Mineralogy in Philadelphia, 1780–1820." American Philosophical Society, *Proceedings*, CXIII (Aug. 1969), 283–295.

Guralnick, Stanley M. "Geology and Religion before Darwin: The Case of Edward Hitchcock, Theologian and Geologist (1793–1864)." *Isis*, LXIII (Dec. 1972), 529–543.

Hale, William H. "Early Years of the American Association.' *Appleton's Popular Science Monthly*, XLIX (1896), 501–507.

Hall, Michael G. "The Introduction of Modern Science into Seventeenth-Century New England: Increase Mather." *Ithaca: Proceedings of the Tenth International Congress of the History of Science . . .*, I (1962), 261–263.

Hardy, Kenneth R. "Social Origins of American Scientists and Scholars." *Science*, CLXXXV (9 Aug. 1974), 497–506.

Harris, P. M. G. "The Social Origins of American Leaders: The Demographic Foundations." *Perspectives in American History*, III (1969), 157–344.

Hendrickson, Walter B. "Naturphilosophie in the United States." *Ithaca: Proceedings of the Tenth International Congress of the History of Science . . .*, II (1962), 977–980.

——. "Nineteenth-Century State Geological Surveys: Early Government Support of Science." *Isis*, LII (Sept. 1961), 357–371.

——. "Science and Culture in the American Middle West." *Isis*, LXIV (Sept. 1973), 326–340.

——. "The Western Academy of Natural Sciences of Cincinnati." *Isis*, XXXVII (July 1947), 138–145.

Holmfeld, John D. "From Amateurs to Professionals in American Science: The Controversy over the Proceedings of an 1853 Scientific Meeting." American Philosophical Society, *Proceedings*, CXIV (16 Feb. 1970), 22–36.

Kallerup, Harry R. "Bibliography of American Academies of Sciences." Kansas Academy of Sciences, *Transactions*, LXVII (Summer 1963), 274–281.

Kohlstedt, Sally Gregory. "The Geologists' Model for National Science, 1840–1847." American Philosophical Society, *Proceedings*, CXVIII (Apr. 1974), 179–195.

——. "A Step toward Scientific Self-Identity in the United States: The Failure of the National Institute, 1844." *Isis*, LXII (Fall 1971), 339–362.

Lane, N. Gary. "New Harmony and Pioneer Geology." *Geotimes*, LX Sept. 1966), 18–22.

Lingelbach, William D. "The American Philosophical Society Library from 1842 to 1952 with a Survey of Its Historical Background." American Philosophical Society, *Proceedings*, XCVII (1953), 471–492.

Livingston, Burton E. "Members of the American Association for the Advancement of Science per Million of Population in the United States." *Science*, LX (July–Dec. 1924), 467–469.

Lurie, Edward. "Louis Agassiz and the Races of Man." *Isis*, XLV (May 1954), 227–242.

———. "Nineteenth-Century American Science: Insights from Four Manuscripts." Rockefeller Institute, *Review*, II, no. 1 (Feb. 1964), 11–19.

———. "Science in American Thought." *Journal of World History*, VIII (1964–65), 638–664.

Mather, Kirtley F. "Geology, Geologists and the AAAS." *Science*, XXIX (24 Apr. 1959), 1106–11.

Mendelsohn, Everett. "The Emergence of Science as a Profession in Nineteenth-Century Europe." In *The Management of Science*. Boston, 1964. Pp. 3–48.

Miles, Wyndham D. "Public Lectures on Chemistry in the United States." *Ambix: The Journal of the Society for the Study of Alchemy and Early Chemistry*, XV (Oct. 1968), 129–153.

Moulton, F. R. "The American Association for the Advancement of Science: A Brief Historical Sketch." *Science*, CVIII (Sept. 1948), 217–218.

Nash, Gerald. "The Conflict between Pure and Applied Science in Nineteenth-Century Public Policy: The California State Geological Survey, 1860–1874." *Isis*, LIV (Sept. 1963), 217–228.

Orange, A. D. "The British Association for the Advancement of Science: The Provincial Background." *Science Studies*, I (1917), 315–329.

Panghorn, C. W., Jr. "A History of the Popularization of Geology in America." Washington Academy of Sciences, *Journal*, XLIX (July 1959), 224–227.

Pickard, Madge D. "Government and Science in the United States: Historical Backgrounds." *Journal of the History of Medicine and Allied Sciences*, I (July 1946), 265–289.

Price, Derek J. DeSolla. "Is Technology Historically Independent of Science? A Study in Statistical Historiography." *Technology and Culture*, VI (Fall 1965), 553–568.

Reingold, Nathan. "Alexander Dallas Bache: Science and Technology in the American Idiom." *Technology and Culture*, XI (Apr. 1970), 163–177.

———. "American Indifference to Basic Science during the Nineteenth Century." In George Daniels, ed., *Nineteenth-Century American Science: A Reappraisal*. Evanston, Ill., 1972. Pp. 38–62.

———. "The Anatomy of a Collection: The Rhees Papers." *American Archivist* (Apr. 1964), pp. 251–259.

———. "Babbage and Moll on the State of Science in Great Britain: A Note on a Document." *British Journal for the History of Science*, IV (1968), 58–64.

Bibliography

Revai, Elizabeth. "Le voyage d'Alexandre Vattamare au Canada: 1840–1875." *Revue d'histoire de l'Amerique française,* XXII (Sept. 1968), 257–299.

Rezneck, Samuel. "The Emergence of a Scientific Community in New York State a Century Ago." *New York History,* LXIII (July 1962), 211–238.

———. "A Traveling School of Science on the Erie Canal in 1826." *New York History,* XL (1959), 255–269.

Ritchie, John, Jr. "The American Association for the Advancement of Science." *New England Magazine,* XVIII (Aug. 1898), 639–661.

Rossiter, Margaret W. "Benjamin Silliman and the Lowell Institute: The Popularization of Science in Nineteenth Century America." *New England Quarterly,* LXIV (Dec. 1971), 602–626.

———. "Women Scientists in America before 1920." *American Scientist,* LXII (May–June 1974), 312–323.

Rostow, Walt W. "The Take-Off into Self-Sustained Growth." *Economic Journal,* LXVI (Mar. 1965), 25–48.

Rudwick, M. J. S. "The Foundation of the Geological Society of London: Its Scheme for Co-operative Research and Its Struggle for Independence." *British Journal for the History of Science,* I, (Dec. 1963), 325–355.

Schneer, Cecil J. "Ebenezer Emmons and the Foundations of American Geology." *Isis,* LX (Winter 1969), 439–450.

Shapin, Steven, and Arnold Thackray. "Prosopography as a Research Tool in the History of Science: The British Scientific Community, 1700–1900." *History of Science,* XII (1974), 1–28.

Sherwood, Morgan B. "Genesis, Evolution, and Geology in America before Darwin: The Dana-Lewis Controversy, 1856–1857." In Cecil J. Schneer, ed., *Toward a History of Geology.* Cambridge, Mass., 1969. Pp. 304–316.

Shryock, Richard. "American Indifference to Basic Science during the Nineteenth Century." *Journal of World History,* II (1948–49), 50–65.

Silverman, Robert, and Mark Beach. "A National University for Upstate New York." *American Quarterly,* XXII (Fall 1970), 701–713.

Stone, Lawrence. "Prosopography." In Felix Gilbert and Stephen Graubard, eds., *Historical Studies Today.* New York, 1971. Pp. 107–140.

Storer, Norman. "The Sociology of Science." In Talcott Parsons, ed., *American Sociology: Perspectives, Problems, Methods.* New York, 1968. Pp. 199–213.

Sydnor, Charles S. "State Geological Surveys in the Old South." In David Kelly Jackson, ed., *American Studies in Honor of William Kenneth Boyd.* Durham, N.C., 1940. Pp. 86–109.

Waller, Adolph E. "Dr. John Locke, Early Ohio Scientist (1792–1856)." *Ohio State Archeological and Historical Quarterly,* LV (Jan.–Mar. 1946), 346–373.

Williams, L. Pearce. "The Royal Society and the Founding of the British Association for the Advancement of Science." *Royal Society of London, Notes and Records,* XVI (Nov. 1961), 221–233.

.

Index

Appendix

APPENDIX

BIOGRAPHICAL DIRECTORY OF THE AAAS, 1848-60

This appendix lists all members and participating non members of the American Association for the Advancement of Science during its formative years from 1848 to 1860. The subjects are grouped according to the analysis subdivisions suggested in Chapter VIII.

A: Leadership

Included in the following list of 337 men are all members who participated in the AAAS by publishing one or more papers or by serving on one or more committees, excepting local committee membership, and were thus discussed as leaders in the foregoing analysis. In each case the material relates to that portion of the subject's life most relevant up to and including the chronological limits of this study. The term of membership and level of leadership are first noted: "primary" if the subject published two or more articles in different years, held two or more offices, was a member of a Standing Committee, or had a combination of publications and offices, with all others listed as "secondary." When available, information on birth and death dates, place of birth, educational experience, occupation, areas of scientific interest, and place of residence during membership included.

Each sketch concludes with a bibliographical citation. When biographical information was derived primarily from the Dictionary of American Biography and sources mentioned therein and no later studies are known to the author, only the DAB is referenced; the only exception is when a longer study was also used extensively by the author. If the individual returned a questionnaire to Spencer F. Baird which is retained in the volume of "Scientific Addresses," that fact is indicated by the initials SFB. The following abbreviations or short forms are used for frequently cited works:

AAA&S, Proc.	American Academy of Arts and Sciences, Proceedings
AJS	American Journal of Science and Arts.
CAB	Appleton's Cyclopedia of American Biography
DAB	Dictionary of American Biography
DNB	Dictionary of National Biography
DSB	Dictionary of Scientific Biography
K&B, Med. Biog.	Howard A. Kelly and Walter L. Burrage, American Medical Biographies
Herringshaw	Herringshaw's Encyclopedia of American Biography of the Nineteenth Century
NAS, Biog. Mem.	National Academy of Sciences, Biographical Memoirs
NCAB	National Cyclopedia of American Biography

Adams, Charles Baker (1814-1853) Primary, 1848-1853.
    Born in Dorchester, Massachusetts, Adams attended Yale but transferred to graduate at the head of his class at Amherst in 1834. He assisted Edward Hitchcock on the Massachusetts survey and was himself state geologist of Vermont, becoming professor of chemistry and natural history at Middlebury, 1838-1847, and then professor of astronomy and zoology at Amherst, 1847-1853. His special interest was zoology, particularly mollusks, and while on a research trip in the West Indies in 1853 he contracted yellow fever and died. DAB.

Adamson, J. Constantine. Secondary, 1857-1859.
    A New York City resident, Adamson presented papers at Montreal in 1857.

Agassiz, Louis (1807-1873). Primary, 1848--.
    Swiss-born and educated in Germany and France with a Ph.D. from Erlangen and Munich, Agassiz arrived in the United States in 1846. Accepting a position essentially created for him at the Lawrence Scientific School, he began to write a monumental study of the natural history of the United States and worked for the establishment of the Museum of Comparative Zoology at Harvard. During the 1850's the zoologist-geologist was everywhere in popular demand and influenced not only public thinking about science but also the emerging scientific community. DAB; Edward Lurie, Louis Agassiz: A Life in Science (Chicago,1960); SFB.

Aiken, William E. A. (1807-1888). Secondary, 1848--(inactive 1851-1858).
Aiken, born in New York state, possibly Onondaga, attended Rensselaer
Polytechnic Institute, 1828-1830, then studied medicine and was licensed by
the New York State Medical Society. He taught at the University of Maryland
and was also a practicing chemist with an interest in physics and geology.
Eugene F. Cordell, Historical Sketches of the University of Maryland School
of Medicine (1807-1890) (Baltimore, 1891), p.144; Henry B. Nason, ed.,
Biographical Record of the Officers and Graduates of Rensselaer Polytechnical
School, 1824-1886 (Troy, New York, 1887), p.550.

Alexander, John Henry (1812-1867). Primary, 1848-1851.
Alexander's merchant father of Annapolis, Maryland, was able to provide
him with a classical education at St. John's College, where he graduated in
1826. His early interest in topography, when he worked on local railroad
surveys, was transposed to the problem of standardization of weights and measures.
He taught mining and engineering at the University of Pennsylvania and then
moved to Baltimore to teach natural history at St. James' College and later
natural philosophy at the University of Maryland. DAB; NAS, Biog. Mem., I
(1877).

Alexander, Stephen (1806-1883). Primary, 1848--.
Born in Schenectady, New York, and a graduate of Union College in 1824,
Alexander followed his cousin Joseph Henry to Princeton where he taught first
in the theological seminary but later in the college. Interested in mathematics and astronomy, Alexander became an expert on star clusters and was
popular as a teacher at Princeton to his retirement in 1877. DAB; NAS, Biog.
Mem., II (1886); SFB.

Alger, Francis (1804-1863). Secondary, 1848-1854.
With only a common school education and a full-time business, Alger,
born in Bridgewater, Massachusetts, found time to do extensive work in
mineralogy and helped found the Boston Society of Natural History. AAA&S,
Proc., VI.

Antisell, Thomas (1817-1893). Secondary, 1850-1854.
Irish-born, Antisell was a graduate of Trinity College, Dublin; he
subsequently studied medicine in Dublin and London and then studied chemistry
in Paris and Berlin. Coming to New York in 1848, he practiced medicine until
1854, when he spent two years as geologist on the Pacific railroad survey.
His later years were primarily spent in Washington, D.C. William H. Goetzmann,
Army Exploration in the American West, 1803-1863 (New Haven, 1959), p.317;
K&B, Med. Biog.

Ayres, William O. (1817-1887). Secondary, 1848-1855.
Although born in New Canaan, Connecticut, and a graduate of Yale, 1837,
Ayres moved to Boston and from 1848 to 1852 was curator for the Boston Natural
History Society. In 1853 he was studying medicine in Boston and in 1854 took
an M.D. from Yale, leaving shortly thereafter for San Francisco. His special
interest was ichthyology. Connecticut State Medical Society, Proceedings,
No. 4 (1884-1887), 174-176; SFB.

Bache, Alexander Dallas (1806-1867). Primary, 1848--.
The great-grandson of Benjamin Franklin, Bache was born in Philadelphia
and graduated from West Point in 1825 with high honors. In 1843 he assumed
control of the Coast Survey and effectively used that post to further basic
research in several aspects of physical science. An adroit administrator,
he was a key figure in the scientific community of the 1850's and prime mover
behind establishment of the National Academy of Sciences. Merle Odgers,
Alexander Dallas Bache, Scientist and Educator, 1806-1867 (Philadelphia, 1947);
Nathan Reingold, "Alexander Dallas Bache: Science and Technology in the
American Idiom," Technology and Culture, XI (April, 1970), 163-177; SFB.

Bache, Franklin (1792-1864). Secondary, 1848--.
Born in Philadelphia, Bache took an A.B. from the University of
Pennsylvania in 1810 and an M.D. in 1814. Professor of chemistry at the
Philadelphia College of Pharmacy and after 1841 at Jefferson Medical School,
he was also a practicing physician. He was especially active in the American
Philosophical Society. Edgar Fahs Smith, Franklin Bache, Chemist (Philadelphia,
1922); SFB.

Bachman, John (1790-1874). Primary, 1848-1854.
Bachman, born in Dutchess County, New York, attended Williams College and then went to Philadelphia where he was acquainted with Alexander Wilson and was also licensed to preach by the Lutheran Church. In 1815 he accepted a pastorate in Charleston, South Carolina. In his free time he prepared a catalogue of local plants and ferns, wrote on the race question, and helped found several scientific groups in Charleston. DAB.

Bailey, Jacob Whitman (1811-1857). Primary, 1848-1857.
Born in Ward (now Auburn), Massachusetts, Bailey worked in Providence before receiving an appointment to West Point, where he graduated in 1832. After 1834 he taught chemistry, mineralogy and geology at West Point but was distinguished among his scientific colleagues for his microscopic technique and his work on algae. DAB; AAAS, Proceedings, XI (1858), 1-8; SFB.

Baird, Spencer Fullerton (1823-1888). Primary, 1848--.
Born in Reading, Pennsylvania, Baird graduated from Dickinson College in 1836 and returned to teach botany and zoology there in 1846. An active zoologist, he became assistant secretary of the Smithsonian Institution in 1850 and secretary on Henry's death in 1878, and was largely responsible for the development of its natural history museum. William H. Dall, Spencer F. Baird, a Biography, including Selections from his Correspondence with Audubon, Agassiz, Dana and Others (Philadelphia, 1915); DSB.

Barnard, Frederick A.P. (1809-1889). Primary, 1853--.
Primarily an educator, Barbard, born in Sheffield, Massachusetts, graduated from Yale in 1828 and taught variously at Yale, the University of Alabama, and the University of Mississippi. In 1856 he became president of the University of Mississippi and after 1864 at Columbia in New York. He was generally interested in science and held positions related to astronomy, although in 1853 he listed his primary interests as organic chemistry and photography. William J. Chute, "The Life of Frederick A.P. Barnard to his Election as President of Columbia College in 1864" (Ph.D. dissertation, Columbia University, 1951); DAB; SFB.

Barringer, Daniel M. (1806-1873). Secondary, 1858-1860.
Primarily a lawyer and politician, Barringer was born near Concord, North Carolina, and graduated from his state university in 1826. He was a leading citizen of Charlotte, North Carolina. DAB.

Bartlett, William H. C. (1809-1893). Primary, 1855--.
A graduate of West Point in 1826, Bartlett served in the army engineers until his appointment as professor of natural and experimental philosophy at West Point in 1836. Born in Lancaster, Pennsylvania, and raised in Missouri, he remained in the New York area as a mathematician and astronomer; after 1871 he was an insurance actuary in New York City. DAB.

Basnett, Thomas. Secondary, 1854--.
A resident of Ottawa, Illinois, Basnett was apparently a farmer who was avocationally interested in meteorology. The Past and Present of La Salle County (Chicago, 1877, p.357.

Batchelder, John M. (1811-1892). Secondary, 1854--.
Born in New Ipswich, New Hampshire, Batchelder studied first at Brunswick and later studied engineering at Harvard. A practicing civil engineer in Maine and Massachusetts, after 1842 he resided in Cambridge and worked also under Bache for the Coast Survey, for whom he invented and improved various kinds of machinery. AAA&S, Proc., XXVIII (1892-1893), 305-310.

Beck, Lewis Caleb (1798-1853). Secondary, 1848-1853.
Born in Schenectady, New York, Beck received his A.B. from Union College in 1817 and was licensed as an M.D. in 1816. He practiced medicine and also taught chemistry at various places including Albany Medical College and Rutgers in New Jersey and was also mineralogist on the New York survey. At New Brunswick in 1853, he listed his interests as chemistry and mineralogy. DAB; SFB.

Beck, Theodric Romeyn (1791-1855). Primary, 1848-1855.
Brother of Lewis, Theodoric also attended Union College, graduating in 1807, and then studied medicine in New York City. From 1817 to 1853 he was

principal of Albany Academy and also practiced some medicine and taught at
upstate medical schools. A local philanthropist, Beck is best known for his
volume on medical jurisprudence. DAB; SFB.

Blake, William Phipps (1825-1910). Primary, 1849-1860.
       Born in New York City, Blake attended private schools and graduated
from Sheffield Scientific School in 1852. He worked briefly as industrial
chemist and mineralogist in New Jersey and then joined the Pacific railroad
survey as geologist. Subsequently he was a mining engineer in North Carolina,
Japan, and California. His AAAS address, however, was generally given as
Washington, D.C. DAB.

Blodgett, Lorin (1823-1901). Primary, 1853-1858.
       Born in Chautauqua County, New York, Blodgett attended Geneva College.
An early meteorological volunteer observer for the Smithsonian Institution,
in 1851 he was placed in charge of researches on climatology, publishing the
accumulated data in 1855 and 1857. He also worked for the Pacific railroad
survey, and his best work was in statistics and meteorology. DAB; SFB.

Bolton, Richard. Secondary, 1848-1854.
       Bolton, who lived in Pontotoc, Mississippi, was an agent for the New
York and Mississippi Land Company. MSS of company at Wisconsin State
Historical Society, Madison.

Bond, George Phillips (1825-1865). Primary, 1849-1856.
       Born in Dorchester, Massachusetts, to astronomer William Bond, George
Bond graduated from Harvard in 1845, assisted his father at the Harvard
Observatory, and in 1859 succeeded him as director. His careful observations
were useful and his study of the Conati comet of 1858 helped to establish
photographic astronomy. DAB; SFB.

Bond, William Cranch (1789-1859). Secondary, 1849-1856.
       Born in Falmouth, Maine, Bond moved with his family to Boston but had
to leave school before finishing, becoming a clock-maker and instrument-maker.
Avocationally he was an astronomical observer and from 1839 to 1859 was first
director at Harvard's Observatory. He withdrew from the AAAS in 1856. DAB;
SFB.

Bowditch, Henry Ingersoll (1808-1892). Secondary, 1849--.
       An ardent abolitionist, Bowditch, born in Salem, was a practicing
physician in Boston. With an 1828 Harvard degree he continued for an M.D.,
1832, and then studied in Paris and in England. Concerned about public
hygiene, he was also interested in microscopical anatomy and physiology.
DAB; SFB.

Boye', Martin Hans (1812-1909). Primary, 1848-1851.
       Born in Copenhagen, Boye' received his college degree from the
university there in 1832, before coming in 1836 to the United States where
he settled in Philadelphia and assisted Robert Hare. In 1844 he received an
M.D. from the University of Pennsylvania and subsequently taught at Central
High School. His most significant work was in the application of chemistry
to agriculture and industry. DAB; Edgar Fahs Smith, Martin Hans Boye',
1812-1909 (Philadelphia, 1924).

Brainerd, Jehu. Primary, 1851-1857.
       An amateur geologist, Brainerd was one of the founding members of the
Cleveland Academy of Natural Science and his paper at the Cleveland meeting
was a source of professional controversy. In 1853 he was professor of
mineralogy and medical botany at the Western College of Hemocopxiria Medicine
and interested in geology, so he wrote Baird. John D. Holmfield, "From
Amateurs to Professionals in American Science: The Controversy over the
Proceedings of an 1853 Scientific Meeting", American Philosophical Society,
Proceedings, CXIV (February 16,1970), p.28; SFB

Breed, Daniel. Secondary, 1849-1851.
       Breed attended New York University Medical College in 1847; he was
apparently a practicing physician and also was a patent examiner. His
address during the period of membership was New York City, but he was later
a surgeon in the Army and a trustee in Washington, D.C. Negro schools.
General Alumni Catalogue of New York University, 1833-1907: Medical Alumni
(New York, 1908), p.16.

Brevoort, James Carson (1818-1887). Secondary, 1848--(inactive 1854-1856).
     Born in Brooklyn to a wealthy landowner, Brevoort was privately
tutored and then went to Paris, 1835-1838, to study civil engineering. He
described his occupation as "leisure", and was active in scientific, literary
and artistic associations. His particular scientific interests were geology
and physiology. CAB; SFB.

Brewer, William Henry (1828-1910). Secondary, 1850-1854.
     Born in Poughkeepsie, New York, Brewer was interested in plants and
minerals and so went to the new Sheffield Scientific School at Yale, where
he graduated in 1852; he subsequently went to Germany to study chemistry
under Bunsen at Heidelberg and under Liebig at Munich. After his return in
1860 he worked on the geological survey of California until accepting an
appointment to teach agriculture at Sheffield, 1864-1907. DAB; SFB.

Brocklesby, John (1811-1889). Primary, 1850--.
     Although born in England, Brocklesby came to America in 1820 and
graduated from Yale in 1835, then studied law and was admitted to the bar in
1838. In 1842 he accepted a chair of mathematics at Trinity College, Hartford,
Connecticut, later adding astronomy and natural philosophy. He was also
interested in meteorology and microscopy. NCAB, XII, 287; SFB.

Brooks, Charles (1795-1872). Secondary, 1857-1859.
     A congregational minister and active in various reform movements,
Brooks was born in Medford, Massachusetts, and graduated from Harvard in 1816.
In 1838 he accepted a post in natural history at New York University and the
following year went to Europe to study zoology. Failing eyesight, however,
sent him back into the education movement, although he did publish on ornithology.
CAB; John Albree, Charles Brooks and his Work for Normal Schools, Medford
Historical Society, 1906.

Brown, Andrew (1793-1871). Primary, 1848--.
     Brown was born in Scotland and attended the University of Edinburgh,
1807-1810. He came to the United States in the 1820's and settled in Natchez,
Mississippi, where he operated a sawmill. Interested in physical science,
broadly construed, he wrote "The Philosophy of Physics . . ." (1854). John
H. Moore, Andrew Brown and Cyprus Lumbering in the Old Southwest (Baton Rouge,
Louisiana, 1967): SFB.

Browne, Peter Arrell (1782-1860). Primary, 1848-1855.
     Born apparently in Philadelphia, Browne was a practicing lawyer there
and was eclectically interested in science, first botany and geology and
later ethnology. He was a founder of the Franklin Institute and an active
member of the Academy of Natural Sciences. Several of his published local
lectures are at the Historical Society of Pennsylvania, Philadelphia;
William Stanton, The Leopard's Spots: Scientific Attitudes toward Race in
America, 1815-1859 (Chicago, 1960), pp. 149-155.

Brunnow, Franz Freidrich Ernst (1821-1891). Secondary, 1856-1860.
     German-born, Brunnow received a Ph.D. from the University of Berlin in
1843 and was subsequently invited by Henry Tappan to direct the observatory of
the University of Michigan. He left Ann Arbor when Tappan resigned in 1863.
Royal Astronomical Society, Monthly Notices, L11 (1891-1892), 203-233.

Brush, George Jarvis (1831-1912). Secondary, 1851-- (inactive 1854-1857).
     Born in Brooklyn and privately educated, Brush joined his father's
mercantile business for a short time and then went to New Haven to study
mineralogy, chemistry, and agriculture under John P. Norton and Benjamin
Silliman, graduating from the new Sheffield Scientific School about 1852.
From 1853 to 1855 he studied chemistry and mining in Germany, when he became
professor of mineralogy at Sheffield; from 1872 until 1898 he headed the
school. DAB; NAS, Biog. Mem., XVII (1924).

Buchanan, Robert (1797-1879). Primary, 1849--.
     Born in western Pennsylvania to Scotch-Irish farmers, Buchanan attended
country school and Meadville Academy and subsequently moved to Cincinnati,
where he founded several businesses and became a leading citizen. He listed
his interests as botany, conchology, mineralogy, and paleontology but was
only active on memorializing committees of the AAAS. The Biographical
Encyclopedia of Ohio of the Nineteenth Century (Cincinnati and Philadelphia,
1876), pp. 62-63; SFB.

Bunce, John B. Secondary, 1850-1854.
A New Haven, Connecticut, resident, Bunce presented a paper to the New Haven meeting on the economic value of anthracite coal ashes.

Burgess, Ebenezer (1805-1870). Primary, 1848-1855.
Born in Grafton, Vermont, Burgess graduated from Amherst in 1831 and also attended Andover Theological Seminary. He went to India as a missionary and remained there until 1851 when he served parishes in Centreville, Lanesville, and South Franklin, Massachusetts. His two papers to the AAAS were on the geology of India. CAB.

Burnap, George Washington (1802-1859). Secondary, 1858-1859.
Born in Merrimac, New Hampshire, Burnap graduated from Harvard in 1824 and continued for three years at the Divinity School. His first call was to replace Jared Sparks in a Unitarian pulpit in Baltimore, and he remained there until his death. Belonging to the conservative wing of the Unitarian church, he was generally interested in history and education. DAB.

Burnet, Jacob (1770-1853). Primary, 1850-1854.
Burnet was born in Newark, New Jersey, graduated at Nassau Hall (Princeton) in 1791 and studied law. In 1796 he moved to Cincinnati, where he became a leading lawyer and active politician, participating in such intellectual movements as the Astronomical Society, the Colonization Society, and Cincinnati College. DAB.

Burnett, Waldo Irving (1828-1854). Primary, 1848-1854.
During his few very productive years, Burnett, born in Southborough, Massachusetts, and with a Harvard M.D. in 1849, studied natural history and microscopy in Europe and published numerous articles, including fifteen with the AAAS. At the time of his death, caused by consumption, he was translating the comparative anatomy of Siebold and Stannius. CAB; Boston Society of Natural History, Proceedings, V (1855), 64-74.

Caswell, Alexis (1799-1877). Primary, 1849--.
Born in Taunton, Massachusetts, Caswell graduated at the head of his class at Brown in 1822. After theological study and one year as a Baptist minister, he returned to Brown in 1828 as professor of mathematics and natural philosophy, later adding astronomy. In his later years he concentrated on meteorology. DAB; SFB.

Chace, George Ide (1808-1885). Secondary, 1848--.
Chace, born in Lancaster, Massachusetts, graduated from Brown in 1830. After teaching and also studying under Robert Hare, he became professor of chemistry, geology and physiology by 1836. Deeply religious and an effective lecturer, he published very little but his mineral collection became the nucleus for Brown's geological collection. DAB; James D. Murray, ed., George Ide Chace, L.L.D. A Memorial (Cambridge, 1886); SFB.

Chandler, W.T. Secondary, 1854-1855.
Chandler's single paper related to meteorological observations on the Mexican Boundary Survey. His address was Washington, D.C., in 1854 and Philadelphia the following year.

Chauvenet, William (1820-1870). Primary, 1851--.
Born in Medford, Pennsylvania, Chauvenet attended private schools and graduated from Yale in 1840. He worked with A.D.Bache in Philadelphia and in 1842 became a professor of mathematics for the Navy; he was largely responsible for establishing the U. S. Naval Academy on a firm scientific basis. Interested primarily in astronomy and mathematics, he later taught at Washington University, St. Louis, and was chancellor there, 1862-1869. DAB; SFB.

Cherriman, J.B. Primary, 1855-1859.
A resident of Toronto, Canada, Cherriman indicated an interest in astronomy and mathematics.

Chittenden, Lucius Eugene (1824-1900). Secondary, 1860--.
Born in Vermont, Chittenden was a lawyer and president of the local bank in Burlington. An active Republican, he was appointed Register of the Treasury under Salmon P. Chase and worked with him on the Peace Conference. Invisible Siege: The Journal of Lucius E. Chittenden, April 15, 1861-July 14, 1861 (San Diego, California, 1969).

Christy, David (1802- ?). Secondary, 1851-1855.
    Christy is most remembered as agent for the American Colonization
Society and his Cotton Is King (1855). While travelling in relation to his
anti-slavery involvement, he became interested in geology, corresponding
with John Locke and Verneuil of Paris and publishing several articles on the
subject. DAB.

Clark, Alonzo (1807-1887). Secondary, 1851-1854.
    Born in Chester, Massachusetts, Clark graduated from Williams in 1828
and took an M.D. from New York College of Physicians and Surgeons. He was an
active physician, publishing, teaching first physiology and pathology, 1848-
1855, and then pathology and practical medicine, 1855-1885, at the College
of Physicians and Surgeons, while actively participating in local and national
medical societies. CAB.

Clark, Alvan (1804-1887). Secondary, 1851-1860.
    Born in Ashfield, Massachusetts, Clark attended local schools and
learned to engrave and draw. At the age of forty he began to study optics
and with his sons produced unexcelled refracting telescopes in their factory
at Cambridgeport, Massachusetts. DAB; DSB; Deborah J. Warner, Alvan Clark &
Sons: Artists in Optics (Washington, D.C., 1968).

Clark, Henry James (1826-1873). Primary, 1859-1860.
    Clark was born in Easton, Massachusetts, graduated from the University
of the City of New York in 1848, went to Cambridge to assist Asa Gray and
graduated from Lawrence Scientific School in 1854. He then assisted Louis
Agassiz on the anatomical and embryological portions of Contributions to the
Natural History of the United States until the two men quarreled in 1863.
Clark was particularly adept with a microscope. DAB.

Clark, Meriwether Lewis. Primary, 1851--.
    Clark was an Army major stationed in St. Louis, Missouri.

Clark, William Warner (1826-1869). Secondary, 1850-1854.
    Born in Bristol, Vermont, Clark received a B.A. from Wesleyan in
1848. He taught at Gouverneau Wesleyan Seminary in New York, becoming
principal in 1850, and after 1853 taught at Genesee Wesleyan Seminary, Lima,
New York; his residence on the AAAS membership rolls was given as Albany.
Alumni Record of Wesleyan University, Middleton,Connecticut (New Haven, 1911),
p. 88.

Coakley, George W. (ca. 1814-1893). Primary, 1851-1856.
    Born in the West Indies, Coakley apparently graduated from Rutgers
College. He subsequently taught at the College of St. James in Maryland and
resided in Hagerstown. His interest was in physical science for he listed
mathematics, astronomy, meteorology and electricity on Baird's questionnaire.
AJS, 3rd Series, XLVI (1893), 484; SFB.

Coffin, James Henry (1806-1873). Primary, 1848--.
    Coming from a poor family on Martha's Vineyard, Coffin graduated
somewhat late from Amherst in 1828. He taught and simultaneously worked on
both theory and instrumentation of meteorology. In 1846 he accepted a chair
of mathematics and natural philosophy at Lafayette College, Easton, Pennsylvania,
and continued his work on wind currents. DAB; SFB.

Coffin, John Huntington Crane (1815-1890). Primary, 1848--.
    Born in Wiscasset, Maine, Coffin graduated from Bowdoin in 1834 and
was appointed professor of mathematics in the Navy; in 1843 he took charge
of the mural circle at the United States naval observatory. After 1853
he taught first mathematics and then astronomy and navigation at the United
States Naval Academy in Annapolis with his special interest remaining mathematics and physical sciences, especially applied. CAB; SFB.

Cook, George Hammell (1818-1889). Secondary, 1848-1860 (inactive various
years).
    Cook was born in Hanover, New Jersey,and graduated from Rensselaer
Polytechnic Institute in 1838. He taught there and at Albany Academy until
becoming professor of chemistry and natural science at Rutgers. He assisted
on the New Jersey survey after 1854 and in 1864 became state geologist. DAB;
SFB.

Cooke, Josiah Parsons (1827-1894). Primary, 1849-1855.
        Son of a wealthy Boston lawyer, Cooke graduated from Harvard in 1848
and then studied in Europe at the chemical laboratories of Regnault and Dumas.
On his return he was elected Erving Professor of Chemistry and Mineralogy,
where he promoted chemistry as a requirement for undergraduates and used the
laboratory for participation as well as demonstration for students. DAB;
NAS, Biog. Mem., IV (1902)

Couper, James Hamilton (1794-1866). Secondary, 1848--.
        A Yale graduate of 1844, Couper studied Dutch water control and
established an experimental rice plantation near Darien, Georgia, with an
elaborate diking and drainage system. He also introduced olive trees and
Bermuda grass to the region and operated two cottonseed oil mills, both of
which were destroyed by the Civil War. Couper was a competent conchologist,
interested in geology and infusoria as well. DAB; SFB.

Courtney, Edward Henry (1803-1853). Secondary, 1853 only.
        Born in Maryland, Courtney was valedictorian at West Point in 1821
and was variously with the Army engineers and professor at West Point and
the University of Pennsylvania. In 1843 he accepted the chair of mathematics
at the University of Virginia until his death. He was also interested in
astronomy. CAB; Manuscript obituary account in Bache MSS, LC; SFB.

Craw, William Jarvis (1830-1897). Primary, 1850-1854.
        Born in Connecticut, Craw graduated with the first class of the
Sheffield Scientific School in 1852 and assisted there in applied chemistry
the following year. Although interested in chemistry as applied to geology
and mineralogy, he was eventually isolated from science as a paint businessman in Springfield, Massachusetts. SFB; Yale Catalogue of Officers and
Graduates, 1701-1924 (New Haven, 1924), pp. 50 and 347; letters from Craw,
dated 29 May 1853 and 17 February 1854, in Brewer Family MSS, YUA.

Culmann, R. Primary, 1850-1855.
        Culmann, whose address was given as Bavaria, was apparently interested
in civil engineering.

Curley, James (1796-1889). Primary, 1854-1860.
        Irish-born, Curley entered the priesthood, becoming professor of
natural philosophy and astronomy at Georgetown College. He was also
interested in botany. SFB; Woodstock College, Woodstock Letters: A Record
of Current Events and Historical Notes Connected with the Colleges and
Missions of the Society of Jesus, XVIII (1889), 381-384.

Dana, James Dwight (1813-1895). Primary, 1848--.
        Born in Utica, Dana's interest in science led him to Yale where he
graduated in 1833; in 1836 he accompanied the Wilkes Expedition as zoologist.
In 1846 he became an editor of the AJS and in 1850 professor of natural
history at Yale. His geological observations on the Pacific complemented
those of Darwin and his Manual of Mineralogy, revised, appeared in its
seventeenth edition in 1959. DSB; Daniel C. Gilman, The Life of James Dwight
Dana (New York, 1899); SFB.

Daniels, Edward. Primary, 1853-1856.
        Daniels, state geologist of Wisconsin in the 1850's, indicated broad
interests in geology, chemistry, zoology and meteorology. He also participated in the abolition movement and served during the Civil War. His
Association address was Ceresco, Wisconsin. Daniels MSS, Wisconsin State
Historical Society, Madison; SFB.

Darby, John (1804-1877). Secondary, 1851-1855.
        Darby, born in North Adams, Massachusetts, worked his way through
Williams College in 1831. He taught at various female colleges in the South
and in 1853 was head of Sigourney Institute at Culloden, Georgia. He
expressed an interest in "all branches" of science but published primarily
in botany and chemistry. DAB; SFB.

Dascomb, James (1808-1880). Secondary, 1853-1860.
        Born in Wilton, New Hampshire, Dascomb received an M.D. from
Dartmouth in 1833. From 1834 to 1878 he taught, at various times, chemistry,
botany, anatomy and physiology at Oberlin College; he also taught briefly

at Ohio Agricultural College and Hillsdale College. General Catalogue of Dartmouth College and the Affiliated Schools, 1769-1925 (Hanover, New Hampshire, 1905); General Catalogue of Oberlin College, 1833-1908 (Oberlin, Ohio, 1909), p. 141; SFB.

Davis, Charles Henry (1807-1877). Primary, 1848-1855.
    Brought up in Boston by a father interested in science, Davis spent two years at Harvard and then joined the Navy; he returned in 1841 to take his degree. After 1842 he worked for the Coast Survey and in 1849 assumed responsibility for the new American Ephemeris and Nautical Almanac, which he supervised until 1855 and again from 1859 to 1862. He was interested primarily in hydrography, astronomy and tidal geology. DAB; SFB.

Dawson, John William (1820-1899). Primary, 1845--.
    Born and educated in Pictou, Nova Scotia, Dawson received an M.S. from Edinburgh University in 1841 and the following year performed a geological survey of his province's coalfields. An active educator, he became principal of McGill University in Montreal in 1855 where he remained until 1893, also teaching various science courses. His major scientific contributions were in geology and paleontology, and he also wrote several books harmonizing science and religion. DSB; Charles F. O'Brien, "The Word and the Work: A Study of Sir William Dawson and Nineteenth Century Controversies between Religion and Science" (Ph.D. dissertation, Brown University, 1968).

Delafield, Joseph (1790-1875). Secondary, 1848--.
    A lifetime resident of New York City, Delafield graduated from Yale in 1808 and in 1811 was admitted to the New York bar. Basically a gentleman of leisure, his interest in mineralogy led to his collections becoming one of the largest private ones in the United States and to his presidency of the New York Academy of Sciences, 1827-1866. CAB; Herman Fairchild, History of the New York Academy of Sciences (New York, 1887), pp. 64-68; SFB.

Desor, Pierre Jean Edouard (1811-1882). Primary, 1848-1855.
    While a German law student Desor was hired by Agassiz to supervise the printing of his work and Desor came to Cambridge in 1847. Later some difficulties between the two men resulted in Desor's return to Europe where he subsequently became professor of geology in the Academie de Neuchatel. AAA&S, Proc., XVII (1882), 422-424; Carl C. Vogt, Eduard Desor: Lebensbild eines Naturforschers (Breslau, 1883).

Dickeson, Montroville Wilson (1810-1882). Secondary, 1848-1851.
    Born in Philadelphia, Dickeson was locally educated and then studied medicine in Philadelphia and affiliated with the dispensary there. From 1837 to 1844 he traveled in the Ohio and Mississippi river valleys collecting archeological remains. Although unsuccessful in opening his own museum with the relics, his collection was later sold to the University of Pennsylvania. Steward Culin, "The Dickeson College of American Antiquities," University of Pennsylvania, University Museum Bulletin, II (1900), 113-168.

Eastwood, George (-1889). Secondary, 1859--.
    Eastwood, a graduate of Lawrence Scientific School in 1859, presented a mathematical paper to the AAAS that year. Harvard University, Quincentennial Catalogue of the Officers and Graduates, 1636-1925 (Cambridge, Massachusetts, 1925), p. 534.

Eaton, Asahel Knowlton (1822-1906). Secondary, 1857-1859.
    Coming from Westmoreland, Eaton graduated from Hamilton College in 1843 and also received an M.D. there. He published a textbook on agricultural chemistry in 1847 and lived in New York City at the time of his membership. The Hamilton College Bulletin: Complete Alumni Register, 1812-1932, XVI (November, 1932), p. 55.

Elliott, Ezekiel Brown (1823-1888). Primary, 1856--.
    Born in Monroe County, New York, Elliott graduated from Hamilton College in 1844. After working briefly on the telegraph, he moved to Boston where he was actuary for a life-insurance company; in 1861 he took a similar position with the U.S. Sanitary Commission and later in federal government agencies. His statistical reports on coinage, weights, and measures surpassed his work in mathematical physics. CAB.

Elwyn, Alfred Langdon (1804-1884). Primary, 1848--.
    Elwyn, who had a degree from Harvard, 1823, and an M.D. from the
University of Pennsylvania, 1831, was a philanthropist interested in history
and literature. Although born in Portsmouth, New Hampshire, he became a
leading citizen in Philadelphia, participating in social and reform activities
and editing materials in history and philology. DAB.

Emerson, George Barrell (1797-1881). Primary, 1848--.
    A Harvard graduate of 1817, Emerson, born in Wells, Maine, headed a
private school in Boston from 1828 to 1855. He was first president of the
Boston Society of Natural History and prepared an important report on the
trees and shrubs of Massachusetts. Interested in education, Emerson helped
found the Boston Mechanics' Institute and after the Civil War worked for the
education of freedmen in the South. DAB.

Emmons, Ebenezer (1799-1863). Primary, 1848-1854.
    Born in Middlefield, Massachusetts, Emmons graduated from Williams
College in 1818 and then studied science. He worked on the New York survey
but had subsequent difficulties over his Taconic System and after 1851 taught
and was state geologist of North Carolina. DAB; Cecil J. Schneer, "Ebenezer
Emmons and the Foundations of American Geology", Isis, LX (Winter, 1969),
439-450; DSB.

Emory, William Helmsey (1811-1887). Primary, 1851-1858.
    Son of a Queen Anne's County, Maryland, planter, Emory graduated
from West Point in 1831. After 1838 he served in the Army topographical
engineers and was chief astronomer for the Mexican boundary commission, 1848
to 1853. His interests in astronomy and geodesy were reflected in his report
of 1856. DAB; William Goetzmann, Army Explorations in the American West,
1803-1863 (New Haven, 1965), pp. 128-129, 195; SFB.

Engelmann, George (1809-1884). Primary, 1848-- (inactive 1854-1856).
    German-born Engelmann received an M.S. from Wurzburg in 1831 and came
to the United States the following year. After a brief stay in Illinois he
established a practice in St. Louis, Missouri. His meteorological observations
began in 1836 but his chief interest was in botany and his collections made
his home a stopping point for interested western explorers. He had an extensive correspondence and helped found the St. Louis Academy of Science.
DAB; SFB.

Erni, Henri (1833-1885). Secondary, 1850-1854.
    Born near Zurich, Switzerland, and educated at the university there,
Erni immigrated to the United States in 1849 and worked for about two years
under Silliman at Yale. He subsequently taught at East Tennessee University
in Knoxville and at the University of Vermont, 1854-1857. Going South again,
he was detained there during the early years of the Civil War but moved to
Washington and subsequently worked as a chemist in various offices and as a
consul to Basel. Wyndham Miles and Louis Kuslan, "Washington's First Consulting Chemist, Henri Erni," Columbia Historical Society of Washington,
D.C., Records, LXVI-LXVII (1966-1968), 154-166.

Espy, James Pollard (1785-1860). Secondary, 1848-1854.
    Born in Pennsylvania, Espy moved westward with his family, graduated
from Transylvania University in 1808, and studied law in Ohio. At both the
Franklin Institute and the Smithsonian Institution he promoted the gathering
of meteorological data and the development of weather systems; his tireless
propagandizing resulted in his being called by contemporaries "The Storm
King." DAB; Nathan Reingold, ed., Science in Nineteenth-Century America;
A Documentary History (New York, 1964), p. 93.

Everett, Edward (1794-1865). Primary, 1849--.
    Born in Dorchester, Massachusetts, Everett graduated from Harvard in
1814, accepted a Unitarian parish call, taught briefly at Harvard and then
went to Gottingen where he received his Ph.D. in 1817. On his return he
edited the North American Review and became active in state and federal
politics and well-known for his magnetic public speaking ability. His
interests were in moral and political "science." DAB; SFB.

Ewing, Thomas (1789-1871). Secondary, 1851--.
    Ewing, born near West Liberty, Virginia, graduated from Ohio University
at Athens in 1815 and then studied law and was admitted to the bar in 1816.
A resident of Lancaster, he served in Congress and held several cabinet posts

during Whig administrations but urged compromise during the sectional crisis and acted with the Democrats during Reconstruction. DAB.

Farmer, Moses Gerrish (1820-1893). Secondary, 1855-1860.
     An ingenious inventor interested in electrical apparatus, Farmer, born in Bowcasen, New Hampshire, worked in Boston for a telegraph company and on a fire-alarm system. His paper before the AAAS was on his discovery concerning duplex and quarduplex telegraph usage and in 1858-1859 he invented the incandescent light. DAB.

Farrington, A. C. Primary, 1850-1854.
     A resident of Newark, New Jersey, Farrington's papers presented at Albany suggest an interest in mining technology.

Felton, John Brooks (1827-1877). Secondary, 1849-1854.
     Felton was born in Saugus, Massachusetts, graduated from Harvard in 1847, taught, traveled in Europe, and returned to take a law degree from Harvard in 1853. He then went to San Francisco where he was eminent at the bar and in politics, also helped to found the University of California. CAB.

Ferrel, William (1817-1891). Secondary, 1857-1861.
     Born in Fulton, Pennsylvania, Ferrel had limited schooling and worked before graduating from Bethany College in 1844. He then taught and in 1854 opened a school in Nashville, Tennessee, and in 1856 published an essay on winds and ocean currents. As a result of this work he was appointed to the American Ephemeris and Nautical Almanac in 1857 and later continued his early interest in meteorology and geodesy working for the Coast and Geodetic Survey. DAB.

Feuchtwanger, Lewis (1805-1875). Secondary, 1848-- (inactive 1850-1857).
     Raised in Bavaria by his mineralogist father, Feuchtwanger received an M.D. from the University of Jena and then came to the United States in 1829, where he opened a pharmacy and practiced medicine. He later concentrated on chemistry and mineralogy, manufacturing rare chemicals and collecting minerals which were displayed at the London Exhibition in 1851. CAB.

Field, Roswell Marton (1807-1859). Secondary, 1859 only.
     Field came from Newfane, Vermont, to graduate from Middlebury College, Vermont, in 1822 and practiced law and was active in politics in his home town until he moved to St. Louis in 1839. His AAAS address was given as Greenfield, Massachusetts. Catalogue of Students and Officers of Middlebury College in Middlebury, Vermont . . .1800-1900 (Middlebury, Vermont, 1901), p.70.

Foote, John Parsons (1783-1865). Primary, 1850-1854.
     Born in Guilford, Connecticut, Foote moved to Ohio about 1820. He worked as a bookseller and typesetter and then organized the Cincinnati water works. A prominent citizen, he helped organize several literary, scientific, and educational organizations. Vital Statistics File at the Cincinnati Historical Society, Cincinnati, Ohio.

Force, Peter 1790-1868). Primary, 1850-1854.
     Born near Passaic Falls, New Jersey, Force moved to Washington in 1815 and published a strongly Whig newspaper, 1824 to 1831. With general and diverse interests, Force readily promoted such enterprises as the National Inssitute, but his personal forte was historical and statistical editing. DAB; Newman F. McGirr, "The Activities of Peter Force," Columbia Historical Society of Washington, D.C., Records, XLII-XLIII (1942-1943), 35-82.

Forshey, Caleb G. (1812-1881). Secondary, 1853-1859 inactive 1855-1857).
     Born in Somerset County, Pennsylvania, Forshey attended Kenyon College and also West Point, graduating from neither. Interested in "natural and mechanical philosophy," he worked as a civil engineer in the Mississippi Delta and in Texas. Operating from Rutersville, Texas, he also helped to found the New Orleans Academy of Sciences. CAB; SFB.

Fosgate, Blanchard. Secondary, 1854--.
     Apparently a practicing physician, Fosgate stated his interest as ethnology; his AAAS address was Auburn, New York. SFB.

Foster, John (1811-1897). Secondary, 1848-1860 (inactive 1854-1856).
    Graduating from Union College in 1845, Foster spent the rest of his
life teaching mathematics and natural philosophy at his alma mater. Stanley
M. Guralnick, "Science and the American College: 1828-1860" Ph.D.
dissertation, University of Pennsylvania, 1969), pp. 263-264; SFB.

Foster, John Wells (1815-1873). Primary, 1848--.
    Born in Petersham, Massachusetts, Foster graduated from Wesleyan and
then studied law and was admitted to the bar in Zanesville, Ohio. His
interest shifted to civil and mining engineering, and he worked on the state
geological survey and on the survey of Lake Superior under Charles T.
Jackson. He returned to Massachusetts in the mid-1850s but in 1858 went
west again, settling in Chicago to work for a railroad and then teaching.
Herringshaw; Marcus Benjamin, "The Early Presidents of the American
Association," AAAS, Proceedings, XLVIII August, 1899), 435-436.

Frazer, John Fries (1812-1872). Primary, 1848-- (inactive 1855-1857).
    A Philadelphia resident throughout his life, Frazer graduated from
the University of Pennsylvania in 1830, assisted Alexander Bache, studied
medicine, transferred to law, and finally accepted a position under Henry
Rogers on the geological survey of the state. He then taught at the
Philadelphia high school until 1844 when he succeeded his mentor Bache as
professor of chemistry and natural philosophy at the University of Pennsylvania.
He was especially interested in chemistry and geodesy and was an active member
in the Franklin Institute, editing its Journal from 1850 to 1866.
DAB; NAS, Biog. Mem., I (1877).

Friedlander, Eduard Julius Theodore (1813-1884). Secondary, 1853-1855.
He was especially interested in mathematics and numismatics. Allemeine
deutsche Biographie, XLVIII (Leipzig, 1904), 780-785.

Gaillard, Peter Cordes (1815-1859). Secondary, 1849-1854.
    Born and educated in Charleston, Gaillard took his degree from the
College of South Carolina in 1834 and also a medical degree. He went to
Paris, studying and practicing there for several years, but then returned
to Charleston to practice. He specialized in hygiene and sanitary science
but also studied natural history and comparative anatomy. CAB; SFB.

Gale, Leonard Dunnell (1800-1883). Secondary, 1854--.
    Massachusetts-born Gale received an M.D. from the New York College
of Physicians and Surgeons in 1830. For several years he taught chemistry
at the University of New York and from 1846 to 1857 was a patent examiner.
Thereafter he worked privately as a patent lawyer. AJS, 3rd Series, XXVI
(1884), 490; Smithsonian Institution, Annual Report for 1883.

Germain, Lewis J. Secondary, 1848-1854.
    Although his residence in 1848 was Burlington, New Jersey, when
Germain wrote to Baird in 1853 he was a civil engineer in Galena, Illinois,
and interested in geology and botany. SFB.

Gibbes, Lewis Reeves (1810-1894). Primary, 1848--.
    A lifetime resident of Charleston, Gibbes received his degree from
South Carolina College in 1829 and his M.D. in 1836 from the Medical College
of South Carolina before going to Paris to study at the Sorbonne and the
Jardin des Plantes. The somewhat fastidious Gibbes never practiced medicine
but taught mathematics and later astronomy and physics, at the College of
Charleston, 1838 to 1892. K&B, Med. Biog.; Thomas C. Johnson, Scientific
Interests in the Old South (New York, 1936), pp. 133-134; SFB.

Gibbes, Robert Wilson (1809-1866). Primary, 1848-1860 (inactive 1854-1856).
    Born in Charleston, Gibbes graduated from South Carolina College in
1827 and received an M.D. from the Medical College of South Carolina in 1830.
A practicing physician, he also built an extensive natural history collection,
was twice mayor in Columbia, and edited the South Carolinian there. DAB.

Gibbon, John Heysham, Primary, 1850--.
    Gibbon, who apparently had an M.D. was assayer at the U. S. Mint at
Charlotte, North Carolina. He was interested in natural history as well as
physical and moral philosophy and also promoted uniform weights and measures.
SFB; Don Taxay, The United States Mint and Coinage: An Illustrated History
from 1776 to the Present (New York, 1966).

Gibbs, Oliver Wolcott (1822-1908). Primary, 1848--.
     Born in New York City, Gibbs demonstrated an early interest in
chemistry. He received a degree from Columbia College in 1841, took an
M.D. from the College of Physicians and Surgeons, and then studied for three
years in Germany and Paris. From 1849 to 1863 he taught at New York City
College (Free Academy), and then became Rumford professor at Harvard. Gibbs
was a young and precocious member of the Lazzaroni. DAB; NAS, Biog. Mem., Vii
(1913); Milton H. Thomas, "The Gibbs Affair at Columbia in 1854" (M.S. thesis,
Columbia University, 1942).

Gillespie, William Mitchell (1816-1868). Primary, 1856--.
     Born in New York City, Gillespie graduated from Columbia in 1834 and
then traveled and studied in Europe. From 1845 to 1868 he taught engineering
at Union College and helped shape a four-year curriculum in engineering not
only at Union but at other schools as well. DAB; SFB.

Gilliss, James Melville (1811-1864). Primary, 1848-1855.
     Gilliss, born in Georgetown, D.C., entered the navy as midshipman in
1831; he studied at the University of Virginia for two years but then returned
to duty. In 1835 he was assigned to the Depot of Charts and Instruments and
planned and equipped the naval observatory, 1842-1844, only to be displaced
by the appointment of Matthew Maury as its head; he returned to head the
observatory in 1861. During the 1850s he worked primarily on astronomical
observations in South America, listing wide-ranging interests also including
magnetism, meteorology, and paleontology. DAB; NAS, Biog. Mem, I (1877); SFB.

Gilmore, Quincy A., Jr. (1825-1888). Primary, 1859--.
     Born in Ohio, Gilmore graduated from West Point in 1849 and was
assigned to the engineers. From 1853 to 1856 he was assistant instructor of
practical military engineering and then worked for an ordinance division in
New York. He wrote several volumes on civil engineering. George Cullum,
Biographical Register of the Officers and Graduates of the U.S. Military
Academy at West Point, New York, II (New York, 1891), 367-370.

Girard, Charles Francis (1822-1895). Primary, 1849-1859 inactive 1855-1857).
     Born in upper Alsace, Girard studied under Aggasiz at Neuchatel and
followed his teacher to Cambridge in 1847. From 1850 to 1860 he joined
S. F. Baird at the Smithsonian where he published extensive reports on the
zoology of America, many based on specimens returned by exploring expeditions.
In 1854 he became a citizen and received an M.D. from Georgetown College in
1856, but at the close of the Civil War he returned to Paris. DAB; SFB.

Glynn, James (1801-1871). Secondary, 1848--.
     At the very early age Glynn left his home in Philadelphia to become a
seaman during the War of 1812, and in 1815 he was appointed midshipman. By
1841 he was a commander and served in the Pacific, actively promoting trade
relations with Japan. His residence even before retirement in 1862 was New
Haven, Connecticut. DAB.

Goadby, Henry. Secondary, 1851-1854.
     Goadby, who signed the register as "doctor," seems to have been
interested in applied chemistry and published A Textbook of Vegetable and
Animal Physiology (1859).

Gould, Augustus Addison (1805-1866). Primary, 1857--.
     Born in New Ipswich, Gould graduated from Harvard in 1825 and took an
M.D. there in 1830. A practicing surgeon, Gould also did important work as
a naturalist, especially in conchology, and was a central figure in the Boston
Society of Natural History. DAB; Richard I. Johnson, The Recent Mollusca of
A. A. Gould; Illustrations of the Types Described by Gould with a Bibliography
and Catalogue of His Species (Washington, D.C., 1964); SFB.

Gould, Benjamin Apthorp (1824-1896). Primary, 1849--.
     Born in Boston, Gould graduated from Harvard in 1844 and went to study
at Berlin and Gottingen where he expanded his mathematical and astronomical
knowledge acquired under Benjamin Peirce. He returned home to found
Astronomical Journal and from 1852 to 1867 affiliated with the longitude
department of the Coast Survey, also directing the Dudley Observatory in
Albany, 1855 to 1859. Involvement in controversy perhaps inhibited the early
promise shown by Gould, but his work on stars in the southern hemisphere was
an important contribution. DAB; NAS, Biog. Mem., XVII (1924).

Graham, George (1798-1881). Secondary, 1851-1856.
		Graham went to Cincinnati from Somerset County, Pennsylvania, and he
successfully made his fortune in the riverboat and steamboat business. With
a limited education, he was competently involved in local social and reform
activities, and he served one term in Congress as a Democrat. Cincinnati,
Past and Present (Cincinnati, 1872), pp. 44-50.

Graham, James Duncan (1799-1865). Secondary, 1848--.
		Born in Prince William County Graham graduated from West Point in
1817. Beginning with his assignment to Stephen Long's survey, he distinguished
himself on other national border surveys conducted by the Corps of Topographical Engineers. His discovery of a lunar tide on the Great Lakes was a
major result of his interest in astronomy and geodesy. DAB; SFB.

Gray, Asa (1810-1888). Primary, 1848--.
		With an M.D. from Fairfield Medical School, Gray, born in Oneida
County, New York, taught school while actively collecting plants and minerals.
His excellent work with John Torrey earned him the Fisher Professorship of
Natural History at Harvard in 1842, the same year in which his standard Botanical
Text Book was published. He was the acknowledged leader of American botanists
and a pioneer in the field of plant geography. DAB; A. Hunter Dupree, Asa
Gray, 1810-1888 (Cambridge, 1959); SFB.

Green, Traill (1813-1897). Secondary, 1848--.
		Born in Easton, Pennsylvania, Green received an M.D. from the
University of Pennsylvania in 1835. After practicing medicine briefly, he
taught first chemistry and then natural science at Lafayette College. His
broad interests included chemistry, engineering and geodesy and he worked
for medical reform, temperance, and higher education for women. K&B, Med.
Biog.; SFB.

Greene, Thomas A. Secondary, 1856-1860.
		Greene, of New Bedford, Massachusetts, presented a single paper
relating to a new maximum thermometer.

Guyot, Arnold Henry (1807-1884). Primary, 1848--.
		Swiss-born, Guyot received a Ph.D. from Berlin and worked with
Agassiz while teaching at the Academy of Neuchatel. He came to the United
States in 1849 to deliver lectures at the Lowell Institute and stayed on, accepting
a chair of physical geography and geology at Princeton. He was active in
perfecting research methods in meteorology as well, working in conjunction
with the Smithsonian Institution. DAB.

Hackley, Charles William (1809-1861). Primary, 1850-1861.
		Born in Herkimer County, New York, Hackley graduated from West Point
in 1829, taught, and was ordained an Episcopal priest. In 1843 Columbia
College appointed him professor of mathematics and astronomy. He worked for
the establishment of an astronomical observatory in New York City and wrote
several mathematical textbooks. CAB; SFB.

Haldeman, Samuel Stehman (1812-1880). Primary, 1848--.
		Born in Lancaster County, Pennsylvania, Haldeman attended Dickinson
College for two years and was then sent by his father to manage a sawmill.
Returning to Columbia, Pennsylvania, he engaged in iron manufacturing but
also taught natural history at the University of Pennsylvania, 1851-1855, and
later geology and chemistry at Pennsylvania Agricultural College. He was
also interested in ethnology and after 1868 taught philology at the University
of Pennsylvania. DAB; SFB.

Hall, James (1811-1898). Primary, 1848--.
		Hall, born and raised in Hingham, Massachusetts, graduated from
Rensselaer Polytechnic Institute, studied with Amos Eaton, and was initially
a close friend of Ebenezer Emmons. His section report for the New York
survey was the best of those submitted and he was subsequently assigned the
report on paleontology, a monumental eight volumes not completed until 1894.
He was an acknowledged American leader in stratigraphic geology and inverte-
brate paleontology. DAB; SFB; John M. Clarke, James Hall of Albany, Geologist
and Palaeontologist, 1811-1898 (Albany, 1923).

Hamel, Joseph (1788-1862). Secondary, 1854-1855.
		Russian born, Hamel was named to the Russian Academy of Medicine in
1813. Interested in mechanics and in education, he invented an electrical

machine, was an active participant in world exhibitions, and wrote histories
of steam engines and the electrical telegraph. He spent much of his life in
England and introduced the Lancastrian system of education into Russia. The
Annual Register of World Events: A Review of the Year, CIV (London, 1862), 357.

Hamlin, Augustus Choate (1829-1905). Primary, 1856--.
     A lifetime resident of Bangor, Maine, Hamlin graduated from Bowdoin
College in 1851 and received an M.D. from Harvard in 1855. For the remainder
of his life he practiced medicine in Bangor, except when he served during
the Civil War in the army. General Catalogue of Bowdoin College and the
Medical School of Maine: A Biographical Record of Alumni and Officers,
1794-1905 (Brunswick, Maine, 1950), p.97.

Hare, Robert (1781-1858). Primary, 1848--. (inactive 1854-1857).
     Educated at home in Philadelphia, Hare managed his father's brewery
and studied chemistry, inventing the oxy-hydrogen blow-pipe. He subsequently
taught at William and Mary College and at the University of Pennsylvania.
After 1847 he retired from teaching but continued work in electricity and
also became involved with spiritualism. DAB; Edgar F. Smith, The Life of
Robert Hare: An American Chemist (Philadelphia, 1917).

Harris, Thaddeus William (1795-1856). Secondary, 1848-1856.
     Born in Dorchester, Massachusetts, Harris graduated from Harvard in
1815 and received an M.D. there in 1820. In 1831 Harvard appointed him
librarian and from 1837 to 1842 he also lectured in natural history only
to be succeeded by Asa Gray for the full professorship. He published several
articles and reports on his specialty, entomology. DAB.

Hartwell, Charles (1825-1905). Primary, 1851-1854.
     Born in Lincoln, Massachusetts, Hartwell graduated from Amherst in
1849, taught for a year at Westford Academy in Amherst, and then attended
Hartford Theological Seminary, 1850 to 1852. In the latter year he sailed
as a missionary to China. Amherst College, Biographical Record of the
Graduates and Non-Graduates of the Classes of 1822-1862, Inclusive (Amherst,
1963), p.59.

Harvey, William Henry (1811-1866). Secondary, 1850-1859.
     An Irishman, Harvey travelled throughout the British Isles and also
in Africa, making extensive zoological and botanical collections. In 1849
he came to the United States and presented lectures on botany and on his
specialty, algae, to the Lowell Institute and the Smithsonian Institution.
From 1844 to his death he was curator of the herbarium at Trinity College,
Dublin. AJS, XCII (December, 1866), 129, 273-277; DNB; SFB.

Haupt, Herman (1817-1905). Primary, 1851-1857.
     Born in Philadelphia, Haupt graduated from West Point in 1835 but
resigned his commission to work for a Pennsylvania railroad. A prolific
writer on technical subjects and known for his civil engineering ability,
Haupt constructed the Hoosac tunnel. He improved the pneumatic drill and
encouraged technology in mining. DAB; SFB.

Hayes, Augustus Allen (1806-1882). Secondary, 1848-1851.
     Hayes, born in Windsor, Vermont, attended a military academy and
Dartmouth College but moved to Boston in 1828 where he continued his interest
in chemistry by applying it to industrial processes for textiles. He subsequently did studies in fuel economy and manufacturing processes for iron
and copper. DAB.

Hayes, Isaac Israel (1832-1881). Secondary, 1858--.
     With his M.D. from the University of Pennsylvania in 1853, Philadelphia-
born Hayes sailed as surgeon on the second Arctic expedition under Elisha Kent
Kane in that year. Convinced by his experience that there was an open polar
sea, Hayes commanded a similar expedition in 1860. After serving as a surgeon
during the Civil War he retired to New York City to write and lecture about
his theory and his experiences. DAB.

Hays, Isaac (1796-1879). Secondary, 1848-1854.
     Hays was born in Philadelphia, took his degree from the University
of Pennsylvania in 1816, and an M.D. there in 1820; in subsequent practice
he specialized in diseases of the eyes and ears. His interest in natural
history led him to edit Wilson's ornithology and he also was interested in
geology. DAB; SFB.

Henry, Joseph (1797-1878). Primary, 1848--.
    Born in Albany, Henry was largely self-taught and his precocious
ability was clear in his appointments, first as professor of mathematics at
the Albany Academy, then as professor of natural philosophy at Princeton,
and finally Secretary of the Smithsonian Institution in 1846. His work on
the electromagnet resulted in the inductive unit "henry" being named for him.
At the Smithsonian Institution Henry was a tireless worker on numerous
government projects and a steady supporter of individual scientific research.
DAB; Thomas Coulson, Joseph Henry, His Life and Work (Princeton, 1950); the
Henry Papers are currently being edited by Nathan Reingold.

Herrick, Edward Claudius (1811-1862). Primary, 1848-1858.
    Weak eyesight prevented Herrick, born in New Haven, from taking his
degree at Yale. He worked on Silliman's Journal and became librarian of
Yale in 1843 and treasurer in 1852. His entomological, astronomical, and
meteorological observations, all avocational, were well received. DAB; SFB.

Hilgard, Eugene Woldemar (1833-1916). Secondary, 1857--.
    Born in Bavaria, Hilgard came to the United States in 1836. Leaving
the Midwest, he studied in Philadelphia and then went to Europe where he
took a Ph.D. from Heidelberg in 1853. On his return he directed the
Mississippi geological survey and taught at the University of Missouri. His
most important work was on the geology of the Mississippi Delta and on soil.
DAB; Typescript biography at the Illinois Historical Survey, Champaign-
Urbana; Jenny Hans, E. W. Hilgard and the Birth of Modern Soil Science
(Pisa, 1961).

Hilgard, Julius Erasmus (1825-1891). Primary, 1850--.
    Brother of Eugene, Julius also came with his family to America in
1836, studied civil engineering in Philadelphia, and then accepted a position
under Alexander Bache on the Coast Survey. There he worked on geodetic
reports and also actively promoted acceptance of a metric system. DAB:
NAS, Biog. Mem., III (1895); SFB.

Hilgard, Theodore Charles (1828-1875). Primary, 1854--.
    A brother of Julius and Eugene, Theodore was interested in botany
and worked with his cousin George Engelmann on classifications. He received an
M.D. from Wurzburg in 1852 and also studied at Heidelberg and Zurich. In
1853 he returned to the United States and in 1855 moved to St. Louis where
he established a medical practice until 1870. Popular Science Monthly, III
(March, 1898), 641, and LXXIV (February, 1909), 130-133.

Hill, Thomas (1818-1891). Primary, 1849--.
    Born in New Brunswick, New Jersey, Hill was briefly apprenticed but
graduated from Harvard in 1843. Interested in the relationship between
science and religion, he served a Congregational parish in Waltham while
lecturing on science. He was president of Antioch College, 1859-1862, and
of Harvard, 1862-1868. His major scientific interests were astronomy and
mathematics. DAB; William G. Hand, Thomas Hill, Twentieth President of Harvard
(Cambridge, Massachusetts, 1933); SFB.

Hirzel, Henri Heinrich (1783-1860). Secondary, 1850-1855.
    Joseph Henry read a paper on the numerical vibrations in music
submitted by Hirzel, whose address was Lausanne, Switzerland. British
Museum Catalogue presents the dates.

Hitchcock, Charles Henry (1836-1919). Primary, 1857--.
    Son of Amherst professor Edward Hitchcock, Charles worked with his
father on geological excursions and graduated from Amherst in 1856. While
studying theology at Yale and Andover, he assisted his father on the
geological survey of Vermont. From 1858 to 1866 he was curator of Amherst's
museum; after 1868 he taught at Dartmouth. DAB.

Hitchcock, Edward (1793-1864). Primary, 1848--.
    Born in Deerfield, Massachusetts, Hitchcock was early influenced by
Benjamin Silliman while a student at Yale's theological seminary. He taught
science, first chemistry and natural history and then geology and natural
theology, at Amherst College and was president there from 1845 to 1855.
Known for his work on the fossil footprints in the Connecticut red sandstone,
he conducted the first two geological surveys in Massachusetts and that in
Vermont from 1856 to 1861. DAB; SFB.

Hitchcock, Edward, Jr. (1828-1901). Primary, 1850-- (omitted in 1855).
Hitchcock received his degree from Amherst in 1849, while his father was president there, and received an M.D. from Harvard in 1853. He taught briefly and then went to England to study comparative anatomy. In 1861 he taught in the new Department of Hygiene and Physical Education and established important precedents in the teaching of physical education in the United States and also abroad. DAB; DSB; Edmund Welch, "Edward Hitchcock, M.D., Founder of Physical Education in the College Curriculum" (Ph.D. dissertation, East Carolina College, 1966); SFB.

Hodge, James Thatcher (1816-1871). Secondary, 1848-1851.
Hodge was born in Plymouth and graduated from Harvard in 1836. Interested in geology and mineralogy, he worked under Charles T. Jackson in Maine and Henry Rogers in Pennsylvania as well as on surveys in Ohio and New Hampshire. He was active as a mining engineer and promoted many mechanical inventions. CAB; NCAB.

Holbrook, John Edwards (1794-1871). Primary, 1848-1857 (inactive 1855-1857).
Born in Beaufort, South Carolina, Holbrook graduated from Brown in 1815 and took an M.D. from the University of Pennsylvania in 1818. A practicing physician in Charleston, he helped found the Medical College of South Carolina and taught anatomy there. His most important work was in zoology, especially reptiles and fishes. DAB; NAS, Biog. Mem., V (1905).

Holmes, Francis Simmons (1815-1892). Primary, 1850-1860 (inactive 1854-1858).
Apparently a Charleston businessman, Holmes was interested in geology and paleontology. In 1850 he became curator of the new Museum of Charleston and helped to found the Elliott Society of Natural History in 1853. He was professor of geology and paleontology in the College of Charleston and published on the Pleiocene and post-Pleiocene fossils of South Carolina. Thomas C. Johnson, Scientific Interests in the Old South (New York, 1936); pp. 68, 85, 92, 145-150; SFB.

Holton, Isaac Farwell (1812-1874). Secondary, 1855-1857.
Born in Westminster, Vermont, Holton graduated from Amherst in 1836 and from Union Theological Seminary in 1839. After serving as home missionary in Illinois, he returned to the East to teach at New Jersey School of Pharmacy, then at Rutgers, and, from 1856 to 1857, chemistry and natural history at Middlebury College. Thereafter he was a Presbyterian minister and a newspaper editor. AJS, CVII (June, 1874), 240; Biographical Record of the Alumni of Amherst College During its First Half Century, 1821-1871 (Amherst, 1883), p. 129.

Hopkins, William. Secondary, 1851-1858.
Hopkins graduated from Williams College and taught natural science at Genesee College, 1854 to 1860, subsequently teaching at a private school in Metuchin, New Jersey. He was interested in ornithology and paleontology. Alumni Record and General Catalogue of Syracuse University, 1872-'99, Including Genesee College, 1852-'71 and Geneva Medical College, 1835-'72 (Syracuse, 1899), p. 132; SFB.

Horsford, Eben Norton (1818-1893). Primary, 1848--.
Born in Moscow, New York, Horsford graduated from Rensselaer Polytechnic Institute in 1838 as a civil engineer. In 1844 he went to Germany and studied chemistry for two years with Justus Liebig and in 1847 became the Rumford Professor at Harvard, subsequently transferring to the Lawrence Scientific School. He pioneered in using the laboratory for teaching but resigned in 1863 to become an industrial chemist. DAB; DSB; Samuel Rezneck, "The European Education of an American Chemist and Its Influence in 19th-Century America: Eben Norton Horsford," Technology and Culture, XI (July, 1970), 366-388; SFB.

Hough, Franklin Benjamin (1822-1885). Primary, 1850-1860 (inactive 1851-1858).
Born in Martinsburg, New York, Hough graduated from Union College in 1843, taught and then received an M.D. from Western Reserve Medical College in 1848. Intermittently a practicing physician, he directed the New York census in Albany in 1854 and after the Civil War became actively involved in promoting forest conservation. DAB.

Howell, Robert. Secondary, 1851--.
A resident of Nichols, New York, Howell presented a single paper to the AAAS "On the wheat-fly and its ravages."

Hubbard, Joseph Stillman (1823-1863). Primary, 1848-1855.
        Graduating from Yale in 1843, Hubbard, born in New Haven, went to
Philadelphia to assist Sears C. Walker. In 1845 he was commissioned
professor of mathematics in the navy and assigned to the Naval Observatory
in Washington, where he helped organize a system of zone observations.
Whimsically he informed Baird that his scientific interest was in "star-
gazing and pen-chasing of comets." DAB; NAS, Biog. Mem., I (1877); SFB.

Hubbard, Oliver Payson (1809-1900). Secondary, 1848-- (inactive 1854-1856).
        Born in Pomfret, Connecticut, Hubbard graduated from Yale in 1828,
assisted Benjamin Silliman, his father-in-law, for several years and then took an
M.D. from South Carolina Medical College in 1837. He then taught chemistry,
mineralogy and geology at Dartmouth until 1866, thereafter teaching pharmacy
as well. His publications were in mineralogy and geology. CAB; Science,
New Series, XI (11 May 1900), 742-743; SFB.

Hunt, Edward Bissell (1822-1863). Primary, 1849--.
        Raised in Portage, New York, Hunt graduated from West Point in 1843,
was appointed to the Corps of Engineers, and worked on fortifications throughout New England. During the 1850s he also assisted the Coast Survey. An able
civil engineer, he was also interested in natural philosophy. AAA&S, Proc.,
VI (1862-1865), 301-302; NAS, Biog. Mem., III (1895); SFB.

Hunt, Thomas Sterry (1826-1892). Primary, 1848--.
        Born in Norwich, Connecticut, Hunt attended Yale and assisted
Benjamin Silliman, Jr., until appointed professor of geology at the Massachusetts
Institute of Technology, 1872 to 1878. He also taught chemistry at Lavel
University, Quebec, and at McGill, working especially on organic chemistry.
DAB; NAS, Biog. Mem., XV (1932).

Hurlburt, Jesse Beaufort (1812?-1891). Primary, 1857-1859.
        Raised in Upper Canada, Hurlburt graduated from Wesleyan in Middletown,
Connecticut, in 1835. He taught at an academy and from 1841 to 1847 was
professor of natural science at Victoria College, then opening an academy for
young women in Toronto. His residence at the time of membership was Hamilton,
Ontario. William S. Wallace, The Dictionary of Canadian Biography (Toronto,
1945).

Hyatt, James (1817-1904). Primary, 1856--.
        Hyatt was an active member of the New York Lyceum and founder of the
Torrey Botanical Club. He was a volunteer observer for the weather bureau
and professor of chemistry and toxicology in the New York Women's Medical
College. Science, New Series, XIX (15 April 1904), 635-636.

Jackson, Charles Thomas (1805-1880). Primary, 1848-- (inactive 1854-1856).
        After receiving an M.D. from Harvard in 1829, Jackson, born in
Plymouth, studied at the Sorbonne and the Ecole des Mines in Paris. Active
on surveys in Maine, Massachusetts, Rhode Island, and New Hampshire, he
became involved in several scientific controversies including priority
disputes with S. F. B. Morse on the electric telegraph and W. Morton on use
of ether for surgical anaesthesia. DAB.

Jackson, Samuel (1787-1872). Secondary, 1848-1859.
        Jackson, born in Philadelphia, took an M.D. from the University of
Pennsylvania in 1808 and became involved in his father's drug business. In
1821 he helped to found the Philadelphia College of Pharmacy and after 1827
taught "institutes of medicine" at the University of Pennsylvania. A popular
teacher, his publications were chiefly concerned with disease theory. DAB.

Jenks, John Whipple Potter (1819-1894). Secondary, 1849-1860.
        Born in West Boylston, Massachusetts, Jenks graduated from Brown in
1835 and returned to his home state to become principal of Peirce Academy,
1842 to 1871. Always interested in natural history, especially ornithology,
he was appointed curator of Brown's museum in 1873, combined with the chair of
agricultural zoology. CAB; Reuben Guild, Memorial Address on the late
John Whipple Potter Jenks . . . (Providence, 1895); SFB.

Jewett, Charles Coffin (1816-1868). Secondary, 1849-1857.
        A bibliographer and librarian, Jewett was born in Lebanon, Maine,
graduating from Brown in 1835 and from Andover in 1840. In 1848 he became
librarian at the Smithsonian Institution and collected statistics on American

libraries; when he sought to make the Smithsonian a research library, however, Henry removed him in 1854. From 1858 to 1868 he directed the new Public Library of Boston. DAB; SFB.

Jillson, Benjamin Cutler (1830-1899). Secondary, 1860--.
Born in Willimantic, Connecticut, Jillson attended Amherst in 1850, received a degree in science at Sheffield Scientific School in 1853 and later an M.D. from the University of Nashville. He subsequently became professor of zoology and geology at Cumberland University and also taught at the University of Nashville and the Western University of Pennsylvania. Amherst College, Biographical Record of the Graduates and Non-Graduates of the Classes of 1822-1962, Inclusive (Amherst, 1963), p.74.

Johnson, Walter Rogers (1794-1852). Primary, 1848-1852.
After graduating from Harvard in 1819, Johnson, born in Leominster, Massachusetts, taught for several years. He became affiliated with the Franklin Institute after 1826, then from 1839 to 1843 was professor of physics and chemistry at the University of Pennsylvania. He conducted research on coal for Congress and after 1848 affiliated with the Smithsonian Institution. CAB.

Johnston, John (1806-1879). Secondary, 1848--.
Born in Bristol, Maine, Johnston graduated from Bowdoin in 1832. After 1835 he was lecturer and then professor of natural sciences at Wesleyan in Middletown, Connecticut, and he published two chemical textbooks which went through three editions each. CAB; Nehemiah Cleaveland, History of Bowdoin College with Biographical Sketches of Its Graduates (Boston, 1882), pp. 432-433; SFB.

Jones, George (1800-1870). Primary, 1855--.
Jones was born near York, Pennsylvania, graduated from Yale, and then joined the navy. He helped to establish the United States Naval Academy and taught English and mathematics there. In 1851 he became the first chaplain in the navy when he joined Matthew Perry for the Japanese expedition. Interested in the zodiacal light, in 1856 he went to Quito, Ecuador, where he confirmed his theory that this astronomical phenomenon resulted from a nebulous ring around the earth. DAB.

Jones, William Louis (1827-   ). Secondary, 1850-1854.
Raised in Liberty County, Georgia, by a planter and botanist, Jones took his degree from the University of Georgia, then an M.D. from the New York College of Physicians and Surgeons, and attended Lawrence Scientific School, 1851-1852. From 1851-1872 he was professor of natural sciences at the University of Georgia and also helped establish the Georgia agricultural experiment station. NCAB; SFB.

Judd, Orange (1822-1892). Primary, 1850--.
Born near Niagara Falls, New York, Judd graduated from Wesleyan in 1847, taught and then studied agricultural chemistry at Yale, 1850-1853. With his knowledge of analytical and agricultural chemistry he moved to New York City where he joined in editing the American Agriculturist; his brief, practical articles rapidly increased the magazine's circulation. During the Civil War he worked with the Sanitary Commission but then returned to editing. DAB; William Ogilvie, Pioneer Agricultural Journalists (Chicago, 1927), pp. 29-37; SFB.

Keely, George Washington (-1878). Secondary, 1848--.
Keely was professor of mathematics and natural philosophy at Colby College from 1828 until he resigned in 1852 and was important in maintaining the college through difficult years in the 1830's. He was apparently interested in magnetic and meteorological observations. Third General Catalogue of Colby College, 1820-1908 (Waterville, Maine, 1909), p.12; Ernest C. Marriner, The History of Colby College (Waterville, Maine, 1963), pp. 93-98; SFB.

Kellogg, Orson. Primary, 1848-1854.
Kellogg, whose address was given simply as New York, expressed an interest in the geological sciences. SFB.

Kent, Edward N. Secondary, 1854-1859.
In 1855 Kent presented a summary of his separately published pamphlet on "Apparatus for Separating Gold and other Previous Metals from foreign substances," which listed its author as a chemist on the title page.

King, Henry. Primary, 1848-1854.
    King, interested in geology and natural history, had worked on geology
in Missouri before becoming curator of the National Institute collections in
1841. Possibly a minister and signing himself M.D., King seems to have returned to St. Louis in the 1850's and become involved in local copper mining.
A.Hunter Dupree, Science in the Federal Government; A History of Policies
and Activities to 1940 (Cambridge, Massachusetts, 1959), pp. 72-73; Henry
King, "Report on the Rives, Hench, Bleeding Hill and Blanton Copper Mines,
in the State of Missouri" (St. Louis, 1853); SFB.

Kirkwood, Daniel (1814-1895). Primary, 1853--.
    With only an academy education, Kirkwood, born in Harford County,
Maryland, taught mathematics at several schools in Pennsylvania. Shortly
after he presented his important paper on the rotation of primary planets (1849)
he was appointed to teach mathematics and astronomy at Delaware College in
Newark. He subsequently taught at Indiana University, 1856-1891, and then
at Stanford. DAB; Ronald Numbers, "The Nebular Hypothesis in American Thought"
(Ph.D. dissertation, University of California at Berkeley, 1969); SFB.

Kirtland, Jared (1793-1877). Primary, 1848-1856.
    Born in Wallingford, Connecticut, Kirtland studied at Yale and the
University of Pennsylvania before returning to Yale to take his M.D. in 1815.
He practiced in Connecticut and then moved to Ohio; in 1837 he moved to a
farm near Cleveland and taught at the Medical College of Ohio at Cincinnati.
In 1843 he became a founder of the Cleveland Medical College of Western
Reserve College and taught there until 1864. He was actively interested in
natural history, especially freshwater mollusks, helping found the local
scientific society. DAB; A. R. Gehr, "Jared Potter Kirtland, 1793-1877"
(M.A. thesis, Western Reserve University, 1950).

Kneeland, Samuel (1821-1888). Primary, 1848-1855.
    A lifetime resident of Boston, Kneeland received his A.B. in 1840 and
his M.D. in 1843 at Harvard. He was a practicing physician and briefly a
demonstrator in anatomy at Harvard. When Massachusetts Institute of Technology
opened, he became first professor of zoology and physiology. Active in
scientific circles in Boston and Cambridge, he was especially interested in
comparative anatomy and in ethnology. DAB; SFB.

Lachlan, Robert. Primary, 1857-1860.
    Lachlan seems to have left Montreal, Canada, to go to Cincinnati in
the mid-1850's. He published several papers on meteorology in the Canadian
Journal and presented a paper to the AAAS in 1858 advocating an organization
for doing observations. How Patriotic Services are rewarded in Canada.
Exemplified in the case of Major Lachlan, late of Montreal (Cincinnati, 1856).

Lane, Ebenezer (1793-1866). Secondary, 1851-1856.
    Born in Northampton, Massachusetts, Lane graduated from Harvard in
1811, was admitted to the bar in 1814, and practiced briefly in Connecticut.
He moved to Ohio where he practiced law and was a judge on the state supreme
court from 1837 to 1845. Active in encouraging railroads, he assessed his
scientific interest as a "general but superficial knowledge of everyday
novelties." CAB; SFB.

Lapham, Increase Allen (1811-1875). Primary,1849--.
    Born in Palmyra, New York, Lapham had only a common school education
when he began working on canal construction. He moved to Wisconsin in 1838
and while surveying for canals produced maps and a catalogue of plants in the
area near Milwaukee. Active in local educational activities, his later work
in natural history was criticized. He did, however, contribute to the
establishment of a national weather bureau and in 1873 was made state geologist.
DAB; Graham P. Hawks, "Increase A. Lapham: Wisconsin's First Scientist"
(Ph.D. dissertation, University of Wisconsin, 1960); SFB.

Lea, Isaac (1792-1886). Secondary, 1848--.
    Born in Wilmington, Delaware, Lea entered the wholesale and importing
business of his father and married into the publishing firm of Matthew Carey &
Sons. Avocationally he wrote nearly three hundred papers, many on his
specialty, the fresh-water mussels. He was active in Philadelphia scientific
societies but his single activity in the AAAS was presiding in 1860. DAB.

LeConte, John (1818-1891). Primary, 1850--.
          With his degree from Franklin College, Athens, Georgia, and an M.D.
from the College of Physicians and Surgeons in New York, LeConte left his
father's plantation in Liberty County, Georgia, to practice medicine in
Savannah. In 1846 he began teaching physics and chemistry at Franklin, then
moved to the College of Physicians and Surgeons in 1855, but in 1856 returned
South to teach physics in South Carolina College; in 1869 he went to the
University of California. LeConte, an intimate friend of Alexander Bache
and Benjamin Peirce, was especially interested in physics and in meteorology.
DAB; SFB.

LeConte, John Lawrence (1825-1883). Primary, 1848--.
          Raised in New York by a father interested in botany and zoology,
LeConte graduated from Mount Saint Mary's College in 1842 and took an M.D.
from the College of Physicians and Surgeons in New York in 1846. Freed from
the necessity of earning a living by inherited wealth, he studied the
geographic distribution of species and was preeminent among contemporary
entomologist. DAB; NAS, Biog. Mem., II (1886).

LeConte, Joseph (1823-1901). Primary, 1850--.
          LeConte, following his brother John's course, graduated from Franklin
College in 1841, then studied for an M.D. at the College of Physicians and
Surgeons. Interested in science, he made a western trip with his cousin John
Lawrence LeConte and graduated from Lawrence Scientific School in 1851. He
returned to teach all sciences at Oglethorpe University and then at the
University of Georgia; in 1857 he accepted the chair of geology at the
University of South Carolina, Columbia, and in 1869 went to the University of
California. He was interested in natural history and also wrote on the
relationship between science and religion. DAB: Autobiography of Joseph
LeConte, ed. William Armes (New York, 1903); SFB.

Lefroy, John Henry (1817-1890). Primary, 1851-1854.
          Born in Hampshire, England, and educated at the Royal Military Academy,
Lefroy worked with the magnetic survey in Canada from 1839 to 1854 and helped
to found the Canadian Institute in 1849. He returned to England and after
his retirement from the army in 1870 he served as governor of Bermuda and of
Tasmania. DNB.

Leidy, Joseph (1823-1891). Primary, 1853-1860 (inactive 1855-1858).
          A lifetime Philadelphia resident, Leidy received a classical education
and in 1844 received an M.D. from the University of Pennsylvania. He traveled
in Europe, practiced medicine, and in 1853 became professor of anatomy at
the University of Pennsylvania. Publishing relatively little, he was interested
in paleontology but did his best work in comparative anatomy. DAB; NAS, Biog.
Mem., VII (1913), SFB.

Lesley, (J.) Peter (1819-1903). Primary, 1848--.
          Born in Philadelphia, Lesley took his degree from the University of
Pennsylvania in 1838 and worked for Henry Rogers on the state geological
survey. In 1842 he went to Princeton Theological Seminary and then to Halle,
returning to work for the American Tract Society and to serve a Congregational
church. After 1852 he returned to science, working again on the state survey
and becoming professor of mining at the University of Pennsylvania; from 1873
to 1887 he was state geologist. DAB; Life and Letters of Peter and Susan
Lesley, ed. Mary Lesley Ames (New York, 1909).

Lieber, Oscar Montgomery (1830-1862). Primary, 1854--.
          Born in Boston to Francis Lieber, Oscar Lieber studied at South
Carolina College, at Berlin and at Gottingen. In 1851 he was appointed
assistant professor of geology at the University of Mississippi where he
also worked on the geological survey. Later he accompanied Michael Tuomey
in Alabama and joined the United States Astronomical Expedition in 1860 as
meteorologist; his early promise in petrography ended with his death in the
Civil War. NCAB.

Linck, Christian, Primary, 1849-1854.
          Linck, who signed himself as Dr. and as an assistant at the Cambridge
laboratory in 1849, presented three papers that year on chemistry.

Logan, William Edmond (1798-1875. Primary, 1848--.
          After taking his degree from the University of Edinburgh in 1817
Logan worked briefly in London and then with mining in Wales. In 1842 he

returned home to Montreal, where he headed the geological survey in Canada until 1870 and was a leader in the intellectual community there. Dictionary of National Biography; Bernard J. Harrington, Life of William E. Logan (Montreal, 1883); SFB.

Loomis, Elias (1811-1889). Primary, 1848--.
     Born in Willington, Connecticut, Loomis graduated from Yale in 1830 and briefly attended Andover before returning to teach at Yale. Interested in astronomy and in magnetic variations, he taught at Western Reserve University, 1837 to 1844, then taught mathematics and natural philosophy at the University of the College of New York, and finally succeeded Denison Olmsted at Yale in 1859. DAB; SFB.

Loomis, Silas Lawrence (1822-1896). Secondary, 1853--.
     After taking his degree at Wesleyan in 1844, Loomis, born in Coventry, Connecticut, taught for several years. In 1856 he received an M.D. from Georgetown University and the following year became an astronomer for the Coast Survey. From 1859 to 1867 he was on the Georgetown faculty and later taught physiology at Harvard. CAB; SFB.

Lovering, Joseph (1813-1892). Primary, 1849-- (inactive 1851-1854).
     Lovering spent his life in Cambridge, Massachusetts, graduating from Harvard in 1833 and remaining there until his appointment as Hollis Professor of Natural Science. A respected teacher of mathematics, his strength was in his systematic and careful work for the Coast Survey, as Permanent Secretary of the AAAS from 1854 to 1873, and in the American Academy of Arts and Sciences. DAB; NAS, Biog. Mem., VI (1909).

Lyman, Chester Smith (1814-1890). Secondary, 1850--.
     Born in Manchester, Connecticut, Lyman graduated from Yale in 1837 and then attended Union and Andover theological seminaries. After 1845 he traveled in Hawaii and California but returned to New Haven in 1850. After 1859 he became professor of industrial mechanics and physics at Sheffield Scientific School at Yale and invented and improved several astronomical instruments. DAB.

Lynch, Patrick Neeson (1817-1890). Secondary, 1850--.
     Irish by birth, Lynch emigrated with his parents as an infant. Ordained in Rome in 1840, he was pastor in Charleston after 1845 and in 1858 became bishop there. A popular and successful clergyman, he also taught theology and edited the United States Catholic Miscellany. DAB.

McCay, Charles Francis (1810-1889). Primary, 1850-1859.
     McCay was born in Danville, Pennsylvania, and graduated from Jefferson College in 1829. From 1833 to 1853 he taught mathematics and natural philosophy at the University of Georgia and then for two years at South Carolina College. After serving the latter as president, he left to pursue his interest in mathematics and statistics as a banker and an insurance actuary. DAB; SFB.

McChesney, Joseph Henry (-1876). Secondary, 1857--.
     While a resident of Springfield, McChesney was an assistant in the Illinois state geological survey from 1858 to 1860, an officer in the Natural History Society of Illinois, and later professor of chemistry, mineralogy, geology and agriculture at the first University of Chicago. In the 1860s he was a member of the diplomatic corps in England. AJS, CLII (December, 1876), 244; John Moses and Joseph Kirkland, A History of Chicago (Chicago, 1893), II, 32.

McCulloch, Richard Sears (1818-1894). Secondary, 1848-1859.
     Baltimore-born McCulloch graduated from Princeton in 1836 and then studied chemistry with James C. Booth in Philadelphia. After teaching and working at the Philadelphia mint, he was appointed professor of natural philosophy at Princeton to 1854 and then at Columbia College in New York. In sympathy with the South, he became a consulting chemist for the Confederacy in 1863 and after the war taught in that section. Milton H. Thomas, "Professor McCulloch of Princeton, Columbia and Points South," Princeton University Library Chronicle, IX (November, 1947), 17-29; SFB.

Mallet, John William (1832-1912). Primary, 1856--.
     Irish by birth, Mallet graduated from Trinity College in Dublin and took a Ph.D. at Gottingen under Friedrich Wohler. Coming to America in 1853,

he assisted at Amherst College and then went to Tuscaloosa, Alabama, where he assisted in the state geological survey and was professor of chemistry at the University of Alabama, 1855-1860. He made an important study on cotton which was published by the Royal Society, served as general superintendent of Confederate ordnance laboratories, and then returned to teaching. DAB.

Mantell, Reginald Neville (1827-1857). Primary, 1849-1855.
    The English-born son of Gideon Mantell, Reginald studied law and then became interested in engineering and in geology. He visited the United States in 1849 and the early 1850s, then went to India on an engineering project, and died there of cholera. E. Cecil Curwin, ed., The Journal of Gideon Mantell, Surgeon and Geologist, Covering the Years 1818-1852 (London, 1940), passim; SFB.

Marcy, Oliver (1820-1899). Secondary, 1856--.
    Born in Colrain, Massachusetts, Marcy worked his way through Wesleyan in 1846 to become teacher of mathematics and geology at Wilbraham Academy. In 1862 he accepted the chair of natural science at Northwestern and also served briefly as president. His geological publications frequently had religious overtones. American Geologist, XXIV (December, 1899), 67-72.

Mathews, Joseph W. (1812-1862). Secondary, 1849-1851.
    Born in Huntsville, Alabama, Mathews came to Mississippi on a government survey and decided to stay, settling on a plantation near Salem. He was active in both legislative houses and in 1850 was elected governor of the state. A colonel in the militia, he died on his way to Richmond in 1862. Rowland Dunbar, Mississippi. Comprising Sketches of Counties, Towns, Events, Institutions, and Persons, Arranged in Cyclopedic Form (Atlanta, 1907), p.178.

Maury, Matthew Fontaine (1806-1873). Primary, 1848-1876.
    Born in Fredericksburg, Virginia, and raised in Franklin, Tennessee, Maury joined the navy in 1825. His work in navigation and astronomy led to his appointment as superintendent of the Depot of Charts and Instruments in 1842, which included the new Naval Observatory; there he concentrated on hydrography and meteorology rather than on astronomy. A capable promoter, his influence with the federal government and in commercial circles contributed to American naval exploration in the 1850s, although it alienated several important Washington scientists. DAB; Edward L. Towle, "Science, Commerce and the Navy on the Seafaring Frontier (1842-1861)--The Role of Lieutenant M. F. Maury and the U. S. Navy Hydrographic Office in Naval Exploration, Commercial Expansion and Oceanography before the Civil War" (Ph.D. dissertation, University of Rochester, 1966); SFB.

Meech, Levi Witter (1821-1912). Primary, 1854--.
    Meech was born in Connecticut, graduated from Brown in 1845, and was a sometime resident in Preston, Connecticut. From 1850 to 1860 he was a mathematical assistant in the Census Office and in 1853 also worked on astronomical computations for the Coast Survey. After the Civil War he was an insurance actuary. Historical Catalogue of Brown University, 1764-1904 (Providence, 1905), p. 203.

Michel, William Middleton (1822-1894). Secondary, 1849-1854.
    Although born in Charleston, Michel studied in Paris and graduated from Ecole de Medicine, Paris, 1845, returning to take an M.D. from the Medical College of South Carolina. A practicing physician, he opened a Summer Medical Institute in 1847 and was also active as a medical editor. After the Civil War he was professor of physiology and histology at the Medical College in Charleston. DAB.

Mitchel, Ormsby MacKnight (1809-1862). Primary, 1849-1859.
    Born in Kentucky and raised in Ohio, Mitchel graduated from West Point in 1829, then taught mathematics. In 1831 he moved back to Cincinnati, was admitted to the bar, and in 1835 became professor of mathematics at Cincinnati College. His popular lectures on astronomy led to the establishment of the Cincinnati Observatory; he later directed the Dudley Observatory at Albany. DAB; Frederick A. Mitchel, Ormsby M. Mitchel, Astronomer and General, A Biographical Narrative (Boston, 1887).

Mitchell, William (1u91-1869). Primary, 1849--.
    A businessman and teacher on Nantucket Island, Mitchell was an avid astronomer. He used his several telescopes to observe star positions for the Coast Survey and for the Nantucket whaling fleet. Together with his daughter Maria, he was a close friend of the Bonds at the Harvard Observatory and was a fellow of the American Academy of Arts and Sciences. DAB; SFB.

Morgan, Lewis Henry (1818-1881). Primary, 1856--.
        Morgan was born in Aurora, graduated from Union College in 1840, and
was admitted to the bar in New York in 1844. Moving to Rochester, he
practiced law but became fascinated with the local Iroquois Indians and
researched a number of important works relating to American Indian anthropology and history. DAB; Carl Resek, Lewis Henry Morgan: American Scholar
(Chicago, 1960).

Morris, John Gottleib (1803-1895). Secondary, 1848-- (inactive 1854-1858).
        Morris was born in York, Pennsylvania, graduated from Dickinson College
in 1823, and then studied at Princeton and Gettysburg theological seminaries.
From 1827 to 1860 he was a Lutheran minister in Baltimore. An interested
amateur in natural history, he did excellent work in teomology. DAB; SFB.

Morris, Oran Wilkinson (1798-1877). Secondary, 1848-1854.
        Born in Ames, New York, Morris attended Albany Academy and from 1833
to 1869 was principal at the New York Institute for the Deaf and Dumb. He
was a meteorological observer for the Smithsonian Institution and also
interested in mineralogy and conchology. SFB; Torrey Botanical Club,
Bulletin, VI (1877), 166-168.

Morris, Robert (1818-1888). Secondary, 1849-1854.
        Morris, born near Boston, settled in Oxford, Mississippi, as a
principal of Mount Sulvan Academy. He later moved to Kentucky where he was
president of the Masonic College at LaGrange; his writings on Masonry are
most remembered, but he was also interested in geology, numismatics, and
archeology. DAB; Beulah Malone, He Belongs to the Ages (Tulsa, Oklahoma,
1967).

Morton, Samuel George (1795-1851). Primary, 1848-1851.
        A lifetime Philadelphia resident, Morton received an M.D. from the
University of Pennsylvania in 1820 and from the University of Edinburgh in
1823. An extensive correspondent, he was a clearing house for new material
in anatomy and anthropology, teaching the former at the University of
Pennsylvania from 1839 to 1843. He was an active member of the Philadelphia
scientific societies but evidenced little interest in the AAAS. DAB; Patsy
Ann Gerstner, "The 'Philadelphia School' of Paleontology: 1820-1854"
(Ph.D. dissertation, Case Western Reserve University, 1967), pp. 23 ff.

Moultrie, James (1793-1869). Primary, 1848-1854.
        Born in Charleston, Moultrie received an M.D. from the University
of Pennsylvania and returned to practice at home. He was active in the state
medical society and later in the American Medical Association. Interested in
natural history, he also taught physiology at the Medical College of South
Carolina, 1833 to 1867. CAB.

Napoli, Raphael. Secondary, 1854-1855.
        Listing his residence as Naples, Italy, Napoli gave a single paper
on chemistry before the AAAS in 1854.

Newberry, John Strong (1822-1892). Primary, 1851--.
        Newberry was born in Windsor, Connecticut, but raised in Ohio where
he graduated from Western Reserve College in 1846. After studying medicine
at the Cleveland Medical School in Paris, he returned to accompany a
number of government surveys in the West. An old-style naturalist, he did
his best work on fossil plants; after 1866 he was professor of geology and
paleontology at the School of Mines, Columbia. DAB.

Newcomb, Simon (1835-1909). Secondary, 1859--.
        Born in Nova Scotia, Newcomb ran away from a medical apprenticeship
and through Joseph Henry secured a position with the Nautical Almanac office
and graduated from Lawrence Scientific School in 1858. After 1861 he was
commissioned professor of mathematics in the navy and worked for the Naval
Observatory and the Nautical Almanac until his retirement. Newcomb made
important calculations relating to the moon and planets, and with A.A.
Michelson determined the velocity of light. DAB; Simon Newcomb, Reminiscences
of an Astronomer (Boston, 1903).

Newton, Hubert Anson (1830-1896). Secondary, 1851--.
        Newton was born in Sherburne, New York, graduated from Yale in 1850,
and was appointed professor of mathematics at Yale in 1853. In 1855 he was

granted a leave to study at Paris under geometer Michel Chasles. On his
return to Yale he concentrated also on astronomy, especially meteors, and
was a staunch advocate of the metric system. DAB; NAS, Biog. Mem., IV (1902);
SFB.

Norton, John Pitkin (1822-1852). Primary, 1848-1852.
    Born in Albany, Norton intended to follow his father's occupation of
farming but first studied at Yale under Silliman, then with James Johnston at
Edinburgh, and finally with Gerard Mulder at Utrecht. He returned to Yale
as professor of agricultural chemistry and initiated that study in Sheffield
Scientific School; he also strongly supported the movement for a national
university at Albany. An active publicist of practical chemistry, his early
death was a loss to the new field. DAB; Norton MSS recently acquired by YUA; SFB.

Norton, William Augustus (1810-1883). Primary, 1851--.
    Norton was born in East Bloomfield, New York, and graduated from West
Point in 1831. He taught at various schools before accepting the chair of
civil engineering at Sheffield Scientific School in 1852. His major publications were in molecular physics, terrestrial magnetism, and astronomy.
DAB; SFB.

Nott, Josiah Clark (1804-1873). Secondary, 1850-1859.
    Nott graduated from South Carolina College in 1824 and received an M.D.
from the University of Pennsylvania in 1827, later studying in Paris. In
1836 he moved to Mobile, Alabama, where he practiced medicine, started a
local medical society, and in 1858 helped organize the Medical College of
Alabama. His best work was on disease theory, especially cholera; his
ethnological papers were largely polemical. DAB.

Oliver, James Edward (1829-1895). Secondary, 1849-1854.
    Born in Portland, Maine, Oliver took a degree from Harvard in 1849
and accepted a position as mathematician for the Nautical Almanac. He
remained there until 1871 when he was names assistant professor of mathematics
at Cornell, and in 1873, full professor. CAB; NAS, Biog. Mem., IV (1896).

Olmsted, Denison (1791-1859). Primary, 1848-1859.
    Olmsted was born near East Hartford, Connecticut, and took his degree
from Yale in 1813. In 1817 he became professor of chemistry at the University
of North Carolina and was state geologist and mineralogist from 1822 to 1825.
Then he returned to Yale as professor of mathematics and natural philosophy;
his most important work was in astronomy and meteorology. DAB; SFB.

Olmsted, Alexander Fisher (1822-1853). Secondary, 1850-1853.
    Born at Chapel Hill, North Carolina, a son of Denison Olmsted,
Alexander graduated from Yale in 1844 and was called to the chair of chemistry
at the University of North Carolina. He published a textbook on chemistry
before his early death. CAB; NCAB.

Osborn, A. Secondary, 1848-1859.
    Osborn listed his residence first as Albany and later as Herkimer,
New York. Probably a farmer, he was interested in geology and meteorology
and published Field Notes of Geology (New York, 1858); SFB.

Owen, David Dale (1807-1860). Primary, 1848-1854.
    Owen was born at his father's experimental community of New Lanark,
Scotland, studied in Switzerland and London, but finally took his degree from
Ohio Medical College of Cincinnati in 1836. Thereafter he worked out of New
Harmony as a geologist, first on the state surveys of Tennessee and Indiana,
then on the federal survey in Wisconsin and Iowa, and finally from 1847 to
1852 directed the survey of the Chippewa Land District. He later worked on
surveys in Kentucky, Arkansas, and Indiana. DAB; Walter B. Hendrickson,
David Dale Owen: Pioneer Geologist of the Middle West (Indianapolis, 1943);
SFB.

Page, Charles Grafton (1812-1868). Secondary, 1848-1854.
    Born in Salem, Massachusetts, Page graduated from Harvard in 1832,
studied medicine and opened a practice in Salem. Becoming interested in
electricity, Page improved Joseph Henry's calorimeter and devised a new
self-acting circuit breaker. In 1841 he accepted a position at the Patent
Office, which he held to 1852 and again after 1862. His lectures on
electromagnetism in Washington resulted in a year's appropriation from
Congress to continue his investigation. DAB.

Park, Roswell (1807-1869). Secondary, 1848-1854.
       Park was born in Lebanon, Connecticut, graduated from West Point in
1831, resigned his commission in the corps of engineers in 1836, and taught
chemistry and natural philosophy at the University of Pennsylvania. From
1842 to 1852 he entered the Episcopal ministry but in the latter year went
to Europe to study and returned to head the new Racine College in Wisconsin.
His "general interest in all science" was expressed in his continuing efforts
to expand collegiate and preparatory school curricula to include more
science. DAB; SFB.

Parker, Samuel (1779-1866). Secondary, 1848-1851.
       Born in Ashfield, Massachusetts, Parker graduated from Williams
College in 1806 and from Andover in 1810. He is chiefly remembered as one of
the missionaries joining Marcus Whitman in the trek to Oregon in 1835. When
Parker returned east in 1837, he settled in New York where he taught at a
girls' seminary. DAB; SFB.

Parry, Charles Christopher (1803-1890). Secondary, 1851-1860.
       English-born, Parry came with his family to American where he took a
degree from Union College in 1842 and an M.D. from Columbia in 1846. While
practicing medicine in Davenport, Iowa, he worked on surveys under David Dale
Owen and made numerous personal expeditions to enlarge existing knowledge of
botany in the West. After the Civil War he was briefly in Washington as a
botanist for the Department of Agriculture and at the Smithsonian Institution.
DAB.

Peirce, Benjamin (1809-1880). Primary, 1848--.
       Born in Salem, Massachusetts, Peirce graduated from Harvard, taught
school, and in 1832 received the appointment as Perkins Professor of
Mathematics and Astronomy at Harvard. He consulted with the Nautical Almanac
and the Coast Survey, heading the latter from 1869 to 1874. Generally
acknowledged as the leading mathematician of his generation, Peirce was a
pivotal member of the Lazzaroni but his temperament involved him in squabbles
within the Boston intellectual community. DAB; Victor F. Lenzen, Benjamin
Peirce and the U.S. Coast Survey (San Francisco, 1969); SFB.

Perkins, George Roberts (1812-1876). Secondary, 1848--.
       Largely self-taught, Perkins was born in Otsego County, New York,
and worked on a river survey before becoming a mathematics teacher at Utica
Academy in 1831 and later at the new normal school there. After 1852 he
superintended the construction of the Dudley Observatory in Albany. An active
educator, he wrote several textbooks on mathematics and astronomy. CAB.

Peters, Christian Henry Frederick (1813-1890). Primary, 1854--.
       Born in Schleswig, Peters received a Ph.D. at Berlin in 1836 and also
studied at Gottingen under Karl Gauss. In 1854 he arrived in the United
States and was appointed to the Coast Survey, stationed first at Cambridge
and then at Albany; after 1858 he directed the Hamilton College Observatory
and in 1867 was appointed professor of astronomy there. His special interest
was solar observation and he also undertook preparation of a new edition of
Ptolemy's Almagest. DAB.

Phillips, John S. Secondary, 1848-1850.
       Phillips, a resident of Philadelphia, was appointed to the committee
on the Coast Survey in 1854, after his membership had lapsed.

Piggot, Aaron Snowden (1822-1869). Secondary, 1856-1860.
       Born in Philadelphia, Piggot was professor of anatomy and physics at
the Washington Medical College and of chemistry at Maryland Institute,
Baltimore; he published an early text on chemistry and metallurgy as applied
to dental surgery. Samuel Allibone, A Critical Dictionary of English
Literature and British and American Authors Living and Deceased from the
Earliest Accounts to the Latter Half of the Nineteenth Century, II
(Philadelphia, 1872), 1594.

Poole, Henry Ward (1826). Primary, 1860--.
       Poole was a professor at the National College of Mines in Mexico and
received an honorary M.A. from Yale in 1875. He was particularly interested
in the mathematical laws of musical tones. Catalogue of the Officers and
Graduates of Yale University, 1701-1899 (New Haven,1898), p.260; "William
F. Poole," NCAB.

Pourtales, Louis Francois de (1823-1880). Primary, 1848--.
        Swiss-born, Pourtales studied under and then accompanied Louis
Agassiz to the United States, 1846. In 1848 he joined the Coast Survey, where
he remained until he went to the Museum of Comparative Zoology at Harvard in
1873. His most important work was on deep sea life, especially the corals.
DAB; SFB.

Pugh, Evan (1828-1864). Primary, 1860--.
        Educated in Jordan Banks, Pennsylvania, Pugh then studied in New York
before going to Europe to study first at Leipzig and then at Gottingen, where
he received his Ph.D. in 1856 in chemistry. He next worked in England on
plant growth until called to head the new Agricultural College of Pennsylvania
in Centre County; he also headed the laboratory for chemistry and mineralogy
until his early death. DAB.

Rainey, Thomas. Primary, 1850-1854.
        Rainey, active in organizing the Cincinnati meeting, edited the only
two volumes of the Western Review, 1850-1852, and published an account of his
new improved abacus in 1849. In 1858 he published a treatise arguing for
governmental support of mail lines as a public service.

Ramsay, Andrew Crombie (1814-1891). Primary, 1857-1859.
        Born in Scotland, Ramsay was employed as a clerk when Roderick
Muchison invited him to assist on the British Geological Survey. In 1845
he became local director of the survey in Wales and in 1847 professor of
geology at University College, London. A popular lecturer, he visited North
America in 1857 and reported on the survey and on his interest in stratigraphy
at the Montreal meeting of the AAAS. DNB.

Ravenel, Henry William (1814-1887). Primary, 1850-1854.
        Ravenel was born in Berkeley County, South Carolina, and graduated from
the state college in 1832 but was dissuaded from attending medical school by
his father. A successful planter prior to the Civil War, he was avocationally
interested in botany, especially fungi, and participated in Charleston scientific societies. DAB; Arney R. Childs, ed., The Private Journal of Henry
William Ravenel, 1859-1887 (Columbia, South Carolina, 1947), xiii-xxi.

Ravenel, St. Julian (1819-1882). Primary, 1848-1854.
        Born in Charleston, Ravenel graduated from Charleston Medical College
in 1840 and then studied in Philadelphia and Paris. Dissatisfied with general
practice, he became interested in agricultural chemistry and in 1857 established
the first limestone works in the state; after 1860 he worked with phosphate
fertilizers and experimented with the nutritive values of legumes. DAB.

Ray, Joseph (1807-1865). Secondary, 1850-1854.
        Born in present-day West Virginia, Ray was educated in Ohio and
graduated from Ohio Medical College in 1829. Opening a practice in
Cincinnati, he decided instead to teach mathematics at Woodward High School
and in 1851 was named principal. A close associate of William McGuffey, he
wrote several popular mathematical texts. Jerry Dennis, "Joseph Ray,"
Ohio State Archaeological and Historical Quarterly, LXVI (January, 1937),
42-50; SFB.

Redfield, William Cox (1789-1857). Primary, 1848-1857.
        Redfield was born in Middletown, Connecticut, and apprenticed to a
saddler after his father died. Becoming interested in storm patterns about
1821, Redfield's subsequent inventigations in meteorology, and to a lesser
degree in geology, resulted in his election as president of the AAAS for 1848.
As eclectic as he was enthusiastic, Redfield was also involved in promoting
railroad and steamship activities in New York. DAB.

Reed, Stephen (1801-1877). Secondary, 1848-1854.
        Born in Cornwall, Connecticut, and a graduate of Yale in 1824, Reed
taught school and studied medicine before moving to Pittsfield, Massachusetts,
in 1848, where he opened his agricultural warehouse and seed store; he also
edited a weekly paper, "The Culturalist and Gazette." After his retirement
in 1858 he spent his leisure studying geology. Obituary Records of the Graduates
of Yale College . . . 1877, 2nd series, No. 37 (1877), 290; SFB.

Reid, David Boswell (1805-1863). Secondary, 1856-1859.
        Scottish-born, Reid received an M.D. from the University of Edinburgh
in 1830. He taught chemistry privately in Edinburgh until his work on

ventilation resulted in his appointment to superintend the ventilation in the
new Houses of Parliament in London. In 1855 he came to the United States where
he lectured on ventilation, taught briefly in Wisconsin, and finally moved to
Washington, D.C. DAB.

Riddell, John Leonard (1807-1865). Primary, 1848-1858.
    Born in Leyden, Massachusetts, Riddell graduated from Rensselaer
Polytechnic Institute in 1829, taught in Cincinnati, and received an M.D. from
the medical college there in 1836. Moving to New Orleans as professor of
chemistry in that year, he demonstrated his broad interests by completing a
study of western flora, doing intensive research on yellow fever, and
designing a pioneer binocular microscope. DAB; SFB.

Ritchie, E. S. Secondary, 1856--.
    A Boston resident, Ritchie presented a paper on induction before the
AAAS in 1857.

Robb, James (1815-1861). Primary, 1850--.
    With an M.D. from the University of Edinburgh, Robb was professor of
chemistry and natural science at the University of New Brunswick, Fredericton,
from 1837 until his death. He was especially interested in geology and his
local collections were the beginning of the University museum. AJS, LXXXII
(December, 1861), 150 and 344; SFB.

Roberts, Solomon White (1811-1882). Primary, 1848-1854.
    A lifetime Philadelphia resident, Roberts attended Quaker schools
before joining his uncle's construction firm as an engineer. Returning from a
business trip to England in 1839, he implemented his observations there by
constructing the first anthracite coal furnace in the Lehigh Valley. Active
on the governing boards of several railroads, he was also an active member of
the leading intellectual and social groups of Philadelphia. DAB.

Rogers, Henry Darwin (1808-1866). Primary, 1848-1854.
    Born in Philadelphia, Rogers was taught largely by his father and
taught himself. Intrigued with the utopian ideas of Robert Dale Owen, Rogers
followed him to London where his attention turned to geology; he returned to
the United States to head the New Jersey and Pennsylvania geological surveys
and to teach geology and mineralogy at the University of Pennsylvania. After
moving to Boston in the 1840s and failing to receive a Harvard chair, Rogers
returned to Britain permanently in 1855 where he taught natural history at
the University of Glasgow until his death. DAB; SFB.

Rogers, James Blythe (1802-1852). Secondary, 1848-1852.
    Born in Philadelphia, James was the eldest of the four Rogers brothers
discussed here. After a year's study at William and Mary College, he took an
M.D. from the University of Pennsylvania and practiced medicine for four years;
he subsequently worked as an industrial chemist, participated in his brothers'
geological surveys, and taught at various colleges, the last being the University of Pennsylvania, 1847-1852. DAB; Edgar F. Smith, James Blythe Rogers
1802-1852, Chemist (Philadelphia, 1927).

Rogers, Robert Empie (1813-1884). Primary, 1848-1858.
    Rogers was born in Baltimore and took his M.D. from the University of
Pennsylvania. After working in Robert Hare's chemical laboratory and on the
Pennsylvania survey, in 1842 he became professor of chemistry at the Uni-
versity of Virginia and in 1852 succeeded his brother at the University of
Pennsylvania. His later efforts in applied chemistry in the oil industry
resulted in financial losses; after 1877 he taught chemistry at Jefferson
Medical College. DAB; NAS, Biog. Mem., IV (1905).

Rogers, William Barton (1804-1882). Primary, 1848--.
    Born in Philadelphia, Rogers was educated at William and Mary College
and taught there, 1827-1835, until he was appointed professor of natural
science at the University of Virginia and also headed the Virginia Geological
Survey, the latter until 1842. In 1853 he moved to Boston where he worked for
the establishment of a polytechnic school; his aspirations were realized with
his appointment as head of the new Massachusetts Institute of Technology in
1861. DAB; Emma Rogers, ed., Life and Letters of William Barton Rogers,
2 vols. (Boston, 1896).

Rood, Ogden Nicholas (1831-1902). Secondary, 1853-- (inactive 1856-1859).
Rood was born in Danbury, Connecticut, and graduated from Rutgers in
1852; for two years he studied at Sheffield Scientific School and thereafter
to 1858 he was in Berlin and Munich studying chemistry and physics. Returning
to teach at a small denominational school in Troy, New York, his contributions
to light polarization resulted in his appointment as professor of physics in
1864 at Columbia, where he did excellent work in optics and on instrumentation.
DAB; SFB.

Roosevelt, Clinton (1804-1898). Secondary, 1857--.
Roosevelt was a leading theorist of the Loco-foco movement and he
urged free banking, no small paper money and a lower tariff. In 1835 he was
elected to the New York legislature. Joseph Dorfman, The Economic Mind in
American Civilization, 1606-1918, II (New York, 1948), 652-653.

Ruggles, Samuel Buckley (1800-1881). Secondary, 1851-1855.
Ruggles was born in Milford, Connecticut, graduated from Yale in
1814, and was admitted to the New York bar in 1821. Becoming an active civic
and political leader, he served in the New York legislature and supported canal
and railroad building. His interest in the Albany university scheme of the
early 1850s and his support of Wolcott Gibbs as a Trustee of Columbia won him
the approval of the scientific community; he was a delegate to two international statistical conferences in 1863 and in 1868. DAB; D.G.Brinton
Thompson, Ruggles of New York: A Life of Samuel B. Ruggles (New York, 1946).

Runkle, John Daniel (1822-1902). Primary, 1849--.
Born in New York City, Runkle graduated from Lawrence Scientific
School in 1851. He held various positions with the Nautical Almanac after
1849 and worked on tables of planetary motion. Interested in the new
Massachusetts Institute of Technology, he became professor of mathematics
there, 1865 to 1868 and 1880 to 1902, serving as president in the interim period.
DAB; SFB.

Saemann, Louis (-1866). Secondary, 1850-1855.
Saemann, who gave his address as Berlin, Prussia, was proprietor of
a Paris shop which specialized in the sale of minerals and fossils; he himself
was interested in mineralogy, geology, and paleontology. He visited the
United States in 1847 for about a year. AJS, XCII (December, 1866), 435.

Safford, James Merrill (1822-1907). Primary, 1854-- (inactive 1855-1857).
Born in Zanesville, Ohio, Safford graduated from Ohio University in
1844 and then attended Yale. From 1847 to 1872 he taught chemistry and
natural sciences at Cumberland University, Lebanon, Tennessee, later at the
University of Nashville and at Vanderbilt University. As state geologist,
1854-1860 and 1871-1900, he pioneered investigation of the state. DAB; SFB.

Safford, Truman Henry (1836-1901). Primary, 1859--.
Safford was born in Royalton, Vermont, and graduated from Harvard in
1854, where he remained on the staff of the Harvard Observatory to 1866 and
edited two volumes of its Annals. He subsequently taught at the old
University of Chicago and at Williams College. His astronomical work was in
the older tradition, that is, the computation of positions and orbits of
heavenly bodies. DAB.

St. John, Samuel (1813-1876). Primary, 1848-1854.
Born in New Canaan, Connecticut, St. John headed the 1844 class at
Yale, studied law, and then traveled to Europe. From 1838 to 1851 he was
professor of chemistry, geology, and mineralogy in Western Reserve College,
Hudson, Ohio, then taught at Cleveland Medical College, and after 1857 took
the chair of chemistry at the New York College of Physicians and Surgeons.
K&B, Med. Biog.

Salisbury, James Henry (1823-1905). Primary, 1849-1854.
Salisbury, born in Scott, New York, took a bachelor's degree in
natural science from Albany Medical College in 1850. As chemist on the
New York survey, after 1846 he worked on plant pathology. Later he settled
in Ohio, where he practiced medicine, taught, and worked on the germ theory
of disease. DAB.

Sanford, Solomon N. Secondary, 1851-1855.
Sanford headed the Granville Episcopal Female Seminary from 1848 to

1857. A science teacher, he was especially interested in chemistry and natural science. Henry Bushnell, The History of Granville, Licking County, Ohio (Columbus, Ohio, 1889), p. 254; SFB.

Saxton, Joseph (1799-1873). Secondary, 1848-1859.
     Born in Huntington, Pennsylvania, Saxton attended public schools and was apprenticed to a watchmaker. Interested in scientific instruments, in 1828 he associated with the Adelaide Gallery of Practical Science in London and received several British patents. He returned to the United States as curator at the national mint in Philadelphia, and after 1843 was superintendent of weights and measures for the Coast Survey. DAB; NAS, Biog. Mem., I (1877).

Schaeffer, George Christian (1815-1873). Primary, 1848-1860.
     German-born Schaeffer, interested in chemistry and physics, studied medicine and became professor of natural philosophy and chemistry at Centre College, Danville, Kentucky. He was also librarian at the Patent Office. J. D. Poggendorff, Biographische-literarisches Handwortenbuch zur Geschichte de exacten Wissenschaften (Leipzig, 1898), p. 1177; SFB.

Schofield, John McAllister (1831-1906). Secondary, 1859--.
     Born in New York and raised in the Midwest, Schofield graduated from West Point in 1853 and from 1855 to 1860 returned as assistant professor of natural and experimental philosophy. In 1860 he taught at Washington University in St. Louis and during the war worked with the Missouri Volunteer Infantry. After the war he served as Superintendent at West Point, 1876-1881, and became commanding general of the army in 1888. DAB.

Schott, Arthur Carl Victor (1814-1875). Primary, 1854--.
     Schott, a German protege of John Torrey with a technical school training, was employed as scientist on the Mexican Boundary Survey in 1851; his collections and artistic illustrations were important for the Eastern botanists. His own major interest was in geology and he also was involved with mining enterprises. William H. Goetzmann, Army Exploration in the American West, 1803-1863 (New Haven, 1959), pp. 182, 194-195, 201-205; C. S. Sargenat, Silva of North America, X (1890), p. 18.

Schott, Charles Anthony (1826-1901). Primary, 1854--.
     German-born, Schott immigrated to America in 1848 where he became a computer at the Washington office of the Coast Survey. After 1855 he was division chief and made important contributions in the fields of geodesy and terrestrial magnetism. DAB; NAS, Biog. Mem., VIII (1919).

Seemann, Berthold Carl (1825-1871). Secondary, 1857-1859.
     Although born in Hanover and a graduate of Gottingen in Germany, Seemann studied botany at Kew, England, and accompanied several British expeditions as naturalist after 1847. He was editor of two botanical journals and also published in literary and political journals of London. CAB.

Serrell, Edward Wellman (1826-1906). Secondary, 1859--.
     English-born to American parents, Serrell attended New York schools and then joined his father as an engineer. Known as an expert on railroad and bridge design and construction, he became a chief engineer in the Army during the Civil War. DAB.

Shepard, Charles Upham (1804-1886). Primary, 1850-1860.
     Shepard was born in Little Compton, Rhode Island, attended Brown, but graduated from Amherst in 1824, studying under Amos Eaton. He assisted Benjamin Silliman and lectured at Yale, and in 1847 became full professor of chemistry and natural history at Amherst. As a result of numerous trips abroad and throughout the United States, Shepard had perhaps the finest private mineralogical cabinet in America. DAB.

Shumard, Benjamin Franklin (1820-1869). Secondary, 1848-1859.
     Shumard was born in Lancaster, Pennsylvania, and graduated from Miami of Ohio in 1841. While practicing medicine in Kentucky he became interested in prehistoric remains and this led to his involvement in geological surveys of the Northwest Territory, Kentucky, Missouri and Texas. After serving as governor of Texas, he returned to St. Louis where he practiced medicine and became professor of obstetrics at the University of Missouri. K&B, Med. Biog.; W.K.Ferguson, Geology and Politics in Pioneer Texas. SFB.

Silliman, Benjamin (1779-1864). Primary, 1848--.
    Born in Trumbull, Connecticut, Silliman graduated from Yale in 1796
and while tutoring there passed the bar examination in 1802. Invited to
teach chemistry and natural history at Yale, where he remained until retirement in 1853, Silliman traveled to Philadelphia and later to Europe to teach
himself the rudiments of these fields. A founder of science in America,
from 1819 he edited the American Journal of Science, worked to develop the
science curriculum at Yale and lectured popularly on science from Boston to
New Orleans. DAB; John F. Fulton and E. H. Thomson, Benjamin Silliman
1779-1864, Pathfinder in American Science (New York, 1947).

Silliman, Benjamin, Jr. (1816-1885). Primary, 1848--.
    Son of the Yale chemistry professor, Silliman also graduated from Yale
in 1837 and assisted his father with the laboratory as well as the Journal.
After helping found Sheffield Scientific School, he taught at the University
of Louisville, 1849-1854, writing a text on chemistry. After 1853 he became
professor of chemistry at Yale but his involvement with a California oil
scandal caused him to resign in 1870, although he remained affiliated with
the medical school until 1880. A genial man, Silliman demonstrated more
ability to synthesize than to penetrate in his work in chemistry and
mineralogy. DAB; Gerald T. White, Scientists in Conflict: The Beginnings
of the Oil Industry in California (San Marino, California, 1968).

Smallwood, Charles (1812-1873). Primary, 1853--.
    English by birth, Smallwood came to Canada in 1853 where he again
established a medical practice. Building a meteorological and electrical
laboratory, he began making reports of his observations to the Smithsonian
Institution. In 1858 he became professor of meteorology at McGill College
in Montreal, and later professor of astronomy as well. CAB; SFB.

Smith, Francis Henney (1812-1890). Primary, 1855-1860.
    Born in Norfolk, Virginia, Smith graduated from West Point in 1833 where
he briefly taught geography, history and ethics. From 1840 to 1889 he was
principal of the Virginia Military Institute, where he attempted to follow
the West Point pattern and worked to make the curriculum more practical. DAB.

Smith, Hamilton Lamphere (1819-1903). Secondary, 1851-1859.
    Smith was born in New London and graduated from Yale in 1839. While
a student at Yale he constructed a telescope and made extended observations
on nebulae. From 1853 to 1868 he was professor of natural philosophy and
astronomy in Kenyon College, Gambier, Ohio, and after that at Hobart College.
Although he expressed an early interest in geology, his work was chiefly in
astronomy. CAB; Science, CXVIII (14 August 1903), 221.

Smith, John Lawrence (1818-1883). Primary, 1848-1857.
    Born near Charleston, Smith studied at the University of Virginia for
two years, worked as an engineer, and in 1840 took an M.D. from the Medical
College of South Carolina. Studying in Germany and France, he returned in
1844 as practitioner and lecturer in Charleston and also founded the Charleston
Medical Journal and Review. He worked as a science consultant in Turkey for
three years, then taught at the University of Virginia, and from 1854 to 1866
was professor of medical chemistry and toxicology at the University of Louisville. He was a leading organic chemist and also interested in meteorites.
DAB.

Smith, Lyndon Arnold (1790-1865). Secondary, 1854--.
    Smith was born in Haverhill, New Jersey, and then studied medicine.
He took M.D.s from both Dartmouth and Williams Medical Colleges in 1823. A
member of the New York College of Physicians and Surgeons, he was an active
and publishing physician in Newark, New Jersey. George T. Chapman, Sketches
of the Alumni of Dartmouth College from the First Graduation in 1771 to the
Present Time (Cambridge, 1867), p. 190.

Smith, Sanderson Arnold (1832-1915). Primary, 1855-1860.
    English-born, Smith attended the London School of Mines and was an
active civil engineer in New York and Long Island. He was also interested in
the geological and biological sciences, preparing a study of the mollusca of
Long Island. AJS, CL XXXIX (1915), 685-686; Staten Island Association of Arts
and Sciences, Proceedings, VI (1916), 141-143.

Snell, Ebenezer Strong (1801-1876). Primary, 1849--.
    Snell, born in North Brookfield, attended Williams College but
graduated from Amherst, Massachusetts, in 1824. Interested in experimental
physics and mathematics, he tutored at Amherst and from 1834 to 1872 was
professor of mathematics and astronomy there. A careful scholar, he wrote
several textbooks for academy and college use. Biographical Record of the
Alumni of Amherst College During Its First Half Century, 1821-1871 (Amherst,
1883), p. 10; SFB.

Sparks, Jared (1789-1866). Primary, 1849--.
    Sparks graduated from Harvard in 1815, tutored science while attending
divinity school there, and took a Unitarian church in Baltimore. In 1823
he left the ministry to become editor of the North American Review and in
Boston edited and wrote important historical studies. He taught at Harvard
after 1839 and was elected president in 1849; after his resignation in 1853,
he devoted himself to collecting documents on the Revolution. DAB; Herbert
Baxter Adams, The Life and Writings of Jared Sparks, Comprising Life and
Writings from His Journals and Correspondence (Boston, 1893).

Squier, Ephraim George (1821-1888). Primary, 1848-1854.
    Born in Bethlehem, New York, Squier taught briefly after finishing
his own local schooling. Moving to Ohio, he was a politician and editor and
actively researched the mounds in Onio for archeological remains. After 1849
he served in various Latin American countries as diplomat and consul, and
wrote archeological and ethnological articles on the countries visited.
DAB; SFB.

Steiner, Lewis Henry (1827-1892). Primary, 1853--.
    Steiner was born in Frederick, Maryland, graduated from Marshall
College in 1846, and took an M.D. from the University of Pennsylvania in 1848.
In 1852 he moved to Baltimore, where he practiced medicine and taught at a
private medical institute and then chemistry and botany at the College of St.
James; after 1884 he was librarian of Enoch Pratt Free Library in Baltimore.
His major scientific interests were chemistry and botany; he was also active
in local public health efforts. DAB; SFB.

Stoddard, Orange Nash (1812-1892). Primary, 1853 only.
    Stoddard came from Lisle, New York, to graduate from Union College
in 1836. He taught at various academies and at Hanover College before becoming
professor of natural science at Miami University, 1845-1870; after 1870 he
taught at Wooster University. His early interest was meteorology. General
Catalogue of the Graduates and Former Students of Miami University, 1809-
1909 (Oxford, Ohio, 1909), p. xxiv; Centennial Catalog, 1795-1895, of the
Officers and Alumni of Union College in the City of Schenectady, New York
(Troy, New York, 1895), p. 49.

Stevens, Robert P. Secondary, 1853-1859.
    Apparently a medical doctor, Stevens resided in Ceres, New York, in
1853 but later moved to North Egremont, Massachusetts. He was interested in
"geology and collateral sciences," and presented a paper to the AAAS on the
coal fields of Illinois. SFB.

Swallow, George Clinton (1817-1899). Primary, 1856--.
    Born in Buckfield, Maine, Swallow graduated from Bowdoin in 1843.
After teaching in mine academies, he became professor of geology, chemistry,
and mineralogy at the University of Missouri, Columbia; from 1853 to 1861 he
was also state geologist. Especially interested in mineralogy, after 1865
he became involved as a mining consultant and moved to Montana in 1882. DAB.

Taylor, Richard Cowling (1789-1851). Primary, 1848-1851.
    English-born, Taylor was a close friend of pioneer geologist William
Smith and worked on the ordinance survey of England. Immigrating to America in
1830, he made numerous surveys of the Pennsylvania coal regions and of
many other mineral districts, ranging from Canada to Central America. DAB.

Thomas, W. H. B. Secondary, 1853-1856.
    Thomas presented a single paper before the AAAS when it met in his
home town of Cincinnati in 1853.

Torrey, John (1796-1873). Primary, 1848--.
    Born in New York City, Torrey studied under Amos Eaton at Williams
College and then took an M.D. from the New York College of Physicians and

Surgeons in 1818. From 1829 to 1855 he taught chemistry at the medical school and concurrently taught at Princeton. He was the leading botanist of his generation and did much of the classifying of plant specimens returned by the several Western exploring expeditions in mid-century. DAB; Andrew D. Rodgers, John Torrey: A Story of North American Botany (Princeton 1942).

Totten, Joseph Gilbert (1788-1864). Primary, 1848-1864 (inactive 1850-1851).
    Totten was born in New Haven and graduated from West Point in 1805. After the War of 1812 he worked primarily as an engineer in coast defense and river and harbor improvements. A close friend of A. D. Bache and Joseph Henry through his activity on the lighthouse board and as a regent of the Smithsonian Institution, he was an original incorporator of the National Academy of Sciences. DAB; SFB.

Troost, Gerard (1776-1850). Primary, 1848-1850.
    Dutch-born, Troost received an M.D. from Leyden and a master's in pharmacy at Amsterdam, and also studied in Germany under Abraham Werner. In 1810 he settled in Philadelphia and established a pharmaceutical and chemical laboratory but moved to New Harmony in 1825. From 1828 he was professor of geology and mineralogy at the University of Nashville and after 1831 was state geologist of Tennessee; he was also interested in Indian archeology. DAB.

Trowbridge, William Petit (1828-1892). Primary, 1856--.
    Born in Troy, New York, Trowbridge was valedictorian of West Point in 1848 and then worked on primary triangulation for the Coast Survey. In 1856 he resigned his commission to become professor of mathematics at the University of Michigan, but the following year returned as assistant superintendent of the Coast Survey. After the Civil War he worked privately and then taught at Yale and at Columbia School of Mines. DAB; NAS, Biog. Mem., III (1895).

Tuomey, Michael (1805-1857). Primary, 1848-1857.
    Irish-born, Tuomey came to the United States while quite young, farmed, and in 1835 graduated from Rensselaer Polytechnic Institute. He then taught and worked as a civil engineer before becoming professor of mineralogy, geology and agriculture at the University of Alabama; after 1848 he was also state geologist. His interests were in botany and geology and he also published an account of the fossils of South Carolina with Francis Holmes. DAB; SFB.

Turnbull, Lawrence (1821-1900). Secondary, 1856-1859.
    Scottish by birth, Turnbull graduated from the Philadelphia College of Pharmacy in 1842 and received an M.D. from Jefferson Medical College in 1845. A lecturer in chemistry at the Franklin Institute for several years, he was primarily interested in the practice of medicine and researched diseases of the eye and ear. CAB.

Twining, Alexander Catlin (1801-1884). Secondary, 1851-1854.
    Twining was born in New Haven and graduated from Yale in 1820; he then studied at Andover and at West Point. His early interest in astronomy resulted in his appointment as professor of mathematics and natural philosophy at Middlebury College; thereafter he was a consulting engineer to railroads and remained actively interested in mathematics. DAB; SFB.

Tyson, Philip Thomas (1799-1877). Primary, 1858--.
    Born in Baltimore, Tyson worked with his brothers as a manufacturing chemist; they were active in the manufacture of bichromate of potash and controlled copper mining in the state. He was active in the Maryland Academy of Sciences and after 1858 was state geologist, publishing a map and report of his findings. NCAB.

Vancleve, John Whitten (1801-1858). Secondary, 1848--.
    A lifetime resident of Dayton, Ohio, Vancleve edited the local Journal from 1828 to 1832 and in the latter year was mayor of the city. An active member of local literary and scientific groups, Vancleve expressed a special interest in botany and geology. History of Dayton, Ohio, with Portraits and Biographical Sketches of Some of Its Pioneer and Prominent Citizens (Dayton, 1889), pp. 44, 69, 129, 164, and 546; SFB.

Vaughan, Daniel (1818-1879). Primary, 1851-1857.
Irish-born, Vaughan came to the United States in 1840, where he tutored in Kentucky and Cincinnati and worked independently in mathematics and astronomy. Although a popular lecturer, he was somewhat retiring and taught at various times at medical schools in Cincinnati. He did important early work on the rings of Saturn but died in poverty in Kentucky. DAB.

Vaux, William S. (1811-1882). Secondary, 1848--.
Vaux was an active member of the Academy of Natural Sciences after 1834 and contributed generously to its expansion; his large Etruscan collections were donated to the Pennsylvania museum at his death. Apparently independently wealthy, he indicated no occupation in 1852 but stated an interest in mineralogy. AJS, CXXIII (June, 1882), 498; Academy of Natural Sciences, Proceedings (1882), 111-112; SFB.

Walker, Sears Cook (1805-1853). Primary, 1848-1853.
Born in Wilmington, Massachusetts, Walker graduated from Harvard in 1825 and then taught at Boston and Philadelphia. Interested in astronomy, he founded an observatory at the Philadelphia High School in 1837, worked at the Naval Observatory, 1845-1847, and then went to the Coast Survey. He originated the telegraphing of transits of stars and application of the graphic registration of time-results to the registry of time-observations for general astronomical purposes, which became known as the "American method." DAB.

Ward, James Warner (1816-1897). Secondary, 1851-1854.
Ward was born in Newark, New Jersey, attended Boston public schools, and then went to Cincinnati, where he assisted John Locke. He taught literature at Ohio Female College, 1851-1854, and because of his interest in geology and botany then assisted editing the Horticultural Review and Botanical Magazine. After 1859 he worked in New York City in the customs office and after 1874 was a leader in library science in Buffalo. DAB; SFB.

Warder, John Aston (1812-1883). Secondary, 1850-1859.
Born in Philadelphia, Warder took an M.D. from Jefferson Medical College in 1836 and then practiced medicine in Ohio. Interested in geology, natural history, and botany, he was active in local scientific societies and published early articles on pomology and fruit culture; in the 1870s he was involved in forestry. DAB; SFB.

Warner, Horatio Gates (1801-1876). Primary, 1859--.
Warner was born in Canaan, New York, and graduated from Union College in 1826. He was a lawyer and later a judge in Rochester and traveled West in the 1850s. NCAB.

Warren, John Collins (1778-1856). Secondary, 1848-1856.
A lifetime Boston resident, Warren graduated from Harvard in 1797; he then studied medicine with his father and also in London, Edinburgh, and Paris. In 1809 he became adjunct professor of anatomy and surgery at Harvard Medical School, and full professor, 1815-1847. Interested in geology and paleontology, he was active in the Boston Society of Natural History and was known for his demonstration of ether anesthesia. DAB; Edward Warren, Life of John Collins Warren, M.D., Compiled Chiefly from his Autobiography and Journals, 2 vols. (Boston, 1860); SFB.

Watson, William (1834-1915). Primary, 1858--.
Watson was born in Nantucket and graduated from Lawrence Scientific School in 1857 in engineering; in 1862 he received a Ph.D. from Jena and also studied at Ecole Nationale des Ponts et Chaussees. The information he collected on technical schools in Europe was applied in the new Massachusetts Institute of Technology, where he was first professor of mechanical engineering and descriptive geometry, 1865-1873. DAB.

Webber, Samuel (1797-1880). Primary, 1848-1856.
Born in Cambridge, Webber graduated from Harvard in 1815 and took an M.D. there in 1822. He moved to Charleston, New Hampshire, to practice medicine and write poetry; he expressed an interest in natural philosophy. CAB; SFB.

Webster, Nathan Burnham (1821-1900). Primary, 1853-1860.
Webster was born in Unity, New Hampshire, and attended Norwich University without graduating. He taught at various academies before founding

his own in 1849, the Virginia Collegiate Institute in Norfolk, Virginia; he moved to Ottawa, Canada, in 1862 to found a similar school. He was interested in natural philosophy and especially meteorology. CAB; SFB.

Weinland, David Freidrich (1829-1915). Primary, 1856-1859.
The author was unable to consult the biography by Fritz Berger, David Freidrich Weinland (Blaubeuren, 1967).

Weld, Mason Cogswell (1829-1887). Secondary, 1850-1854.
Born in Philadelphia, Weld assisted Benjamin Silliman in chemistry from 1848 to 1853 but only received a degree in 1858 from Sheffield Scientific School; he also studied in Germany under Liebig and Bunsen. He served in the Civil War and was later known for his expertise in scientific agriculture, especially animal breeding. Yale University, Obituary Record, Series 3, No. 8 (1888), 477; SFB.

Wells, David Ames (1828-1898). Primary, 1849-1858.
Wells was born in Springfield,Massachusetts, then graduated from Williams College in 1847 and from Lawrence Scientific School in 1851. From 1850 to 1866 he published the Annual of Scientific Discovery and also helped to compile handbooks on chemistry, geology and natural philosophy. After 1866 he was an active economist as well, writing books and pamphlets to popularize his particular interpretation of laissez faire. DAB; Herbert Ferleger, David A. Wells and the American Revenue System, 1865-1870 (New York, 1942); SFB.

West, Charles (1809-1900). Primary, 1848--.
Born in Washington, Massachusetts, West graduated from Union College in 1832 and was also admitted to the bar. Interested in raising the level of women's education, he taught at and headed the Rutgers Female Institute, 1839-1851, and then the Buffalo Female Seminary, 1851-1860. Catalogue of the Officers and Alumni of Rutgers College (Trenton, New Jersey, 1909), p. 240; NCAB.

Weyman, George Washington (1822-1864). Secondary, 1851-1859.
Born in Pittsburgh, Weyman graduated from Sheffield Scientific School in 1852 and took a Ph.D. from Gottingen in 1855. He returned to Pittsburgh where he was a pharmacist and analytical chemist. Yale University, Obituary Record, 1st Ser., No. 5 (1865), 150.

Wheidon, William Wilder (1805-1892). Secondary, 1859--.
Born in Boston, Wheidon was apprenticed to a printer. In 1827 he founded the Bunker Hill Aurora in Charlestown, Massachusetts, which he edited until 1870. He was also active in local politics and wrote poetry. CAB.

Whitney, Josiah Dwight (1819-1896). Primary, 1848--.
Whitney was born in Northampton, Massachusetts, and graduated from Yale in 1839; he subsequently studied with Robert Hare and then in Paris, Berlin, and Geissen. After working on the Lake Superior survey with Charles Jackson, he was a mining consultant and mineralogist, becoming state geologist of California, 1860-1874; thereafter he taught at Harvard. DAB; Edwin T. Brewster, Life and Letters of Josiah Dwight Whitney (Boston, 1909).

Whitney, William Dwight (1827-1894). Primary, 1858--.
Brother of Josiah, William Whitney graduated from Williams College in 1845 and then assisted his brother with botany on the Lake Superior survey. Botany and ornithology became subordinate interests to his evident expertise in linguistics and Sanskrit. After 1854 he taught Sanskrit at Yale and helped organize modern languages at Sheffield Scientific School. DAB.

Whittlesey, Charles (1808-1886). Primary, 1848-- (1854-1856).
Whittlesey was born in Southington, Connecticut, but grew up in Ohio and graduated from West Point in 1831. He then practiced law and edited a Whig newspaper in Cleveland, Ohio; locally prominent, he also served a term as mayor. He assisted on the Ohio survey in the 1830s and later on the Lake Superior survey, showing particular interest in mining and mineralogy. Charles C. Baldwin, "Memorial of Colonel Charles Whittlesey," Western Reserve and Northern Ohio Historical Society, Tract 68 (Cleveland,1887).

Wilkes, Charles (1798-1877). Primary, 1848--.
Wilkes, born in New York City, was appointed midshipman in 1818, studied under Ferdinand Hassler, and then headed the Navy's Depot of Charts

and Instruments. From 1838 to 1842 he headed the United States Exploring
Expedition which surveyed the Antarctic Coast, the Pacific Islands, and the
American northwestern coast; he then supervised publication of the Expe-
dition's results in Washington before returning to active duty. DAB;
Daniel M. Henderson, The Hidden Coast; A Biography of Admiral Charles
Wilkes (New York, 1953); SFB.

Williams, William W. Secondary, 1858-1860.
    Williams, a correspondent of William Redfield on meteorology, gave
his address as Keokuk, Iowa, in 1858 but later as New York, New York.

Wilson, Daniel (1816-1892). Primary, 1856--.
    Scottish-born, Wilson studied archaeology at the University of
Edinburgh. After 1853 he was professor of history and English literature in
University College, Toronto, actively researching and publishing on pre-
historic men and animals. CAB.

Winchell, Alexander (1824-1891). Primary, 1850-- (inactive 1859-1860).
    Born in Northeast, New York, Winchell graduated from Wesleyan in 1844.
While teaching at various seminaries in New Jersey, he studied local flora.
After 1855 he taught at the University of Michigan in various areas including
geology, zoology, and botany; he was also director of the Michigan geological
survey. Basically a popularizer, he wrote several works attempting to
reconcile the supposed conflict between science and religion. DAB; SFB.

Winlock, Joseph (1826-1875). Primary, 1851-1858.
    Winlock was born in Shelby County, Kentucky, graduated from Shelby
College, and taught mathematics and astronomy there until 1851. That year
he met Benjamin Peirce at an AAAS meeting and the result was an offer to
become a computer for the American Ephemeris and Nautical Almanac, which he
headed after 1859 until he became director of the Harvard Observatory after
1866. A traditional astronomer himself, he encouraged the use of spectroscopic
studies in the observatory. DAB.

Wood, Alphonso (1810-1881). Secondary, 1853 only.
    Born in Chesterfield, New Hampshire, Wood graduated from Dartmouth
in 1834 and was licensed to preach after a year's study at Andover. Interested
in natural history, he wrote a popular text on botany and made several expeditions to expand his personal collection. After 1853 he was president of
the Ohio Female College in Cincinnati, later teaching in Brooklyn and New
York City. Torrey Botanical Club, Bulletin, VIII (May, 1881), 53-56.

Worthen, Amos Henry (1813-1888). Primary, 1850-- (inactive 1854-1856).
    Born in Bradford, Vermont, Worthen lived in Kentucky, Ohio, and
Illinois variously working in business establishments. When his interest in
conchology elicited some Eastern response, he worked under J. D. Norwood on
the Illinois geological survey, becoming state geologist in 1858. He
completed seven volumes for the Geological Survey of Illinois (1866-1890),
doing his best work on the Lower Carboniferous strata. DAB; NAS, Biog. Mem.,
III (1895); SFB.

Wright, Chauncey (1830-1875). Primary, 1855--.
    Wright was born in Northampton, Massachusetts, and graduated from
Harvard in 1852, where he indicated an interest in mathematics, natural
science, and philosophy. In 1852 he became a computer for the American
Ephemeris and Nautical Almanac; he was commended by Charles Darwin for his
work on genesis of species. Publishing little, he was a central influence
for the group of Cambridge pragmatists. DAB; Edward Madden, Chauncey Wright
and the Foundations of Pragmatism (Seattle, 1963).

Wurtz, Henry (1828-1910). Primary, 1856--.
    Born in Easton, Pennsylvania, Wurtz graduated from Princeton in 1848,
attended Lawrence Scientific School, and assisted Benjamin Silliman at Yale.
From 1854 to 1856 he was chemist and geologist for the New Jersey geological
survey, subsequently exploring in North Carolina. In 1858 he became
professor of chemistry and pharmacy at the National Medical College (now
George Washington University) and was also chemical examiner in the Patent
Office; after 1861 he became a private consulting chemist in New York City.
DAB.

Wurtz, Jacob H. Secondary, 1849-1851.
    Wurtz, whose address was New York City, presided over three section
meetings in the chemical division in 1849.

Wyman, Jeffries (1814-1874). Primary, 1848-1859 (inactive 1856).
    Born in Chelmsford, Massachusetts, Wyman graduated from Harvard in
1833 and received an M.D. from Massachusetts General Hospital in 1837. More
successful as demonstrator than as practitioner, he taught at Hampton-Sidney
College of Richmond, returning in 1847 to Harvard as Hersey Professor
of Anatomy. His best work was in comparative anatomy and he was an active
member of the Boston Society of Natural History, although he did not assume
his office as President of the AAAS in 1857. DAB; SFB.

Wyman, Morrill (1812-1903). Primary
    Brother of Jeffries, Morrill Wyman graduated from Harvard in 1833 and
from Harvard Medical School in 1837. He practiced in Cambridge and was
innovative in surgical techniques. He briefly taught at Harvard, attempted
to found a medical college, and was an active hospital administrator. DAB.

Wynne, James (1814-1871). Primary, 1854--.
    Born in Utica, New York, Wynne graduated from the University of New
York City, studied medicine, and then practiced in Baltimore. He later
moved to New York City where he worked on medical statistics and jurisprudence,
while also publishing on cholera and public hygiene. In 1867 he emigrated to
Guatemala and worked with coffee culture. CAB.

Yandell, Lunsford Pitts (1805-1878). Primary, 1851-1854.
    Born near Hartsville, Tennessee, Yandell received an M.D. from the
University of Maryland in 1825. He established a practice in Nashville and
in 1831 became professor of chemistry and pharmacy at Transylvania University
to 1837, and then taught chemistry and materia medica at the University of
Louisville. Interested in natural science, he made extensive geological and
paleontological explorations near Louisville and was also on the editorial
boards of several western medical journals. DAB; SFB.

                              B: Membership

    The following list of ordinary members of the AAAS provides the name
of a member, his years active in the Association as recorded in the annual
Proceedings, and his place of residence during the majority of his period of
membership. If a member served on a local committee, this fact is recorded
by the abbreviation LC; SFB continues to indicate inclusion in Spencer F.
Baird's volume of "Scientific Addresses." An asterisk indicates the person
died while an active member.

Abbot, Gorham Dummer. 1853-1860. New York, New York. SFB.
Abbott, Samuel Leonard. .1848-1854. Boston, Massachusetts.
Abernethy, William. 1849-1854. Oregon City, California.
Abert, John James. 1848-1859. Washington, D.C. LC.
Abrahams, Woodward. 1858 only. Baltimore, Maryland. LC.
Adams, John G. 1848-1851. New York, New York.
Adams, Pentegast. 1848-1851. Jefferson County, New York.
Adams, Solomon. 1848-1854 (inactive 1850-1851). Boston, Massachusetts
Adamson, William. 1853-1855. Cambridge, Massachusetts.
Aiken, William. 1850-1854. Charleston, South Carolina.
Ainsworth, J. G. 1860--. Barry, Massachusetts.
Albert, Augustus J. 1858--. Baltimore, Maryland. LC.
Albro, John Adams. 1849-1851. Cambridge, Massachusetts.
Alexander, Joseph H. 1854 only. Baltimore, Maryland.
Alexander, R. C. 1849-1854. Bath, England.
Allen, Edward A. H. 1851--. New Bedford, Massachusetts. SFB.
Allen, George N. 1851-1857. Oberlin, Ohio. SFB.
Allen, Ira M. 1848-1851. New York, New York.
Allen, J. Burnet. 1849-1854. Springfield, Massachusetts.
Allen, J. L. 1848-1851. New York, New York.
Allen, Jerome. 1856 only. Dubuque, Iowa.
Allen, John H. 1851-1858. Oxford, Maryland.
Allen, Nathaniel T. 1856-1859. West Newton, Massachusetts.
Allen, Richard Lamb. 1856-1859. Saratoga Springs, New York.
Allston, Robert F. W. 1850-1860. (inactive 1850-1851). Georgetown, South
        Carolina. SFB.

Allyn, Robert. 1855-1857. East Greenwich, Rhode Island.
Alter, David. 1853 only. Freeport, Pennsylvania.
Alvord, Daniel W. 1848-1850. Greenfield, Massachusetts.
Ames, Bernice D. 1856-1858. Fort Edward, New York.
Ames, Nathaniel Peabody. 1848 (deceased AAGN member). Springfield, Massachusetts.
Amory, Jonathan. 1854-1857. Jamaica Plain, Massachusetts.
Anderson, William. 1857 only. Montreal, Canada.
Andrews, Alonzo. 1853-1859. Lewiston, Maine. SFB.
Andrews, E. H. 1850-1858. (inactive 1854-1856). Charlotte, North Carolina. SFB.
Andrews, Edmund. 1853-1855. Ann Arbor, Michigan. SFB.
Andrews, J. W. 1850-1854. Marietta, Ohio.
Andrews, Joel W. 1851-1854. Albany, New York. SFB.
Angell, James Burrill. 1855-1856. Providence, Rhode Island.
Anthony, Charles H. 1851-1860. Albany, New York.
Anthony, Henry Bowen. 1854--. Providence, Rhode Island. LC.
Appleton, Nathan. 1848--. Boston, Massachusetts. LC.
Appleton, Thomas Gold. 1854-1859. Cambridge, Massachusetts.
Arden, Thomas B. 1853-1858. Putnam County, New York. SFB.
Armour, A. H. 1856-1859. Toronto, Canada.
Armsby, John H. 1851-1859. Albany, New York. LC. SFB.
Ashmead, Samuel. 1848-1854. Philadelphia, Pennsylvania.
Astrop, R. F. 1853-1859. Burns County, Virginia. SFB.
Atkinson, Robert. 1858--. Baltimore, Maryland.
Atkinson, T. C. 1848-1854. Cumberland, Maryland.
Atlee, Washington Lemuel. 1848-1851. Lancaster, Pennsylvania.
Atterbury, John G. 1857-1860. New Albany, Indiana.
Austin, Samuel. 1855-1858. Providence, Rhode Island.
Austin, William W. 1853 only. Richmond, Indiana. SFB.
Avery, Charles A. 1851--(inactive 1854-1859). Schenectady, New York.

Baby, G. 1857-1859. Montreal, Canada.
Bachmeister, H. 1849-1851. Schenectady, New York.
Bacon, Austin. 1853 only. Natick, Massachusetts. SFB.
Bacon, J. S. 1848-1855. Washington, D.C. LC.
Bacon, John. 1848--. Boston, Massachusetts.
Bacon, William M. 1853-1858. Richmond, Massachusetts. SFB.
Baer, William. 1854-1857. Sykesville, Maryland.
Bagg, Moses M. 1850-1860. Utica, New York.
Baird, E. Thompson. 1851-1854. Washington College, East Tennessee.
Baird, Thomas D. 1851-1860 (inactive 1854-1858). Baltimore, Maryland.
Baird, William M. 1853-1854. Reading, Pennsylvania. SFB.
Baker, Eben. 1848-1851. Charleston, South Carolina.
Bakewell, Robert. 1848-1854. New Haven, Connecticut.
Baldon, A. S. 1857 only. Jacksonville, Florida.
Baldwin, F. H. 1856-1860. Waverly, New York.
Baldwin, William. 1853-1854. Platte City, Missouri. SFB.
Bancroft, Rev. Canon. 1857 only. Montreal, Canada.
Bannister, Henry. 1851-1854. Cazenovia, New York.
Barbee, William J. 1851-1854. Jackson, Mississippi. SFB.
Barber, Edgar A. 1856-1859. Albany, New York.
Barber, Isaac R. 1848-1851. Worcester, Massachusetts.
Barbour, J. R. 1848-1851. Worcester, Massachusetts.
Bardwell, F. W. 1859--. Yellow Springs, Ohio.
Barker, George Frederick. 1859--. Boston, Massachusetts.
Barker, Sanford. 1850-1854. Charleston, South Carolina. SFB.
Barlow, Thomas. 1853--. Canastota, New York. SFB.
Barnard, Henry. 1850--(inactive 1854-1858). Madison, Wisconsin.
Barnard, John Ross. 1860--. Washington, D.C.
Barnes, James. 1850--. Springfield, Massachusetts. LC.
Barnston, James. 1856-1859. Montreal, Canada.
Barratt, John P. 1849-1857. Barrattsville, South Carolina. SFB.
Barratt, Joseph. 1848--(inactive 1850-1859). Middletown, Connecticut.
Barrows, George B. 1853-1858 (inactive 1854-1857). Fryeburg, Maine. SFB.
Barry, L.F. 1858-1860. Baltimore, Maryland. LC.
Bartlett, John Russell. 1854-1858. Providence, Rhode Island.
Barton, E.H. 1855-1858. New Orleans, Louisiana.
Barton, William. 1851-1854. Troy, New York. SFB.
Baudry, J.H. 1857-1859. Montreal, Canada.
Bayliss, James. 1857. Montreal, Canada.
Beadle, E.R. 1856--. Hartford, Connecticut.
Beadle, Edward L. 1848-1854. New York, New York.

Beale, James. 1857 only. Richmond, Virginia.
Bean, Sidney A. 1855-1857. Waukesha, Wisconsin.
Beardsley, H.C. 1848-1851. Painesville, Ohio.
Beardsley, Rufus G. 1851-1854. Albany, New York. SFB.
Bebb, M.S. 1859--. Rockford, Illinois.
Beck, C.F. 1848--. Philadelphia, Pennsylvania. SFB.
Belknap, George. 1848-1854. Boston, Massachusetts.
Belknap, Henry. 1849-1854. Boston, Massachusetts.
Bell, John. 1848-1851. Philadelphia, Pennsylvania. SFB.
Bell, John G. 1853-1856. New York, New York.
Bell, John James. 1853-1854. Carmel, Maine. SFB.
Bell, Samuel Newell. 1853--. Manchester, New Hampshire. SFB.
Bell, Sanford. 1860 only. Springfield, Illinois.
Belle, Charles E. 1857-1859. Montreal, Canada.
Bellinger, John. 1850-1851. Charleston, South Carolina.
Bellows, Henry Whitney. 1851-1854. New York, New York.
Beman, Nathan Sidney S. 1851-1854. Troy, New York.
Benedict, Erastus Cornelius. 1856-1858. New York, New York.
Benedict, Farrand Northrup. 1848-1859. Parisippany, New York. SFB.
Benedict, Noah Bennet. 1856-1860. New Orleans, Louisiana.
Benedict, Thomas B. 1857-1860. Kirk's Ferry, Louisiana.
Bent, Silas. 1856--. New York, New York.
Bentley, Cyrus. 1859--. Chicago, Illinois.
Benton, Charles. 1857--. Oxbow, New York.
Berdon, Hiram. 1858-1860. New York, New York.
Berezy, William. 1857-1860. Daillehaut, Canada.
Bernard, A. 1857-1859. Montreal, Canada.
Berthoud, Edward L. 1853-1854. Maysville, Kentucky. SFB.
Berton, S. R. 1857 only. Port Gibson, Mississippi.
Bibaud, J. G. 1857 only. Montreal, Canada.
Bidwell, Walter Hilliard. 1857-1860. New York, New York.
Bierce, L. V. 1853 only. Akron, Ohio.
Bigelow, Artemas. 1855--. Newark, New Jersey.
Bigelow, Henry Jacob. 1848-1854. Boston, Massachusetts. SFB.
Bigelow, Jacob. 1849-1854. Boston, Massachusetts. LC.
Biggs, Lansing. 1853 only. Auburn, New York.
Billings, Elkanah. 1857-1859. Montreal, Canada.
Bingham, Joel Foote. 1853-1855. New York, New York.
Binkerd, R. S. 1856 only. Germantown, Ohio.
Binney, Amos. 1848 (deceased AAGN member). Boston, Massachusetts.
Binney, John. 1850-1857. Boston, Massachusetts.
Blackie, George S. 1856-1860. Nashville, Tennessee.
Blackmarr, Henry. 1857-1859. Rochester, New York.
Blair, Franklin Otis. 1859--. Lebanon, Illinois
Blake, C. M. 1857-1859. Chile.
Blake, Eli Whitney. 1848 (inactive 1854-1856). New Haven, Connecticut. SFB.
Blake, J. R. 1856--. La Grange, Tennessee.
Blake, John Lauris. 1848-1851. Boston, Massachusetts.
Blanding, William. 1848-death.
Blaney, James V. Z. 1858--. Chicago, Illinois.
Blasius, Wilhelm. 1853-1855. New York, New York.
Blatchford, Thomas W. 1851-1858 (inactive 1854-1856). Troy, New York.
Bledsoe, Albert Taylor. 1858-1860. Charlottesville, Virginia.
Bliss, George. 1850-1854. Springfield, Massachusetts. LC.
Blumenthal, C. E. 1850-1854. Carlisle, Pennsylvania.
Bolton, James. 1856-1860 (inactive 1857-1858). Richmond, Virginia.
Bomford, George. 1848 only. Washington, D.C.
Bonar, J. B. 1857-1859. Montreal, Canada.
Bond, Henry. 1848-1854. Philadelphia, Pennsylvania. SFB.
Bonnycastle, Charles. 1848-1859. Montreal, Canada.
Booth, James Curtis. 1848-1854. Philadelphia, Pennsylvania. SFB.
Borland, John Nelson. 1855-1857. Boston, Massachusetts.
Botta, Vincenzo. 1855-1860. New York, New York.
Boutelle, C. O. 1850-1854. Washington, D.C. SFB.
Bouve', Thomas Tracy. 1848--. Boston, Massachusetts. SFB.
Bowman, A. H. 1849-1854. Charleston, South Carolina.
Bowman, Francis Caswell. 1858-1860. New York, New York.
Boyd, George W. 1848-1854. Charlottesville, Virginia.
Boyden, Uriah Atherton. 1849-1859. Boston, Massachusetts. SFB.
Boynton, Edward Carlisle. 1859--. Oxford, Mississippi.
Boynton, John F. 1850-1859. Chicago, Illinois.
Brace, John Peirce. 1849-1854. Hartford, Connecticut. SFB.

Bradford, George W. 1856-1858. Homer, New York.
Bradford, Hezekiah. 1853-1855. New York, New York. SFB.
Bradford, Isaac. 1860--. Cambridge, Massachusetts.
Bradford, James C. 1857-1859. Elyria, Ohio.
Bradish, Luther. 1848-1853. New York, New York.
Bradley, Francis. 1850-1854. New Haven, Connecticut.
Brainerd, A. 1853. Norwald, Ohio.
Braithwaite, Joseph. 1857-1859. Chambly, Canada.
Brant, James R. 1855-1858. New York, New York.
Braw, J. B. 1851-1854. Hartford, Connecticut.
Breckenridge, Robert Jefferson. 1851-1854. Lexington, Kentucky.
Brewer, Fist Parsons. 1857--. New Haven, Connecticut.
Bridges, Robert. 1848-1850. Philadelphia, Pennsylvania.
Briggs, A.D. 1859--. Springfield, Massachusetts. LC.
Briggs, Caleb. 1851-1854. Ironton, Ohio.
Briggs, Charles. 1848-1851. Columbus, Ohio.
Briggs, Charles C. 1848-1851. Charlottesville, Virginia.
Briggs, George Ware. 1857 only. Pittsfield, Massachusetts.
Brinsmade, Thomas Clark. 1851-1854. Troy, New York. SFB.
Briston, William. 1857-1859. Montreal, Canada.
Britton, A. A. 1853-1855. Keokuk, Iowa.
Bross, William. 1853--. Chicago, Illinois. SFB.
Brown, B. B. 1848-1851. St. Louis, Missouri. SFB.
Brown, John Carter. 1855-1859. Providence, Rhode Island. LC.
Brown, Richard. 1848--. Sydney, Capte Breton. SFB.
Brown, Robert. 1857--. Cincinnati, Ohio.
Brown, W. Leroy. 1853-1857. Oakland, Mississippi. SFB.
Browne, Daniel Jay. 1848-1853. New York, New York.
Browne, Robert H. 1848-1854. New York, New York.
Brumby, Richard Tapier. 1848-1854. Tuscaloosa, Alabama. SFB.
Brush, George S. 1857-1859. Montreal, Canada.
Buchanan, Joseph Rodes. 1851-1854. Cincinnati, Ohio. SFB.
Buckland, David. 1853-1855. Brandon, Vermont. SFB.
Buckley, Samuel Botsford. 1848-1854. New York, New York.
Budd, B. W. 1848-1851. New York, New York.
Buell, David. 1851-1858. (inactive 1854-1857). Troy, New York.
Bulfinch, Thomas. 1848-1851. Boston, Massachusetts.
Bulkley, John Williams. 1856-1858. Brooklyn, New York.
Bullard, Edward F. 1856-1858. Waterford, New York.
Burden, Thomas Legere. 1850-1854. Charleston, South Carolina.
Burdett, F. 1848-1851. Boston, Massachusetts.
Burr, Enoch Ritch. 1851-1854. Lyme, Connecticut. SFB.
Burton, C. V. W. 1851-1859. (inactive 1854-1857). Lansingburg, New York.
Burwell, George N. 1851-1854. Buffalo, New York. SFB.
Busher, James. 1855--. Worcester, Massachusetts.
Butler, James Davis. 1859--. Madison, Wisconsin.
Butler, Thomas Belden. 1856--. Norwalk, Connecticut.

Cabell, James Lawrence. 1851--. Charlottesville, Virginia. SFB.
Cabot, Edward Clarke. 1848-1854. Boston, Massachusetts.
Cabot, Henry. 1848-1851. Boston, Massachusetts.
Cabot, James Elliott. 1848-1851. Boston, Massachusetts.
Cabot, Samuel. 1848-1851. Boston, Massachusetts.
Cady, Calvin Brainerd. 1857-1859. Alburgh, Vermont.
Cain, D.J.C. 1850-1854. Charleston, South Carolina.
Caldwell, John D. 1851-1854. Cincinnati, Ohio.
Calkars, Alonzo. 1848-1851. New York, New York.
Cameron, Daniel. 1856 only. Johnstown, New York.
Cameron, Hector. 1857 only. Toronto, Canada.
Campbell, A.D. 1857-1859. Montreal, Canada.
Campbell, Archibald Barrington. 1848-1851. Philadelphia, Pennsylvania.
Campbell, George W. 1857-1859. Montreal, Canada.
Campbell, John Nicholson. 1856-1858. New York, New York. LC.
Campbell, Robert. 1848-1854. Pittsfield, Massachusetts.
Campbell, W.D. 1857-1859. Quebec, Canada.
Campbell, William M. 1853 only. Battle Creek, Michigan.
Canning, Edward W.B. 1850-1851. Stockbridge, Massachusetts.
Cappy, T.P. 1850-1854. Montgomery, Alabama.
Capron, David I. 1853--. Annapolis, Maryland. SFB.
Carey, Henry Charles. 1854-1856. Burlington, New Jersey. SFB.
Carleton, James Henry. 1853-1854. Fort Leavenworth, Missouri. SFB.
Carley, S.T. 1851-1859. Cincinnati, Ohio.

Carpenter, George W. 1851-1854. Albany, New York.
Carpenter, J.N. 1858-1860. Washington, D.C.
Carpenter, Thornton. 1853-death. Camden, South Carolina. SFB.
Carpenter, William Marbury. 1848 only. New Orleans, Louisiana.
Carr, Ezra. 1848-- (inactive 1851-1855). Albany, New York.
Carter, Restor C. 1851-1854. Cincinnati, Ohio. SFB.
Cartier, George Etienne. 1857-1859. Montreal, Canada.
Case, William. 1851-1860. Cleveland, Ohio. LC.
Casseday, Samuel Addison. 1851-1854. Louisville, Kentucky. SFB.
Cassells, John Long. 1848-- (inactive 1850-1853). Cleveland, Ohio.
Cassin, John. 1848-1853 (inactive 1851). Philadelphia, Pennsylvania. SFB.
Cavert, Michael P. 1856-1858. Amsterdam, New York.
Chadbourne, Paul Ansel. 1856--. Williamstown, Massachusetts.
Chamberlain, Frank. 1856-1858. Albany, New York.
Chamberlain, Nathan Henry. 1849-1851. Boston, Massachusetts.
Chamberlin, B.C. 1857-1859. Montreal, Canada.
Chandler, John. 1848-1851. Boston, Massachusetts.
Channing, Walter. 1849-1851. Boston, Massachusetts.
Channing, William Francis. 1849-1859. Davenport, Iowa. SFB.
Chapin, Aaron Lucius. 1860--. Beloit, Wisconsin.
Chapin, Lebeus Cornelius. 1857--. New Haven, Connecticut.
Chapman, C.B. 1857--. Madison, Wisconsin.
Chapman, Edward. 1857-1859. Toronto, Canada.
Chapman, Nathaniel. 1848-1853. Philadelphia, Pennsylvania.
Chappellsmith, John. 1853-1855. New Harmony, Indiana. SFB.
Chase, Benjamin. 1853-1859. Natchez, Mississippi. SFB.
Chase, D. 1849-1851. Hanover, New Hampshire.
Chase, Stephen. 1849-1853. Dartmouth, New Hampshire.
Chase, Theodore R. 1853-1856. Cleveland, Ohio, SFB.
Chauveau, Pierre J. O. 1857-1859. Montreal, Canada. LC.
Chauvenet, C. U. 1848-1854. Annapolis, Maryland. (A note on the SFB return
    of William Chauvenet stated that this individual did not exist.)
Cheeseborough, A. 1851-1854. Springfield, Massachusetts.
Chenoweth, B. D. 1850-1854. Winchester, Virginia.
Chesbrough, Ellis Sylvester. 1849--. Chicago, Illinois.
Chester, C. 1854 only. Rochester, New York.
Cheves, C. M. 1850-1854. Charleston, South Carolina.
Chickering, Jesse. 1848-1851. Boston, Massachusetts.
Chilton, J. R. 1848-1851. New York, New York.
Choate, Charles Francis. 1853-1856. Cambridge, Massachusetts.
Church, Albert E. 1856--. West Point, New York.
Churchill, Franklin H. 1853 only. New York, New York. SFB.
Churchill, Marlborough. 1859--. Sing Sing, New York.
Clapp, Asahel. 1848--. New Albany, Indiana.
Clark, Francis M. 1851-1854. West Point, New York. SFB.
Clark, J. H. 1858-1860. Philadelphia, Pennsylvania.
Clark, James. 1854-1860. Georgetown, D. C. SFB.
Clark, John S. 1857 only. New Orleans, Louisiana.
Clark, Joseph. 1851--. Cincinnati, Ohio. SFB.
Clark, Lester M. 1851-1858 (inactive 1854-1856). Canandaigua, New York. SFB.
Clark, Thomas March. 1855 only. Providence, Rhode Island.
Clark, William P. 1848-1855. Norwalk, Ohio.
Clark, William P. 1853-1855 (inactive 1854). Hillsdale, Michigan.
Clark, William Smith. 1849-1853. East Hampton, Massachusetts.
Clarke, Alpheus Bryant. 1859--. Holyoke, Massachusetts.
Clarke, Francis N. 1850-1854. West Point, New York. SFB.
Clarke, Henry. 1860--. Worcester, Massachusetts.
Clarke, Robert. 1853-1858. Cincinnati, Ohio. SFB.
Clay, Joseph A. 1848-1854. Philadelphia, Pennsylvania.
Cleaveland, A. B. 1849-1856. Cambridge, Massachusetts.
Cleaveland, A. B. 1849-1851. Dedham, Massachusetts.
Cleaveland, C. H. 1855--. Cincinnati, Ohio.
Cleaveland, Parker. 1848-1854. Brunswick, Maine. SFB.
Cleghorn, E. B. 1857-1859. New Orleans, Louisiana.
Clement, H. H. 1855-1857. Providence, Rhode Island.
Clingman, Thomas Lanier. 1848-1854. Asheville, North Carolina.
Clum, Henry A. 1855--. New York, New York.
Coan, Titus. 1848-1855. Hilo, Hawaii.
Coates, Benjamin Horner. 1848-1857. Philadelphia, Pennsylvania.
Coates, Reynell. 1848-1854. Philadelphia, Pennsylvania. SFB.
Cobleigh, Nelson Ebenezer. 1859--. Lebanon, Illinois.
Cochran, J. W. 1857-1859. New York, New York.

Coffin, Charles Carleton, 1859--. Malden, Massachusetts.
Cogswell, Mason Fitch. 1850-1858. Albany, New York. LC.
Cohen, J. J. 1848-1854. Baltimore, Maryland.
Cole, Seth. 1856 only. Albany, New York.
Cole, Thomas. 1848-1860. Salem, Massachusetts.
Coleman, Henry. 1848-death. Boston, Massachusetts.
Collier, G. H. 1857-1859. Wheaton, Illinois.
Collins, D. B. 1857-1859. Carlisle, Pennsylvania.
Collins, George Lewis. 1855-1857. Providence, Rhode Island.
Colton, Oliver. 1849-1851. Salem, Massachusetts.
Colton, Willis S. 1854-1857. New Haven, Connecticut.
Comfort, George Fisk. 1856-1858. Middletown, Connecticut.
Comstock, Cyrus Ballou. 1860--. West Point, New York.
Comstock, J. C. 1857-1859. Hartford, Connecticut.
Comstock, John Lee. 1851-1854. Hartford, Connecticut. SFB.
Conant, Marshall. 1853--. South Bridgewater, Massachusetts. SFB.
Congdon, Charles Taber. 1848-1860. New York, New York.
Conger, A. B. 1856 only. Haverstown, New York.
Conkey, J. S. 1851-1855. Jefferson County, New York.
Conkling, Frederick Augustus. 1857--. New York, New York.
Conrad, Timothy Abbott. 1848-1850. Philadelphia, Pennsulvania.
Convers, Charles C. 1851-1854. Zanesville, Ohio. SFB.
Conway, Moncure Daniels. 1860--.Cincinnati, Ohio.
Cooke, George Hammel. 1848--. Troy, New York,
Cooke, Robert L. 1853-1856. Essex County, New Jersey. SFB.
Cooley, James E. 1856-1858. New York, New York.
Cooper, George F. 1853 only. Houston County, Georgia. SFB.
Cooper, James Graham. 1856-1858. Orange, New Jersey.
Cooper, William. 1848-1859 (inactive 1854-1855). Hoboken, New Jersey.
Copes, Joseph S. 1857--. New Orleans, Louisiana.
Cordner, John. 1857-1859. Montreal, Canada.
Corless, George Henry. 1855 only. Providence, Rhode Island. LC.
Corning, Erastus. 1851--. Albany, New York.
Coryell, Thomas D. 1859--. Madison, Wisconsin.
Cotting, B. C. 1848-1854. Georgia.
Cottle, Thomas J. 1856-1860. Woodstock, Canada.
Couch, Carius Nash. 1854-1856. Washington, D.C.
Coulton, Oliver. 1849-1851. Massachusetts.
Couthouy, Joseph P. 1848-1854. New York, New York.
Cowan, J. P. 1857-1859. Montreal, Canada.
Cox, Christopher C. 1858-1860. Easton, Maryland.
Cox, Samuel Hanson. 1856 only. New York, New York.
Cozzens, Issachar. 1848-1854. New York, New York.
Craig, Hugh. 1853 only. Jefferson, South Carolina. SFB.
Craik, Robert. 1857--. Montreal, Canada.
Cramp, John Mockett. 1857--. Acadia College, Nova Scotia.
Crandall, Pardon S. 1856-1858. Troy, New York.
Cresson, John Chapman. 1848-1854. Philadelphia, Pennsylvania.
Croft, Henry. 1857-1859. Toronto, Canada.
Crosby, Alpheus. 1856--. Hanover, New Hampshire.
Crissley, Richard. 1849-1851. Boston, Massachusetts.
Croswell, Edwin. 1851-1858 (inactive 1854-1856). Albany, New York.
Cruikshank, James. 1856-1858. Albany, New York.
Cummings, Joseph. 1859--. Middletown, Connecticut.
Cummings, William M. 1857-1859. New York, New York.
Cunynghame, Thurlow. 1857-1859. Montreal, Canada.
Curry, W. F. 1857--. Geneva, New York.
Curtis, James L. 1851-1854.
Curtis, Jasper. 1857-1859. St. Albans, Vermont.
Curtis, Josiah. 1848-1854. Lowell, Massachusetts. SFB.
Curtis, L. W. 1849-1854. Unionville, South Carolina.
Curtis, Moses Ashley. 1849-1851. Society Hill, South Carolina.
Curtiss, Charles W. 1859--. Evanston, Illinois.
Cutts, Richard D. 1853-1855. San Francisco, California. SFB.
Cyr, Narcisse. 1857 only. Montreal, Canada.

Dahlgren, John A. B. 1853-1855. Washington, D.C. LC.
Dakin, Francis E. 1856 only. Freeport, Illinois.
Dale, L. D. 1848-1851. New York, New York.
Dalrymple, E. A. 1857--. Baltimore, Maryland.
Dalson, Augustus. 1851-1854. Cambridge, Massachusetts.
Dalton, John Call. 1857-1859. New York, New York.

Dana, J. J. 1848-1851. South Adam, Massachusetts.
Dana, Samuel Luther. 1848-1850. Lowell, Massachusetts.
Danforth, Edward. 1857--. Grand Rapids, Michigan.
Daniels, W. C. 1848-1854. Savannah, Georgia. SFB.
Darracott, G. B. 1849-1851. Boston, Massachusetts.
Davenport, James R. 1851-1858 (inactive 1854-1856). Albany, New York.
David, A. H. 1857-1859. Montreal, Canada.
Davidson, Robert. 1856-1859. New Brunswick, New Jersey.
Davies, Charles. 1856--. Fishkill Landing, New York.
Davies, W. H. 1857-1859. Toronto, Canada.
Davies, W. H. A. 1857-1860. Montreal,Canada.
Davis, Charles. 1850-1854. Charleston, South Carolina.
Davis, Emerson. 1851-1854. Westfield, Massachusetts. SFB.
Davis, James. 1848--. Boston, Massachusetts.
Davis, Noah Knowles. 1855-1859. Marion, Alabama.
Day, Charles D. 1857-1859. Montreal, Canada.
Day, Jeremiah. 1850-1854. New Haven, Connecticut.
Day, John J. 1857-1859. Montreal, Canada. LC.
Dayton, Aaron Ogden. 1850-1857. Washington, D. C. LC. SFB.
Dayton, Edwin A. 1853-1859. Lawrence County, New York. SFB.
Dean, Amos. 1851--. Albany, New York. LC.
Dean, Philotus. 1853-1857. Alleghany, Pennsylvania. SFB.
Deane, James. 1848-1851. Greenfield, Massachusetts.
Dearborn, Henry A. S. 1848-1851. Roxbury, Massachusetts.
Dearborn, William L. 1848-1854. (inactive 1851-1853). Roxbury, Massachusetts.
Deems, Charles F. 1853. Greensboro, North Carolina. SFB.
DeKay, James Ellsworth. 1848-1851. New York, New York.
Delano, Joseph C. 1851--. New Bedford, Massachusetts. SFB.
Delavan, Edward Cornelius. 1856-1858. Albany, New York.
Denson, Claudius B. 1858--. Duplin County, North Carolina.
DeSaussure, H. A. 1850-1854. Charleston, South Carolina.
DeSaussure, H. W, 1850-1854. Charleston, South Carolina.
Devol, Charles. 1856-1859. Albany, New York.
Dewey, Chester. 1848--. Rochester, New York. SFB.
DeWolf, John. 1854-1857. Bristol, Rhode Island.
Dexter, G. M. 1857--. Boston, Massachusetts.
Dexter, George. 1856. Albany, New York.
Dexter, James. 1851--. (inactive 1854-1856). Albany, New York.
Dickinson, John Woodbridge. 1856-1858. Westfield, Massachusetts.
Dickson, Cyrus. 1858-1860. Baltimore, Maryland.
Dickson, Samuel Henry. 1848-1851. New York, New York.
Diehl, Israel S. 1856-1858. Sacramento, California.
Dilke, C. Wentworth. 1853-1855. London, England.
Dillaway, Charles Knapp. 1848-1851. Roxbury, Massachusetts.
Dinwiddie, Robert. 1848--. New York, New York, SFB.
Disturnell, John. 1857-1859. New York, New York.
Dixwell, Epes Sargent. 1848--. Cambridge, Massachusetts.
Donaldson, Francis. 1858-1860. Baltimore, Maryland.
Doolittle, L. 1857-1859. Lenoxville, Canada.
D'Orbigny, M. Alcide. 1848-1855. Paris, France.
Doremus, Robert Ogden. 1856-1860. New York, New York.
Dorr, Robert L. 1857. Danville, New York.
Douglass, C. C. 1848-1854. Detroit, Michigan.
Douglass, Silas Hamilton, 1851-1858 (inactive 1854-1856). Ann Arbor.
    Michigan. SFB.
Dow, George W. 1859--. Chicago, Illinois.
Dowie, J. Muir. 1857-1859. Liverpool, England.
Downes, John. 1856--. Washington, D.C.
Doyle, John P. 1857-1859. Montreal, Canada.
Drowne, Charles. 1851--. (inactive 1854-1857). Troy, New York, SFB.
Ducatel, Julius Timoleon. 1848-1849. Baltimore, Maryland.
Duffield, George. 1856-1859. Detroit, Michigan.
Dumont, A. H. 1860--. Rhode Island.
Duncan, Lucius Campbell. 1856-death. New Orleans, Louisiana.
Dunglison, Robley. 1854-1860. Philadelphia, Pennsylvania.
Dunkin, Christopher. 1857-1859. Montreal, Canada.
Dunn, R. P. 1860--. Providence, Rhode Island.
Dunn, T.C. 1860--. Newport, Rhode Island.
Dupuy, Charles H. 1858-1860. Baltimore, Maryland.
Dutton, Thomas R. 1848-1854. (inactive 1850-1851). New Haven, Connecticut. SFB.
Dwight, James M. B. 1851-1854. New Haven, Connecticut. SFB.

Dwight, Theodore William. 1851-1854. Clinton, New York.
Dwight, Timothy. 1850-1854. New Haven, Connecticut.
Dwinelle, John H. 1857-1860. Rochester, New York.
Dwinelle, William H. 1856--. New York, New York.
Dyer, David. 1856-1859. Albany, New York.
Dyer, Elisha. 1855--. Providence, Rhode Island.

Easter, John Day. 1851-1860. Athens, Georgia.
Eastman, Seth. 1853. Washington, D.C. LC.
Eastob, Norman. 1860--. Fall River, Massachusetts.
Eaton, Daniel Cady. 1859--. New Haven, Connecticut.
Eaton, Horace. 1853. Middlebury, Vermont. SFB.
Edgerton, Alonzo Jay. 1853. Grenada, Mississippi. SFB.
Edmondstone, William 1857-1859. Montreal, Canada.
Edmundson, Thomas. 1856-1858. Baltimore, Maryland.
Edwards, Richard. 1856-1858. Salem, Massachusetts.
Eggleston, N. H. 1859--. Madison, Wisconsin.
Eights, James. 1848-1851. Albany, New York.
Ela, Joseph. 1851-1854. Concord, New Hampshire.
Elderhorst, William. 1856-1858. Troy, New York.
Elin, Frederick. 1857-1859. London, England.
Eliot, Charles William. 1854-1855. Cambridge, Massachusetts.
Ellet, William Henry. 1848-1853. New York, New York.
Elliott, William. 1848-1854. Beaufort, South Carolina.
Ellsworth, Henry Leavitt. 1848-1851. Lafayette, Indiana.
Ely, Charles Arthur. 1850-1858. Elyria, Ohio.
Ely, George H. 1859--. Rochester, New York.
Ely, James W. C. 1855-1857. Providence, Rhode Island.
Emerson, Alfred. 1853-1855. Hudson, Ohio.
Emerson, George Barrell. 1848--. Boston, Massachusetts.
Emerson, George D. 1853-1856. (inactive 1854-1855). Cleveland, Ohio.
Engstrom, A. B. 1848--. Burlington, New Jersey. SFB.
Erwin, John S. 1850-1854. Marian, North Carolina. SFB.
Estabrook, Joseph. 1848-1854. Nashbille, Tennessee.
Este, D. K. 1850-1854. New Haven, Connecticut.
Estes, D. C. 1856-1858. Albany, New York.
Eustis, Henry Lawrence. 1849--. Cambridge, Massachusetts. LC SFB.
Evans, Alexander C, 1850-1854. Newark, New Jersey.
Evans, John. 1851--. Chicago, Illinois.
Evans, John. 1853-1854. Tadnor, Pennsylvania. SFB.
Eve, Paul Fitzsimmons. 1858-1860. Nashville, Tennessee.
Everett, Frank. 1848-1851. Canajoharie, New York.
Everett, Franklin. 1848-1851. Boston, Massachusetts.
Everett, J. D. 1860. Winsor, Nova Scotia.
Ewbank, Thomas. 1854-1856. Washington, D.C.
Ewendberg, L. C. 1853-1854. New Weid, Texas.
Ewing, Thomas. 1851--. Lancaster, Ohio.

Fair, Samuel. 1857 only. Columbia, South Carolina.
Fairbanks, Henry. 1860--. Hanover, New Hampshire.
Fairchild, Charles. 1857-1859. Madison, Wisconsin.
Fairchild, James Harris. 1851-1858. Oberlin, Ohio. SFB.
Fairfield, J. W. 1856-1859. Hudson, New York.
Fairhaven, James. 1857 only. Montreal, Canada.
Fairie, James. 1853-1859. Morehouse Parish, Louisiana. SFB.
Fairly, David. 1857-1859. Glasgow, Scotland.
Farnham, J. E. 1853-1859. (inactive 1855-1858). Georgetown, Kentucky.
Farquhar, W. H. 1858-1860. Montgomerie County, Maryland.
Fay, Charles. 1857-1859. St. Albans, Vermont.
Fearing, E. P. 1854-1859. Nantucket, Massachusetts.
Fellows, Joseph. 1856-1859. Albany, New York.
Felton, Cornelius Conway. 1849-1854. Cambridge, Massachusetts.
Felton, Frank Elliot. 1850-1854. Cambridge, Massachusetts. SFB.
Felton, Samuel Morse. 1849-1854. Charlestown, Massachusetts. SFB.
Ferland, John A. B. 1857-1859. Quebec, Canada.
Ferris, Isaac. 1851--. New York, New York.
Fillmore, Millard. 1853--. Buffalo, New York.
Finley, William P. 1851-1854. Charleston, South Carolina.
Fisher, Mark. 1856--. Trenton, New Jersey.
Fisher, George Park. 1857--. New Haven, Connecticut.
Fisher, Nathaniel Augustus. 1855-1859. Norwich, Connecticut.
Fisher, Robert A. 1855-1857. Providence, Rhode Island.

Fisher, Thomas. 1850-1857. (inactive 1851-1854). Philadelphia, Pennsylvania.
Fisk, L. R. 1856-1858. Ypsilanti, Michigan.
Fitch, Alexander. 1848-death (ca.1860). Hartford, Connecticut. SFB.
Fitch, Asa. 1848-1851. Salem, New York.
Fitch, Enoch. 1851-1854. Lyme, Connecticut.
Fitch, Orramel H. 1853--. Ashtabula, Ohio. SFB.
Flagg, Henry Collins. 1848-1854. New Haven, Connecticut.
Folsom, George. 1857--. New York, New York.
Foote, Elisha. 1856-1860. Seneca Falls, New York.
Foote, Thomas Moses. 1851-1854. Albany, New York.
Forbes, Charles Edwin. 1848-1851. Northampton, Massachusetts.
Forbes, R. W. 1849-1857. New York, New York.
Forbes, Robert Watson. 1848-1851. New Haven, Connecticut.
Force, Charles F. 1858--. Washington, D.C.
Ford, Richard. 1858-1860. Columbia, South Carolina.
Foreman, Edward. 1848-1851. Baltimore, Maryland.
Forrest, John. 1850-1854. Charleston, South Carolina.
Foster, Joel. 1848-1851. New York, New York.
Fowle, William Bentley. 1848--. Boston, Massachusetts. SFB.
Fowler, Asa. 1856 only. Concord, New Hampshire.
Fowler, William Chauncey. 1851-1854. Amherst, Massachusetts.
Fox, Charles. 1853-death (ca. 1858). Grosse Isle, Michigan. SFB.
Francford, E. 1855-1858. Middletown, Connecticut.
Franklin, Danforth. 1857-1859. Buffalo, New York.
Fraser, Hugh. 1857-1859. Montreal, Canada.
Fraaer, W. 1857-1859. Montreal, Canada.
Freeman, Samuel H. 1856-1858. Albany, New York.
Fremont, John Charles. 1848-1854. Washington, D.C.
French, Benjamin Franklin. 1848-1851. Philadelphia, Pennsylvania.
French, J. H. 1858-1860. Syracuse, New York.
French, John William. 1857-1859. West Point, New York.
Frick, George. 1848-1851. Baltimore, Maryland.
Fristoe, Edward T. 1856-1860. Washington, D.C.
Frost, Adolph. 1853-1855. Burlington, New Jersey. SFB.
Frost, Charles Christopher. 1859--. Brattleboro, Vermont.
Frost, H. R. 1850-1851. Charleston, South Carolina.
Frothingham, Frederick. 1857--. Portland, Maine.
Frothingham, George H. 1857-1859. Montreal, Canada.
Frothingham, Washington. 1856--. Albany, New York.

Gale, Samuel. 1857-1859. Montreal, Canada.
Gardin, Benjamin. 1850-1854. Charleston, South Carolina. SFB.
Gardner, James S. 1856-1859. Whitestown, New York.
Garland, L. C. 1848-1854. Tuscaloosa, Alabama.
Garrigue, Rudolph. 1853-1855. New York, New York. SFB.
Garrigues, S. S. 1856-1860. Philadelphia, Pennsylvania.
Garth, Charles. 1857-1859. Montreal, Canada.
Gavit, John E. 1848--. Albany, New York. LC.
Gay, Austin Milton. 1856-1859. Charlestown, Massachusetts.
Gay, Martin. 1848-1850. Boston, Massachusetts.
Gebhard, John. 1848-1854. Schoharie, New York.
Gebhard, John. 1848-1858 (inactive 1853). Schoharie, New York.
Geddes, Charles. 1857-1859. Montreal, Canada.
Geddings, Eli. 1850-1860 (inactive 1854-1858). Charleston, South Carolina.
Gerolt, Baron A. Von. 1848-1854. Washington, D.C.
Gibbons, Henry. 1848-1851. Philadelphia, Pennsylvania.
Gibbs, George. 1848-1851. New York, New York.
Gibbs, J. Campbell. 1857-1859. Montreal, Canada.
Gibbs, Josiah Willard. 1851-1859. Montreal, Canada.
Gibbs, Josiah Willard. 1851-1854. New Haven, Connecticut. SFB.
Gifford, J.P.S. 1856-1859. Albany, New York.
Gill, Charles. 1853 only. New York, New York. SFB.
Gilham, William. 1851-1854. Lexington, Virginia. SFB.
Gillett, J. 1851-1854. Cleveland, Ohio.
Gilman, Charles. 1858-1860. Baltimore, Maryland.
Gilman, Daniel Coit. 1853-- (inactive 1854-1855). New Haven, Connecticut.
Gilman, Judson. 1858-1860. Baltimore, Mayland.
Gilmor, Robert. 1848-death. Baltimore, Mayland.
Ginder, Henry. 1857 only. New Orleans, Louisiana.
Gladstone, T.H. 1856-1859. London, England.
Gluck, Isidor. 1856-1859. New York, New York.

Goddard, Paul Beck. 1848-1854 (inactive 1853). Philadelphia, Pennsylvania.
Goddard, William. 1855 only. Providence, Rhode Island. LC.
Gold, Stephen A. 1859--. New Haven, Connecticut.
Gold, Theodore Sedgwick. 1850-0. West Cornwall, Connecticut. SFB.
Goldmark. 1850-1855. Vienna.
Goodrich, Charles Rush. 1856 only. New York, New York.
Goodsell, Thomas. 1851-1854. Utica, New York.
Goodwin, William Frederick. 1856--. Concord, New Hampshire.
Gould, Benjamin Apthorp. 1849-1859. Boston, Massachusetts.
Gourdin, Henry. 1850-1854. Charleston, South Carolina.
Gourdin, Robert N. 1850-1854. Charleston, South Carolina.
Grant, John Mason. 1857-1859. Montreal, Canada.
Grant, Seth Hastings. 1857-1860. New York, New York.
Gray, Alonzo. 1848-- (inactive 1851-1858). Brooklyn, New York.
Gray, Henry. 1851-1854. Springfield, Massachusetts.
Gray, Henry C. 1858-1860. Washington, D.C.
Gray, James Harrison. 1851-death 1856. Springfield, Massachusetts. SFB.
Green, A. Thomas. 1848-1851. New Bedford, Massachusetts.
Green, Everett W. 1856 only. Madison, New Jersey.
Green, Horace. 1856--. New York, New York.
Green, James. 1848-1854. Baltimore, Maryland.
Green, John W. 1855-1859. New York, New York.
Greene, Benjamin D. 1848--. Boston, Massachusetts.
Greene, B. Franklin. 1848-1859 (inactive 1854-1858). Troy, New York. LC.
Greene, David B. 1856-1859. New York, New York.
Greene, Everett W. 1856-1859. Madison, New Jersey.
Greene, Francis Coles. 1857--. East Hampton, Massachusetts.
Greene, S.D. 1848-1851. Boston, Massachusetts.
Greene, Samuel. 1854--. Woonsocket, Rhode Island.
Greene, Samuel Stillman. 1855-1859. Providence, Rhode Island.
Greenshields, J. B. 1857 only. Montreal, Canada.
Griffin, Nathaniel Herrick. 1856-1859. Williamstown, Massachusetts.
Griffith, Robert Eglesfeld. 1848-1850. Philadelphia, Pennsylvania.
Grigg, William. 1850-1854. Charleston, South Carolina.
Grinnan, A.G. 1853--. Madison Court House, Virginia.
Griscom, John. 1848-1854. Burlington, New Jersey.
Groneweg, Lewis. 1853-1860. Germantown, Ohio. SFB.
Grosvenor, A. C. 1851-1859. Cincinnati, Ohio. SFB.
Grosvenor, William. 1855-1859. Providence, Rhode Island. LC.
Grundy, John. 1853--. Cincinnati, Ohio. SFB.
Gruvern, Julius. 1856-1859. New Haven, Connecticut.
Guerin, Thomas. 1857-1859. Albany, New York.
Guest, William E. 1851-1859. Ogdensburg, New York. SFB.
Guilford, Nathan. 1850-1854. Cincinnati, Ohio. LC.
Gulick, John Thomas. 1856-1860. Williamstown, Massachusetts.
Gummere, Samuel James. 1853--. (inactive 1856). Burlington, New Jersey. SFB.
Guy, S.S. 1848-1851. Brooklyn, New York.

Hadley, Amos. 1856 only. Concord, New Hampshire.
Hager, Albert David. 1857--. Proctorsville, Vermont.
Hague, John M. 1851-1857. Newark, New Jersey.
Hague, William W. 1851-1857. Newark, New Jersey.
Haines, William S. 1855--. Providence, Rhode Island.
Hale, Albert W. 1856-1859. Albany, New York.
Hale, Enoch. 1848 only. Boston, Massachusetts.
Hale, Josiah. 1853-only. New Orleans, Louisiana.
Hall, Archibald. 1856-1859. Montreal, Canada.
Hall, Edward. 1848-1854. San Francisco, California.
Hall, Frederick. 1848 (deceased AAGN member). Washington, D.C.
Hall, Joel. 1853--. Athens, Illinois. SFB.
Hall, Nathan Kelsey. 1853-1859. Buffalo, New York.
Hall, William. 1850-1855. Mobile, Alabama.
Halleck, Henry Wager. 1848-1854. San Francisco, California.
Hallowell, Benjamin. 1853-1858. Alexandria, Virginia. LC. SFB.
Ham, L. J. 1856 only. Williamsville, New York.
Hamilron (Dr.) 1850-1854. Mobile, Alabama.
Hammond, Charles. 1859--. Groton, Massachusetts.
Hammond, George T. 1860--. Newport, Rhode Island.
Hammond, John Fox. 1853-1855. Pensacola, Florida. SFB.
Hammond, Ogden. 1850-1859. Charleston, South Carolina.
Hamnett, J. 1853-1855. Meadville, Pennsylvania.

Hance, Ebenezer. 1853--. Morrisville, Pennsylvania. SFB.
Hand, Thomas Jennings. 1858--. Baltimore, Maryland.
Hand, Isaac. W. K. 1856--. Portsmouth, Virginia.
Handy, Truman P. 1851-1854. Cleveland, Ohio. LC.
Hanover, M.D. 1859--. Springfield, Tennessee.
Hardcastle, Edmund L.F. 1853-1857. Washington, D.C. LC. SFB.
Hardy (Dr.). 1848-1853. Asheville, North Carolina.
Harlan, Joseph Gibbons. 1854-1857. Javerford, Pennsylvania.
Harlan, Richard. 1848 (deceased AAGN member). Philadelphia, Pennsylvania.
Harman, Henry M. 1848--. Baltimore, Maryland.
Harrequi, Jose' Salazar. 1856-1858. Minerra, Mexico.
Harris, C. Townsend. 1851-1854. New York, New York.
Harris, Chapin Aaron. 1858-1860. Baltimore, Maryland.
Harris, Charles T. 1850-1854. New York, New York.
Harris, Ira. 1857-1859. Albany, New York. LC.
Harris, J.O. 1853-1854. Ottawa, Illinois. SFB.
Harris, William Logan. 1853-1855. Delaware, Ohio. SFB.
Harrison, B. F. 1857--. Wallingsford, Connecticut.
Harrison, Edwin. 1857--. St. Louis, Missouri.
Harrison, Joseph. 1858-1860. Philadelphia, Pennsylvania.
Hart, Simeon. 1858-1863.* Farmington, Connecticut.
Hart, Theodore. 1857-1859. Montreal, Canada.
Harte, Rufus E. 1853-1856. Columbus, Ohio. SFB.
Hartley, M. B. 1848-1854. Montreal, Canada.
Hartshorne, Henry. 1858--. Philadelphia, Pennsylvania.
Hartshorn, O.N. 1853 only. Mt. Union, Ohio.
Harvey, Matthew. 1848-1860. Concord, New Hampshire, SFB.
Haskins, R.W. 1851-1854. Buffalo, New York. SFB.
Hason, Henry B. 1855--. Worcester, Massachusetts.
Hathaway, Charles. 1856-1860. Delhi, New York.
Haven, Samuel Forster. 1855--. Worcester, Massachusetts.
Haven, Simon Z. 1848-1851. Utica, New York.
Hawkesworth, W. 1850-1854. Charleston, South Carolina.
Hay, George. 1850-1854. Barnwell, South Carolina.
Hayden, C. B. 1848-1854. Smithfield, Virginia.
Hayden, Ferdinand Vandiveer. 1858-1860. Washington, D.C.
Hayden, Horace H. 1848 (deceased AAGN member). Baltimore, Maryland.
Hayes, John Lord. 1848-1854. Portsmouth, New Hampshire.
Haynesworth (Dr.). 1848-1854. Sumpterville, South Carolina.
Hayward, James. 1848--. Boston, Massachusetts. SFB.
Hazard, Rowland Gibson. 1855--. Peace Dale, Rhode Island.
Headlam, William. 1857-1860. Albany, New York.
Headley, William S. 1853 only. Albany, New York.
Hedrick, B. S. 1854--(inactive 1856). New York, New York.
Heffron, Daniel S. 1856-1859. Utica, New York.
Helme, W. H. 1855-1859. Providence, Rhode Island.
Henderson, Andrew Augustus. 1848-1851. Huntington, Pennsylvania.
Henderson, Isaac J. 1857 only. New Orleans, Louisiana.
Heneker, William R. 1857-1859. Sherbrooke, Canada.
Herbert, Alfred. 1854-1856. Washington, D.C.
Herrick, F.C. 1853 only. Bowling Green, Kentucky. SFB.
Herring, William A. Duplin. 1858-1860. Sampson County, North Carolina.
Hext, G. 1856 only. Oxford, England.
Hickok, M.J. 1857--. Scranton, Pennsylvania.
Hickok, William C. 1848-1851. New York, New York.
Hicks, Levi I. 1856-1859. Walworth, New York.
Hiester, John P. 1853-1854. Reading, Pennsylvania.
Higgins, James. 1858-1860. Baltimore, Maryland.
Hildreth, Samuel Prescott. 1848-1854. Marietta, Ohio.
Hill, B. L. 1853-1855. Baresville, Ohio.
Hill, Benjamin. 1857-1859. Montreal, Canada.
Hill, Nathaniel Peter. 1856-1859. Providence, Rhode Island. L.C.
Hill, Nicholas. 1851-1859. Albany, New York.
Hill, S. W. 1851--. Eagle Harbor, Michigan. SFB.
Hillard, George Stillman. 1849-1854. Boston, Massachusetts.
Hillhouse, William. 1850-1854. New Haven, Connecticut. LC.
Hincks, William. 1848-- (inactive 1856). Toronto, Canada.
Hingston, W.P. 1856-1859. Montreal, Canada. LC.
Hiscox, Gardner D. 1850-1854. New York, New York. SFB.
Hoadley, E. S. 1859--. East Hampton, Massachusetts.
Hoblitzell, J. H. 1858-1860. Baltimore, Maryland.

Hodgins, George. 1857-1859. Toronto, Canada.
Hodgson,William Brown. 1856--. Savannah, Georgia.
Holcomb, Amasa. 1853 only. Southwark, Massachusetts.
Holland, Joseph B. 1855--. Monson, Massachusetts.
Hollowell, W. E. 1855 only. Alabama.
Holmes, Andrew Fernardo. 1857--. Montreal, Canada.
Holmes, Benjamin. 1857--. Montreal, Canada.
Holton, David P. 1851-1854. New York, New York. SFB.
Holton, Luther Hamilton. 1857--. Montreal, Canada.
Holwell, W. A. 1857-1859. Quebec, Canada.
Holyoke, F. E. 1850-1854. Cambridge, Massachusetts.
Holyoke, Francis. 1851-1854. Salem, Massachusetts. SFB.
Homans, J.S. 1858-1860. New York, New York.
Homans, Sheppard. 1856-1860. New York, New York.
Homes, Henry Augustus. 1857--. Albany, New York.
Hopkins, Albert. 1848-1851. Williamstown, Massachusetts.
Hopkins, J. A. 1854-1856. Milwaukee, Wisconsin.
Hopkins, James G. 1856-1859. Ogdensburg, New York.
Hopkins, John Henry. 1857-1859. New York, New York.
Hopkins, T. O. 1856-1859. Williamsville, New York.
Hopkins, W. Fenn. 1853-1857. Annapolis, Maryland. SFB.
Horan, Eduard John. 1856-1859. Quebec, Canada.
Hord, Kellis. 1853-1856.
Horlbeck, E. 1850-1854. Charleston, South Carolina.
Horsford, Benjamin F. 1859--. Haverhill, Massachusetts.
Horton, C. V. R. 1856-1859. Chaumont, New York.
Horton, William. 1848 (deceased AAGN member). Orange County, New York.
Hotchkiss, Jebediah. 1851-1855. Bridgewater, Virginia. SFB.
Houghton, Douglass. 1848 (deceased AAGN member). Detroit, Michigan.
Houghton, George Frederick. 1857--. St. Albans, Vermont.
Howard, Benjamin. 1859--. Williamstown, Massachusetts.
Howard, R. P. 1857-1859. Montreal, Canada.
Howe, Henry A. 1857-1859. Montreal, Canada.
Howe, Samuel Gridley. 1849-1851. Boston, Massachusetts.
Hoy, Philo R. 1853-1857. Racine, Wisconsin. SFB.
Hoyt, John Wesley. 1854-1857. Cincinnati, Ohio.
Hubbard, Bela. 1848-1851. Detroit, Michigan.
Hulburd, Calvin T. 1857 only. Brasher Falls, New York.
Hume, William. 1849-1854. Charleston, South Carolina.
Humphrey, Frederic. 1857-1859. Iowa City, Iowa.
Humphrey, S. Dwight. 1854-1858 (inactive 1855-1856). New York, New York.
Humphrey, William F. 1853-1855. New Haven, Connecticut.
Humphreys, Andres Atkinson. 1857-1859. Washington, D.C.
Humphreys, Hector. 1848-1851. Annapolis, Maryland.
Hun, Thomas. 1850-1857. Albany, New York. LC.
Hunt, Freeman. 1857-1858. *New York, New York.
Hunt, George. 1855--. Providence, Rhode Island.
Hunt, Washington. 1851-1854. Albany, New York.
Hunter, Andrew W. 1857-1859. New Orleans, Louisiana.
Hunter, George W. 1857--. New Orleans, Louisiana.
Hunton, Ariel. 1850-1854. Hyde Park, Maryland.
Hurd, Isaac N. 1859--. Corning, New York.
Husband, J. J. 1853 only. Cleveland, Ohio.
Huse, Caleb. 1859--. West Point, New York.
Huson, Calvin. 1857-1859. Rochester, New York.

Ingalls, Thomas R. 1848-1854. Greenwich, New York. SFB.
Ingham, Albert C. 1856 only. Madison, Wisconsin.
Inglis, David. 1857-1859. Hamilton, Canada.
Irwin (Dr.). 1848-1851. Morgantown, North Carolina.
Ives, Eli. 1850-1854. New Haven, Connecticut.
Ives, Moses Brown. 1855-1859. Providence, Rhode Island. LC.
Ives, Nathan Beers. 1850-1854. New Haven, Connecticut.
Ives, Thomas P. 1856--. Providence, Rhode Island.

Jack, W. B. 1857--. Frederickton, New Brunswick.
Jackson, John B. S. 1848-1851. Boston, Massachusetts.
Jackson, John. 1853 only. Darby, Pennsylvania. SFB.
Jackson, Robert M.S. 1848-1854. Alexandria, Pennsylvania.
Jacobs, Michael. 1854-1856. Gettysburg, Pennsylvania.
James, Charles S. 1856-1860. Lewisburg, Pennsylvania.

James, John. 1856-1859. Alton, Illinois.
James, U. P. 1851-1860. Cincinnati, Ohio. SFB.
Jamison, D. F. 1850-1851. Orangeburg, South Carolina.
Janes, D. P. 1857-1859. Montreal, Canada.
Jay, John Clarkson. 1848-1854. New York, New York. SFB.
Jayne, J. W. 1851-1854. Cincinnati, Ohio. SFB.
Jeffers, William Nicholson. 1851-1854. Bridgeton, New Jersey. SFB.
Jeffrey, R. W. 1853-1854. Pensacola, Florida.
Jenkins, John F. 1851-1854. New Salem, New York. SFB.
Jenkins, Thornton A. 1853--. Washington, D.C. LC.
Jenner, Solomon. 1850-1856. New York, New York. SFB.
Jennings, N. R. 1850-- (inactive 1855). New Orleans, Louisiana.
Jervey, J. P. 1850-1854. Charleston, South Carolina.
Jewett, George Baker. 1853 only. Amherst, Massachusetts. SFB.
John, Samuel F. 1849-1851. Cambridge, Massachusetts.
Johnson (Lt.). 1848-1851. Washita, Texas.
Johnson, Alexander Smith. 1851-1854. New York, New York. SFB.
Johnson, Benjamin Pierce. 1856 only. Albany, New York. LC.
Johnson, Christopher. 1856-1860 (inactive 1857). Baltimore, Mayland. LC.
Johnson, Hosmer A. 1853-1855. Chicago, Illinois. SFB.
Johnson, James. 1850-1854. Charleston, South Carolina.
Johnson, Joseph. 1850-1854. Charleston, South Carolina. SFB.
Johnson, Lyman H. 1859--. Rockford, Illinois.
Johnson, Samuel. 1853-1855. New Haven, Connecticut. SFB.
Johnson, Samuel William. 1850-1854. New Haven, Connecticut. LC. SFB.
Johnson, Sydney L. 1848-1850. New Orleans, Louisiana.
Johnson, William. 1853-1855. Tuscaloosa, Alabama. SFB.
Johnson, William C. 1851-1860. Utica, New York.
Johnston, Algernon Knox. 1859--. Platteville, Wisconsin.
Johnston, Stephen. 1853-1854. Platte City, Missouri. SFB.
Jones, Catesby. 1854--. Washington, D.C.
Jones, Edward. 1850-1851. Charleston, South Carolina.
Jones, James. 1853 only. New Orleans, Louisiana.
Jones, Thomas P. 1848-1851. Stockbridge, Massachusetts.
Jones, Thomas Walter. 1856-1859. Montreal, Canada.
Joseph, J. H. 1857-1860. Montreal, Canada.
Joslin, Benjamin F. 1856-1859. New York, New York.
Joy, Charles Arad. 1854--. New York, New York.

Kain, John Henry. 1848-1851. New Haven, Connecticut.
Keefer, Samuel. 1857--. Brockville, Canada.
Keefer, Thomas C. 1857-1860. Hamilton, Canada.
Keep, Nathan Cooley. 1859--. Boston, Massachusetts.
Keith, Reuel. 1851-1855. Washington, D.C. LC. SFB.
Keller, William. 1848-1851. Philadelphia, Pennsylvania.
Kelley, Edwin. 1853-1859. Elyria, Ohio. SFB.
Kemp, Alexander. 1857-1859. Montreal, Canada.
Kempton, John. 1850-1854. New York, New York.
Kendall, David. 1854-1856. Rochester, New York.
Kendall, Edward Dwight. 1850-1854. Cambridge, Massachusetts. SFB.
Kendall, Ezra Otis. 1848-1854. Philadelphia, Pennsylvania.
Kendall, Joshua. 1856-1860. Meadville, Pennsylvania.
Kendrick, Henry Lane. 1848-1854. West Point, New York.
Kennedy, Alfred L. 1853-1857. Philadelphia, Pennsylvania. SFB.
Kenney, William B. 1848-1851. Newark, New Jersey.
Kennicott, Robert. 1858-1860. West Northfield, Illinois.
Kerr, George H. 1851-1854. Franklin, New York. SFB.
Kerr, Robert. 1857 only. New Orleans, Louisiana.
Kerr, Washington Caruthers. 1856--. Charlotte, North Carolina.
Kierskowski, A. E. 1857-1859. Montreal, Canada.
Kimball, D. A. 1851-1854. Warren, Massachusetts.
Kimmel, Anthony. 1858-1860. Linganore, Maryland.
King, Alfred T. 1848-1851. Greensburg, Pennsylvania.
King, Charles. 1851-1854. New York, New York.
King, David. 1859--. Newport, Rhode Island. LC.
King, Mitchell. 1850-1860. Charleston, South Carolina.
Kingsley, A. 1851-1856. Meadville, Pennsylvania.
Kingston, G. T. 1857--. Toronto, Canada.
Kinnear, David. 1857-1859. Montreal, Canada. LC.
Kirkpatrick, James A. 1853--. Philadelphia, Pennsylvania. SFB.
Kirkpatrick, John. 1853-1857. Ohio City, Ohio. SFB.

Kitchell, William. 1851-1859. Newark, New Jersey. SFB.
Kite, Thomas. 1851--. Cincinnati, Ohio. SFB.
Kittredge, Josiah. 1857--. South Hadley, Massachusetts.
Klipstein, Louis Frederick. 1850-1854. St. James, South Carolina.
Kloman, W. C. 1858-1860. Baltimore, Maryland.
Knox, J. P. 1857--. Newtown, Long Island, New York.
Kurtz, John. 1850-1860 (inactive 1855-1856). Bucksport, Maine.

Laflamme, G. 1857-1859. Montreal, Canada.
Laflamme, T. A. Radolphe. 1857-1859. Montreal, Canada.
Lamb, James. 1850-1851. Charleston, South Carolina.
Lamoreaux, M. W. 1851-1854. Schenectady, New York.
Lane, Charles W. 1850-1854. Milledgeville, Georgia. SFB.
Langdon, William Chauncy. 1851-1854. Washington, D. C. LC. SFB.
Lansing, Gerritt Y. 1851-1859. Albany, New York. LC.
LaRoche, Charles Pucy. 1858-1860. Philadelphia, Pennsylvania.
LaRoche, Rene. 1858--. Philadelphia, Pennsylvania.
Lasel, Edward. 1848-death. Williamstown, Massachusetts.
Latham, Richard P. 1854-1856. Richmond, Virginia.
Lathrop, Stephen P. 1853-1854. Beloit, Wisconsin. SFB.
Latour, Louis A. Huguet. 1857-1859. Montreal, Canada. LC.
Lauderdale, John V. 1856--. Geneseo, New York.
Lawford, Frederick. 1857-1859. Montreal, Canada.
Lawrence, Amos. 1849-1851. Boston, Massachusetts.
Lawrence, Amos Adams. 1849-1851. Boston, Massachusetts.
Lawrence, Edward A. 1859--. East Windsor, Connecticut.
Lawrence, George Newbold. 1853--. New York, New York. SFB.
Lawrence, William. 1849-1854. Boston, Massachusetts.
Lea, Robert M. 1858--. Aransas, Texas.
Leavenworth, Melines Conklin. 1848-1856. Waterbury, Connecticut.
Leckie, Robert. 1857-1859. Montreal, Canada.
LeConte, John Eatton. 1851-1857. New York, New York.
Lederer, Baron von. 1848 (deceased AAGN member). Washington, D.C.
Lee, Charles Alfred. 1848-1851. New York, New York.
Lee, John C. 1848-1854. Salem, Massachusetts.
Lee, Thomas J. 1851-1860. Ellangowan, Maryland. LC.
Leeds, Stephen P. 1851-1854. Brooklyn, New York. SFB.
Lefferts, John. 1853 only. Seneca County, New York. SFB.
Leffingwell, Edward Henry. 1851-1854. Columbia, Missouri.
Leonard, F.B. 1851-1854. Lansingburgh, New York.
Leonard, Frederick Baldwin. 1848-1854. Washington, New York.
Leonard, Levi Washburn. 1851-1854. Dublin, New Hampshire. SFB.
Lesene, Henry D. 1850-1854. Charleston, South Carolina.
Lesley, Joseph. 1854--. Philadelphia, Pennsylvania.
Lesley, William W. 1854-1856. Monticello, Missouri.
Lesquereux, Leo. 1853 only. Columbus, Ohio. SFB.
Lettsom, W. J. 1848-1851. Washington, D.C.
Lewis, Robert C. 1850-1851. Shelbyville, Kentucky.
Lieber, Francis. 1849-1856 (inactive 1855). Charleston, South Carolina.
Lincklaen, Ledyard. 1848-death (ca. 1860). Cazenovia, New York. SFB.
Lincoln, N.S. 1858-1860. Washington, D.C.
Lindsay, William B. 1857 only. New Orleans, Louisiana.
Lindsley, James H. 1848-death. Stafford, Connecticut.
Lindsley, John Berrien. 1848--. Nashville, Tennessee. SFB.
Lintner, Joseph Albert. 1851-1854. Schoharie, New York. SFB.
Lippitt, Edward Spaulding. 1851-1854. Cincinnati, Ohio. SFB.
Lischka, Emile. 1848-1854. Washington, D.C.
Little, Weare C. 1857-1859. Albany, New York.
Litton, A. 1858--. St. Louis, Missouri.
Livermore, Abiel Abbott. 1851-1857. Cincinnati, Ohio.
Locke, J. H. 1848-1854. Nashua, New Hampshire.
Locke, John. 1849 only. Cincinnati, Ohio.
Locke, Joseph M. 1859--. Cincinnati, Ohio.
Locke, Luther Franklin. 1853--. Nashua, New Hampshire. SFB.
Lockwood, Moses B. 1855-1859. Providence, Rhode Island.
Loomis, Charles Lafayette. 1855-- (inactive 1857-1859). Wheeling, West Virginia.
Loomis, Justin Rudolph. 1856-1860. Lewisburg, Pennsylvania.
Loosey, Charles F. 1858--. New York, New York.
Loranger, Thomas J. J. 1857 only. Montreal, Canada.
Lord, Asa Dearborn. 1853-1855. Columbus, Ohio. SFB.
Lord, John. 1859--. Stamford, Connecticut.

Louson, John. 1857 only. Montreal, Canada.
Lovelace, P. E. H. 1857-1859. Fort Adams, Mississippi.
Lunn, William. 1857--. Montreal, Canada.
Lusher, Robert M. 1856-1859. New Orleans, Louisiana.
Lyell, Charles. 1848-1855. London, England.
Lyman, Addison. 1841-1854. Geneseo, Illinois. SFB.
Lyman, Henry. 1857--. Montreal, Canada. LC.
Lyman, Josiah. 1851-1854. Lenox, Massachusetts. SFB.
Lyman, Theodore. 1857-1859. Montreal, Canada.
Lyon, Caleb. 1854-1860. Lyonsdale, New York.
Lyon, Merrick. 1857-1859. Providence, Rhode Island.

M'Cabe, James Dabney. 1858-1860. Mount Washington, Maryland.
McAlpine, William Jarvis. 1851-1859. Albany, New York.
McCall, George Archibald. 1853-1859. Philadelphia, Pennsylvania.
McCall, John. 1856--. Utica, New York.
MacCallum, Daniel Craig. 1857-1859. Montreal, Canada.
McCauley, James. 1858-1860. Cecil County, Maryland.
McClory, Henry. 1857-1859. Warehouse Point, Connecticut.
McCord, J. L. 1857--. Montreal, Canada.
McCormick, Richard Cunningham. 1857-1859. New York, New York.
McCoy, Amasa. 1856-1859. Albany, New York.
McCron, John. 1858-1860. Baltimore, Maryland.
McDonald, Marshall. 1853-1857. New Creek Depot, Virginia. SFB.
McElroy, James. 1853-1855. Delaware, Ohio. SFB.
McElroy, John C. 1857-1859. Albany, New York.
McEuen, Thomas. 1848-1854. Philadelphia, Pennsylvania.
McFarlan, Henry. 1854-1860. Dover, New Jersey.
Machin, Thomas. 1856-1859. Albany, New York.
McIlvaline, Joshua. 1857-1859. Rochester, New York.
Mackay, Joseph. 1857-1859. Montreal, Canada.
Mackay, T. P. 1857-1859. Montreal, Canada.
McKew, Dennis. 1858-1860. Baltimore, Maryland.
McKinley, Alexander. 1848-1851. Philadelphia, Pennsylvania.
Maclean, George M. 1851-1860. Pittsburgh, Pennsylvania. SFB.
McLean, James. 1853 only. Hudson, Ohio.
McMahon, Mathew. 1856-1859. Albany, New York.
McMinn, J.M. 1853-1855. Fleming, Pennsylvania. SFB.
McMurtie, Henry. 1848-1851. Philadelphia, Pennsylvania.
McNaughton, James. 1850-1859. Albany, New York.
McNaughton, Peter. 1856-1859. Albany, New York.
Macomber, David O. 1856-1859. Middletown, Connecticut.
M'Conihe, Isaac. 1850--. Troy, New York. SFB.
Macrae, Archibald. 1854-1856. North Carolina.
McRae, John. 1850--. Camden, South Carolina.
MacWhorter, Alexander. 1850-1854. New Haven, Connecticut.
Macy, Alfred. 1859--. Nantucket, Massachusetts.
Maffitt, John Newland. 1850-1854. United States Navy.
Mahan, Dennis Hart. 1855--. West Point, New York.
Major, James. 1853-1855. Washington, D.C. LC. SFB.
Malone. 1850-1851. Athens, Georgia.
Mandeville, Henry. 1851-1854. Albany, New York.
Manigault, Gabriel Edward. 1850-1854. Charleston, South Carolina.
Mansfield, Edward Deering. 1850-1854. Cincinnati, Ohio. LC.
March, Alden. 1850-1854. Albany, New York. LC.
Marissal, F.V. 1860--. Fall River, Massachusetts.
Markham, Jesse. 1853-1855. Salem, Ohio. SFB.
Markoe, Francis. 1848-1854. Washington, D.C. LC.
Marsh, Dexter. 1848-1853. Greenfield, Massachusetts.
Marsh, James E. 1856-1859. Roxbury, Massachusetts.
Marshall, Charles. 1850-1854. Indiana.
Martin, Benjamin Nicholas. 1851-1854. Albany, New York.
Martin, W. J. 1858--. Chapel Hill, North Carolina.
Mason, Charles. 1857-1859. Burlington, Iowa. LC.
Mason, Cyrus. 1851-1854. New York, New York.
Mason, Francis. 1853-1855. Maulmain, India.
Mason, Isaac N. 1853-1855. Cleveland, Ohio.
Mason, Owen. 1848-1859 (inactive 1851-1855). Providence, Rhode Island. LC.
Mason, R. C. 1855-1857. Madison, Wisconsin.
Mather, William Williams. 1848-1859.* Columbus, Ohio. SFB.
Mathews, T. J. 1851-1854. Oxford, Ohio.

Matheison, Alexander. 1857-1859. Montreal, Canada.
Mathoit, George. 1854-1859. Washington, D. C.
Mattison, Hiram. 1853-1855. New York, New York. SFB.
Maupin, S. 1856-1860. Charlottesville, Virginia.
Mauran, Joseph. 1849--. Providence, Rhode Island. LC.
Maverick, Augustus. 1850-1854. New York, New York.
Mayhew, D. P. 1859--. Ypsilanti, Michigan.
Maynard, Alleyne. 1853-1855. Cleveland, Ohio.
Mead, Orlando. 1851-1859 (inactive 1855-1857). Albany, New York.
Mead, Samuel Burnam. 1853 only. Augusta, Illinois. SFB.
Means, Alexander. 1851--. Oxford, Georgia. SFB.
Medbury, E. 1851-1854. Columbus, Ohio.
Meek, Fielding Bradford. 1851--. Albany, New York.
Meigs, H. 1848-1851. New York, New York.
Meigs, James Aitken. 1858--. Philadelphia, Pennsylvania.
Meilleur, Jean Baptiste. 1857-1860. Montreal, Canada.
Mensheimer, Frederick E. 1848-1854. Dover, New York.
Meredith, Edmund P. 1857-1859. Toronto, Canada.
Meriwether, Charles I. 1853 only. Albermarle County, Virginia. SFB.
Merrick, Frederick. 1848-1854. Athens, Ohio.
Merrick, Samuel Vaughan. 1848-1857. Philadelphia, Pennsylvania.
Merrill, Hubert H. 1859--. Labanon, Tennessee.
Merrill, J. W. 1856 only. Concord, New Hampshire.
Merrill, Stephen. 1857-1859. Charlestown, Massachusetts.
Merrrman, M. 1858-1860. Baltimore, Mayland.
Metcalfe, Samuel Luther. 1848-1851. Kentucky.
Meyers, A. M. 1858-1860. Baltimore, Maryland.
Michelotti, M. J. 1848-1855. Turin, Italy.
Mighels, J. W. 1848-1851. Cincinnati, Ohio.
Miles, Francis T. 1840-1854. Charleston, South Carolina.
Miles, Henry H. 1857--. Lennoxville, Canada.
Millar, G. M. 1857 only. Montreal, Canada.
Miller, H. B. 1848-1854. New York, New York.
Miller, Samuel. 1860--. New Haven, Connecticut.
Miller, William A. 1851-1854. Albany, New York. SFB.
Millington, John. 1848-1859. Oxford, Mississippi. SFB.
Mills, B. F. 1853-1855. Sauk County, Illinois.
Mills, Charles C. 1857 only. Richmond, Virginia.
Minifie, William. 1858--. Baltimore, Maryland.
Mitchell, Elisha. 1848-1854. Chapel Hill, North Carolina.
Mitchell, Henry. 1856-1860. Nantucket, Massachusetts.
Mitchell, J. B. 1853-1855. Johnson City, Tennessee. SFB.
Mitchell, John Kearsley. 1848-1854. Philadelphia, Pennsylvania. LC.
Mitchell, Maria. 1850--. Nantucket, Massachusetts. SFB.
Mittag, J. F. G. 1858-1860. Hagerstown, Maryland.
Moffat, A. G. 1853 only. Choctaw Nation, Arkansas.
Moffatt, George. 1857-1860. Montreal, Canada. LC.
Moffatt, J. O. 1857-1859. Montreal, Canada.
Molinard, J. 1851-1854. Albany, New York.
Mondelet, Charles J. E. 1857-1859. Montreal, Canada.
Monroe, Nathan. 1851-1854. Bradford, Massachusetts.
Monson, Alfred. 1850-1854. New Haven, Connecticut.
Montague, Theodore L. 1859--. Pomeroy, Ohio.
Montgomerie, Hugh E. 1857-1859. London, England.
Moody, L. A. 1851-1855. Chicopee, Massachusetts.
Moore, George Henry. 1854-1860. New York, New York.
Moore, Thomas V. 1853-1855. Richmond, Virginia. SFB.
Morange, William D. 1856-1859. Albany, New York.
Mordecai, Alfred. 1853-1857. Washington, D.C. LC. SFB.
Morfit, Campbell. 1853-1859 (inactive 1856). New York, New York.
Morgan, DeWitt C. 1860--. Baltimore, Maryland.
Morris, DeWitt. 1860--. Ellingham, Connecticut.
Morris, J. R. 1857--. Houston, Texas.
Morris, Margaretta H. 1850-1859. Germantown, Pennsylvania.
Morrison, Benjamin F. 1859--. Nantucket, Massachusetts.
Morrow, R. G. 1857-1860. Cambridge, Massachusetts.
Morse, Charles M. 1856--. Waterville, Maine.
Morse, M. L. 1856-1858. Dover, New Hampshire.
Morton, Sketchley. 1853 only. Delaware County, Pennsylvania. SFB.
Moss, Theodore F. 1849-1854. Philadelphia, Pennsylvania. SFB.
Moultrie, William L. 1850-1854. Charleston, South Carolina.
Mowat, Oliver. 1857-1859. Toronto, Canada.

Munger, George Gourdy. 1856--. Rochester, New York.
Munro, William. 1857-1859. Quebec, Canada. LC.
Munroe, Nathan. 1855--. Bradford, Massachusetts. SFB.
Munsell, Joel. 1851-1859. Albany, New York. SFB.
Murdock, Charles N. 1858-1860. Baltimore, Maryland.
Murdock, John. 1858-1860. Baltimore, Maryland.
Murphy, Edward. 1857-1859. Montreal, Canada.
Murphy, John W. 1851-1854. Troy, New York. SFB.
Murphy, Alexander. 1857-1859. Woodstock, Canada.
Murray, David. 1857--. Albany, New York.
Mussey, D. 1856 only. Albany, New York.
Mussey, Reuben Dimond. 1856 only. Cincinnati, Ohio.
Mutter, Thomas Dent. 1848-1854. Philadelphia, Pennsylvania. SFB.
Muzzey, John. 1850-1851. Portland, Maine.
Myers, Gustavus A. 1857--. Richmond, Virginia.

Nash, John Adama. 1851-1854. Amherst, Massachusetts. SFB.
Nason, Elias. 1857-1859. Natuck, Massachusetts.
Nason, Henry Bradford. 1855-- (inactive 1856-1858). Worcester, Massachusetts.
Nault, J. Y. 1857-1859. Quebec, Canada.
Nelles, Samuel Sobieske. 1857-1859. Victoria College, Nova Scotia.
Nelson, Cleland K. 1858--. Annapolis, Maryland.
Nelson, J. P. 1853-1860. Lewisburg, North Carolina. SFB.
Newberry, Samuel. 1853 only. Cleveland, Ohio.
Newcomb, Wesley. 1856-1860. Albany, New York.
Newell, William Augustus. 1857-1859. Trenton, New Jersey.
Newell, John. 1853-1854. Albany, New York. SFB.
Newman,(Dr.). 1850-1854. Huntsville, Alabama.
Newton, Ephriam. 1848-- (inactive 1855). Cambridge, New York. SFB.
Newton, John. 1853--. Washington County, Florida. SFB.
Nichols (prof.). 1848-1851. Schenectady, New York.
Nichols, Andrew. 1848-1854. Danvers, Massachusetts.
Nichols, James Robinson. 1853--. Haverhill, Massachusetts.
Nichols, John A. 1856-1860. New York, New York.
Nicollet, Joseph Nicolas. 1848 (deceased AAGN ember). Washington D.C.
Niles, W. W. 1853 only. New York, New York.
Noble (Capt.). 1856 only. London, England.
Nolen, George A. 1859--. New Haven, Connecticut.
North, Edward. 1857-1859. Clinton, New York.
Northrop, Richard H. 1851-1854. Albany, New York. SFB.
Norton, Edward. 1859--. Farmington, Connecticut.
Norton, Hiram. 1854-1856. Meadville, Pennsylvania.
Norwood, Joseph Granville. 1848-1854. Madison, Indiana. SFB.

Oakes, William. 1848 only.* Ipswich, Massachusetts.
O'Callaghan, Edmund Bailey. 1856-1859. Albany, New York.
Oeland, John C. 1853-1855. Fort Prince, South Carolina.
Ogier, T. L. 1850-1854. Charleston, South Carolina.
Okie, Abraham H. 1855 only. Providence, Rhode Island
Olcott, Thomas W. 1849-1858 (inactive 1853). Albany, New York. LC.
O'Leary, Charles. 1856-1859. Emmetsburg, Maryland.
Oliver, James Edward. 1853-1860. Lynn, Massachusetts. SFB.
Olmsted, Charles Hyde. 1848-1859. East Hartford, Connecticut. SFB.
Olmsted, Denison, Jr., 1848 (deceased AAFN member). New Haven, Connecticut.
Olmsted, Lemuel G. 1848-1851. New York, New York.
Olney, Stephen Thayer. 1848-1859 (inactive 1853-1854). Providence, Rhode
    Island. LC.
Opdyke, George. 1854-1860. New York, New York.
Ordway, John Morse. 1849--(inactive 1853-1854). Roxbury, Massachusetts.
Ormiston, William. 1856-1859. Hamilton, Canada.
Orvis, Joseph U. 1851-1854. Troy, New York. SFB.
Osborn, Henry Stafford. 1856-1859. Bedford County, Virginia.
Osborne, E. 1851-1854. Sandusky, Ohio.
Osgood, Samuel. 1854-1860. New York, New York.
Osten Sacken, Carl R.R. 1856-1860. Washington, D.C.
Otis, George Alexander. 1856-1860. Springfield, Massachusetts. LC.
Owen, Richard. 1851-1854. New Harmony, Indiana. SFB.
Owen, Robert Dale. 1848-1854. New Harmony, Indiana.

Paine, Cyrus F. 1858--. Rochester, New York.
Paine, H.D. 1851-1854. Albany, New York.
Paine, Robert Treat. 1853 only. Boston, Massachusetts. SFB.

Painter, Minshall. 1853--. Delaware County, Pennsylvania. SFB.
Palmer, Aaron A. 1848-1851. New York, New York.
Palmer, Alonzo Benjamin. 1858-1860. Detroit, Michigan.
Palmer, Charles H. 1851-1854. Rome, Michigan.
Paradis, H.C.A. 1857-1859. Montreal, Canada.
Parker, Amasa Junius. 1851-1860. Albany, New York. LC. SFB.
Parker, Charles. 1850-1854. Charleston, South Carolina.
Parker, Henry E. 1857--. Concord, New Hampshire.
Parker, William Henry. 1856-1860. Middlebury, Vermont.
Parkman, Samuel. 1848-1849.* Boston, Massachusetts.
Parsons, Charles William. 1855-1857. Providence, Rhode Island.
Parsons, Theophilus. 1848-1854. Cambridge, Massachusetts.
Parsons, Usher. 1855-1859. Providence, Rhode Island.
Parvin, Theodore Sutton. 1853-- (inactive 1854-1856). Muscatine, Iowa. SFB.
Paton, George. 1857-1859. Galt, Canada.
Paton, John. 1857-1859. Toronto, Canada.
Patten, D. 1856 only. Concord, New Hampshire.
Patterson, James Willis. 1857--. Hanover, New Hampshire.
Payne, Henry B. 1851-1854. Cleveland, Ohio.
Peabody, Francis. 1848-1854. Salem, Massachusetts.
Peale, Titian Ramsay. 1848-1860. Washington, D.C. LC. SFB.
Pearson, Jonathan. 1848-1851. Schenectady, New York.
Peck, G. W. 1855-1857. New York, New York.
Peck, William G. 1859--. New York, New York.
Peckham, Fenner Harris. 1858-1860. Providence, Rhode Island.
Peet, Harvey Prince. 1851-1854. New York, New York. SFB.
Peirce, David S. 1851-1854. Albany, New York. SFB.
Peirce, James Mills. 1856--. Cambridge, Massachusetts.
Pendergast, John G. 1848-1854. Jefferson County, New York.
Pendleton, A. G. 1853-1857. Washington, D.C. LC. SFB.
Pendleton, Edmund Monroe. 1850-1854. Sparta, Georgia. SFB.
Percival, James Gates. 1848-1851. New Haven, Connecticut.
Perkins, Henry Colt. 1848-1854. Newburyport, Massachusetts. SFB.
Perkins, James Amory. 1857-1859. Montreal, Canada.
Perkins, Justin. 1848-1855. Oroomiah, Persia.
Perkins, Louis. 1857 only. New London, Connecticut.
Perry, Aaron F. 1853-1855. Columbus, Ohio. SFB.
Perry, Horace. 1851-1854. Cleveland, Ohio.
Perry, Matthew Calbraith. 1848-1859* (inactive 1851-1855). New York, New York.
Phelps, Almira H. L. 1859--. Baltimore, Maryland.
Phelps, Charles Edward. 1859--. Baltimore, Maryland.
Phelps, Edward Elisha. 1851-1854. Windsor, Vermont. SFB.
Phelps, Philip, 1856-1859. Hastings, New York.
Phelps, S. L. 1854-1856. Washington, D.C.
Phelps, William Franklin. 1857-1859. Trenton, New Jersey.
Phillips, William. 1853-1854. Augusta, Georgia. SFB.
Pickering, Charles. 1848-1857. Boston, Massachusetts.
Pierce, William F. 1857-1859. Trenton, New Jersey.
Pierrepont, Henry Evelyn. 1860--. Brooklyn, New York.
Pitcher, Zina. 1848--. Detroit, Michigan.
Pitman, Benn. 1856-1859. Cincinnati, Ohio.
Plant, I. C. 1850--. Macon, Georgia.
Pleasants, Thomas S. 1856-1857. Petersburg, Virginia.
Plumb, Ovid. 1855-death. Salisbury, Connecticut.
Pohlman, Henry. 1851-1854. Albany, New York.
Pope, Charles Alexander. 1858--. St. Louis, Missouri.
Porcher, Francis Peyre. 1850-1854. Charleston, South Carolina.
Porcher, Frederick Adolphus. 1850-1854. Charleston, South Carolina.
Porter, Benjamin Rickling. 1850-1854. Charleston, South Carolina.
Porter, Charles Hogeboom. 1859--. Albany, New York.
Porter, Charles T. 1859--. New York, New York.
Porter, Edward D. 1853 only. Newark, Delaware. SFB.
Porter, John Addison, 1850--. New Haven, Connecticut.
Porter, Noah. 1850-1854. New Haven, Connecticut. LC.
Porter, Samuel D. 1856 only. Rochester, New York.
Porter, Thomas Conrad. 1851-- (inactive 1853-1857). Lancaster, Pennsylvania. SFB.
Potter, Elisha Reynolds. 1853 only. Kingston, Rhode Island. SFB.
Potter, H.B. 1851-1859. Albany, New York.
Poulson, Charles A. 1848-1855. Philadelphia, Pennsylvania. SFB.
Powell, Robert I. 1857 only. Illinois.
Powell, Samuel. 1859--. Philadelphia, Pennsylvania. LC.
Powers, Albert E. 1848-1859. Lansingburgh, New York. SFB.

Pratt, J. D. 1858-1860. Baltimore, Maryland. LC.
Prentice, Ezra P. 1849-1860. Albany, New York. LC.
Prescott, William. 1848--. Concord, New Hampshire.
Priest, J. A. 1856-1859. Homer, New York.
Prince, William Robert. 1856--. Flushing, New York.
Prioleau, Thomas. 1850-1854. Charleston, South Carolina.
Probasco, Henry. 1851-1854. Cincinnati, Ohio.
Prout, Hiram Augustus. 1853-1859. St. Louis, Missouri. SFB.
Pruyn, John V. L. 1848--. Albany, New York. LC.
Pruyn, Robert Hewson. 1856-1859. Albany, New York.
Pulte, John Hippolyt. 1853 only. Cleveland, Ohio.
Purcell, John Baptist. 1850-1854. Cincinnati, Ohio. LC.
Putnam, Allen. 1848-1851. Boston, Massachusetts.
Putnam, Frederic Ward. 1856-1860. Cambridge, Massachusetts.
Pybos, Benjamin. 1853-1857. Tuscumbia, Alabama. SFB.
Pynchon, Thomas Ruggles. 1848-1860 (inactive 1855-1856). Hartford,
    Connecticut. SFB.

Quincy, Edmund. 1857--. Dedham, Massachusetts.

Rachmaninow, J. 1859-1860. Kiev, Russia.
Ramsay, J. K. 1857-1859. Montreal, Canada.
Randall, Samuel Sidwell. 1849-1854. Albany, New York.
Randell, John. 1854-1856. New York, New York.
Rankin, Robert Gozman. 1856-1860. New York, New York.
Rauch, John Henry. 1857-1860. Burlington, Iowa.
Ravenel, Edmund. 1848-1859 (inactive 1856). Charleston, South Carolina.
Ravenel, H. 1850-1854. Charleston, South Carolina.
Ray, Isaac. 1855-1857. Providence, Rhode Island. LC.
Raymond, Charles H. 1851-1854. Cincinnati, Ohio.
Read, D. B. 1857-1859. Toronto, Canada.
Read, Matthew C. 1853-1855. Hudson, Ohio. SFB.
Redfield, Charles B. 1857--. Albany, New York.
Redfield, John Howard. 1848--. New York, New York. SFB.
Reed, Lyman. 1858--. Baltimore, Maryland.
Rehfus, Lewis. 1851-1854. Cincinnati, Ohio.
Reid, William Wharry. 1848-1851. Rochester, New York.
Rennie, Alexander N. 1856 only. Montreal, Canada.
Renwich, James. 1848-1859. New York, New York. SFB.
Resor, Jacob. 1856-1860. Cincinnati, Ohio.
Reuben, Levi. 1856-1859. New York, New York.
Rhees, William Jones. 1854-1856. Washington, D.C.
Rhett, James. 1850-1853. Charleston, South Carolina. SFB.
Rice, Clinton. 1857--(inactive 1858-1859). New York, New York.
Rice, DeWitt C. 1853-1856. Albany, New York.
Rice, Henry. 1853--. North Attleborough, Massachusetts. SFB.
Rice, Nathan Lewis. 1850-1854. Cincinnati, Ohio. LC.
Rice, W. A. 1856-1859. Albany, New York.
Richards, Newton. 1857 only. New Orleans, Louisiana.
Richards, William Carey. 1855-1859. Providence, Rhode Island.
Richards, Zalmon. 1853-1859. Washington, D.C. LC.
Richardson, Horace. 1858--. Boston, Massachusetts.
Richmond, A. B. 1854-1856. Meadville, Pennsylvania.
Riddell, William Pitt. 1853-1859 (inactive 1856). New Orleans, Louisiana.
Riell (Lt.). 1850-1854. United States Navy.
Ripley, Hezekiah W. 1851-1859. Harlan, New York. SFB.
Roberts, Algernon S. 1848-1851. Philadelphia, Pennsylvania.
Roberts, William Milnor. 1851-1859 (inactive 1855-1856). Carlisle,
    Pennsylvania.
Roberts, F. M. 1850-1854. Charleston, South Carolina.
Robertson, Thomas D. 1856--. Rockford, Illinois. SFB.
Robertson, W.H.C. 1857-1859. Stamford, Connecticut.
Robinson, George C. 1859--. Cincinnati, Ohio.
Robinson, Horatio Nelson. 1850-1854. Cincinnati, Ohio. LC.
Robinson, J.R. 1851-1854. Mansfield, Ohio.
Robinson, W.B. 1857-1859. Toronto, Canada.
Rockwell, Alfred P. 1856--. Chicago, Illinois.
Rockwell, John. 1857--. Chicago, Illinois.
Rockwell, John Arnold. 1856--. Norwich, Connecticut.
Rodman, William Mitchell. 1855--. Providence, Rhode Island.
Roemer, Frederick. 1848-1855. Berlin, Prussia.

Rogers, Fairman. 1857--. Philadelphia, Pennsylvania.
Rogers, Samuel Towner. 1851-1854. Chestertown, Maryland. SFB.
Rogers, William. 1856-1859. Philadelphia, Pennsylvania.
Roome, Martin R. 1856-1859. New York, New York.
Root, Oren. 1848-- (inactive 1855-1858). Clinton, New York. SFB.
Rose, Henry. 1857-1859. Montreal, Canada.
Ross, Alexander Coffman. 1851-1854. Zanesville, Ohio. SFB.
Rosseter, George R. 1853 only. Buffalo, Virginia. SFB.
Roulston, Andrew. 1853 only. Freeport, Pennsylvania. SFB.
Rousseau, Henry. 1851-1854. Albany, New York.
Roy, Euclide. 1857-1859. Montreal, Canada.
Ruffin, Edmund. 1848-1851. Petersburg, Virginia.
Ruggles, Daniel. 1848-1854. Detroit, Michigan.
Ruggles, William. 1854--. Washington, D.C.
Ruschenberger, William S. W. 1848-1854. Brooklyn, New York.
Russell, Andrew. 1857--. Toronto, Canada.
Russell, Archibald. 1857-1859. New York, New York.
Russell, John Lewis. 1848-1854. Hingham, Massachusetts. SFB.
Rutherford, Lewis Morris. 1859--. Newport, Rhode Island. LC.
Ruttan, Allan. 1857-1859. Newburgh, Canada.
Ryan, Thomas. 1857-1859. Montreal, Canada.
Ryerson, E. 1857-1859. Montreal, Canada.

Sager, Abraham. 1851-1860. Ann Arbor, Michigan. SFB.
Salisbury, Edward Elbridge. 1850-1854. New Haven, Connecticut. LC.SFB.
Sanborn, Francis G. 1859--. Andover, Massachusetts.
Sanders, J. Milton. 1851-1854. Memphis, Tennessee.
Sanderson, J. E. 1857-1859. Montreal, Canada.
Sands, Samuel. 1858-1860. Baltimore, Maryland. LC.
Sanford, Richard R. 1853-1859. Riga, New York. SFB.
Sanger, W.W. 1858-1859. Blackswell Island, New York.
Sargent, Rufus. 1856--. Auburn, New York.
Savage, Thomas Stoughton. 1856--. Pass Christian, Mississippi.
Saville, Henry Martyn. 1855-1857. Boston, Massachusetts.
Sawyer, A. W. 1859--. Acadia College, Nova Scotia.
Saynisch, Lewis. 1848-1854. Blossburg, Pennsylvania.
Scarborough, George. 1849--. Owensburg, Kentucky.
Schaeffer, Peter W. 1850-1853. Pottsville, Pensylvania.
Schaff, Philip. 1858--. Mercersburg, Pennsylvania.
Schanck, John Stillwell. 1850--. Princeton, New Jersey. SFB.
Schlater, Philip Lutley. 1856 only. Oxford, England.
Schnee, Alexander. 1856-1859. Madison, Wisconsin.
Scholfield, Isaac. 1848-1851. Boston, Massachusetts.
Schoolcraft, Henry Rowe. 1853-1858. Washington, D.C. SFB.
Schoville, John. 1851-1854. Salisbury, Massachusetts.
Schreiner, Francis. 1853 only. Crawford County, Pennsylvania. SFB.
Scott, John Witherspoon. 1853-1855. Oxford, Ohio.
Scott, Joseph. 1857-1859. Durham, Canada.
Scott, William. 1857-1859. Durham, Canada.
Scriven, T.P. 1850-1854. Savannah, Georgia.
Scudder, Samuel Hubbard. 1859--. Boston, Massachusetts.
Sears, Phillip Howes. 1849-1851. Cambridge, Massachusetts.
Seely, William A. 1848-1853. New York, New York.
Seely, Charles A. 1851-1854. Rochester, New York. SFB.
Selden, George M. 1851-1859. Troy, New York.
Selden, Silas Richards. 1850-1854. New Haven, Connecticut.
Sellers, George Escol. 1853-1855. Cincinnati, Ohio. SFB.
Seropyan, Christopher D. 1856-1859. New York, New York.
Sessions, John. 1851-1859. Albany, New York. SFB.
Sestini, Benedict. 1854-1860. Washington, D.C. SFB.
Seward, William Henry. 1848--. Auburn, New York.
Seybert, Henry. 1848-1851. Philadelphia, Pennsylvania.
Seymour, M.H. 1857--. Montreal, Canada.
Shaeffer, Peter Wenrich. 1853--. Pottsville, Pennsylvania. SFB.
Shaffer, David H. 1853-1857. Cincinnati, Ohio. SFB.
Shaler, Nathaniel Burger. 1851-1854. Newport, Kentucky.
Shane, J.D. 1853--. Lexington, Kentucky. SFB.
Shattuck, George. 1848-1851. Boston, Massachusetts.
Shaw, Edward. 1855-1857. Washington, D.C.
Shaw, James. 1853-1855. Newburg, Ohio. SFB.
Sheldon, D.H. 1856--. Racine, Wisconsin.

Sheldon, D.S. 1856 only. Davenport, Iowa.
Sheldon, E.E. 1857-1859. Montreal, Canada.
Shepard, G.C. 1850-1854. Charleston, South Carolina.
Shepard, J. Avery. 1853-1854. Scuppernong, North Carolina. SFB.
Shepard, Perkins. 1848-1851. Providence, Rhode Island.
Shepard, Thomas Perkins. 1855 only. Providence, Rhode Island. LC.
Sheppard, William. 1857--. Drummondville, Canada.
Sherman, S.S. 1850-1854. Birmingham, Alabama.
Sherwin, Thomas. 1849-1859 (inactive 1853-1856). Dedham, Massachusetts.
     SFB.
Shippen, William. 1854-1856. Washington, D.C.
Short, Charles Wilkins. 1849-1851. Louisville, Kentucky.
Shotwell, Samuel L. 1857-1859. Macedon, New York.
Shurtleff, Nathaniel Bradstreet. 1848-1854. Boston, Massachusetts.
Sias, Solomon. 1856--. Fort Edwards, New York.
Silisby, Horace. 1853-1855. Blue Hill, Maine. SFB.
Sill, Elisha Noyes. 1851--. (inactive 1855). Guyahoga Falls,Ohio. SFB.
Simmons, Daniel. 1851-1854. Paris Hill, New York.
Sismonda, Eugene. 1848-1855. Turin, Italy.
Skilton, Avery J. 1851-1859. Troy, New York. SFB.
Skinner, A.W. 1853 only. Syracuse, New York.
Skinner, George W. 1856-1859. Little Falls, New York.
Skipwith, P. H. 1857 only. New Orleans, Louisiana.
Slack, Elijah. 1851-1854. Cincinnati, Ohio. SFB.
Slack, John Hamilton. 1858--. Philadelphia, Pennsylvania.
Smead, Morgan J. 1853-1855. Williamsburg, Virginia. SFB.
Smith, Amos Denison. 1860--. Providence, Rhode Island. LC.
Smith, Augustus William. 1851--. Middletown, Connecticut. SFB.
Smith, Erastus. 1848-1855. Hartford, Connecticut. SFB.
Smith, George. 1853-1860. Upper Darby, Pennsylvania.
Smith, George W. L. 1851-1855. Troy, New York. SFB.
Smith, Howard. 1853 only. New Orleans, Louisiana.
Smith, J. Bryant. 1853-1859. New York, New York.
Smith, J. V. 1851-death. Cincinnati, Ohio.
Smith, J.V.C. 1848-1854. Boston, Massachusetts.
Smith, James Youngs. 1855--. Providence, Rhode Island. LC.
Smith, John. 1857-1859. Montreal, Canada.
Smith, John Chappall. 1853-1855. New Harmony, Indiana.
Smith, John Spear. 1857 only. Baltimore, Maryland. LC.
Smith, Metcalf J. 1856-1859. M'Granville, New York.
Smith, Oliver. 1848-1854. New York, New York.
Smith, Peter. 1848-1851. Nashville, Tennessee.
Smith, Spencer. 1857--. St. Louis, Missouri.
Smith, Thomas. 1853-1855. Charleston, South Carolina.
Snodgrass, William. 1857-1859. Montreal, Canada.
Snow, Charles B. 1854-1859. Washington, D.C.
Snow, Edwin Miller. 1855--. Providence, Rhode Island.
Sola, A. 1857-1859. Montreal, Canada.
Sommers (Rev.) 1850-1854. Charleston, South Carolina.
Soule, Richard. 1848-1850. Boston, Massachusetts.
Sowell, J. F. 1849-1854. Athens, Alabama.
Spear, Charles. 1850-1859 (inactive 1854). Pittsfield, Massachusetts. SFB.
Spencer, Charles A. 1851-1855. Canastota, New York. SFB.
Spencer, Thomas. 1854-1856. Philadelphia, Pennsylvania.
Spillman, William. 1849-1851. Columbus, Ohio.
Spink, William. 1857-1860. Toronto, Canada.
Spinner, Francis Elias. 1848-1851. Herkimer, New York.
Spooner, Edward. 1850-1854. Plymouth, Massachusetts.
Sprague, Charles Hill. 1853-1856. Malden, Massachusetts. SFB.
Sprague, Daniel J. 1857-1860. South Orange, New Jersey.
Sprague, Daniel G. 1857-1859. South Orange, New Jersey.
Spring, Charles H. 1859--. Mt. Holyoke, Massachusetts.
Stadtmuller, Ludwig. 1850-1854. Bristol, Connecticut. SFB.
Staley, George L. 1858-1860. Mt. Washington, Maryland.
Stannard, Benjamin A. 1851--. Cleveland, Ohio. LC.
Stansbury, Howard. 1851-1855. Washington, D.C. SFB.
Starr, William. 1856--. Ceresco, Wisconsin.
Stearns, Eben Sperry. 1856-1859. Albany, New York.
Stearns, Edward Josiah. 1851-1854. Annapolis, Maryland. SFB.
Stearns, Josiah Atherton. 1856 only. Boston, Massachusetts.
Stearns, William. 1859--. Oxford, Mississippi.

Stebbins, Richard. 1850-1854. Springfield, Massachusetts.
Stebbins, Rufus Phineas. 1849-1857. Carlisle, Pennsylvania. SFB.
Steele, Samuel. 1856-1859. Albany, New York.
Stephens, A. H. 1848-1854. New York, New York.
Sterling, John Whalen. 1859--. Madison, Wisconsin.
Stetson, Charles. 1850-1858. Cincinnati, Ohio. LC.
Steuart, Alexander. 1851-1854. Lebanon, Tennessee. SFB.
Stevens, J. J. 1851-1854. Washington, D.C.
Stevens, M.C. 1855-1859. Richmond, Indiana.
Stevenson, Charles L. 1860--. Charlestown, Massachusetts.
Stewart, William M. 1853--. Clarksville, Tennessee. SFB.
Stickney, Lyman D. 1853 only. Perry County, Indiana. SFB.
Stillman, C. H. 1854-1856. Plainfield, New Jersey.
Stillman, J.D.B. 1854-1856. New York, New York.
Stillman, Thomas B. 1854-1857. New York, New York.
Stimson, William. 1849--. (inactive 1851-1857). Washington, D.C.
Stoddard, John William. 1851-1854. Oxford, Ohio.
Stodder, Charles. 1848-1855. Boston, Massachusetts. SFB.
Stone, Charles. 1851-1854. Trenton, New Jersey. SFB.
Stone, Edwin Martin. 1855--. Providence, Rhode Island.
Stone, L. 1851-1854. Cleveland, Ohio.
Storer, David Humphreys. 1848--. Boston, Massachusetts.
Storer, Francis Humphreys. 1859--. Boston, Massachusetts.
Street, Alfred Billings. 1856-1859. Albany, New York.
Streeter, Sebastian Ferris. 1857-1859. Baltimore, Maryland. LC. SFB.
Strong, Woodbridge. 1848-1851. Boston, Massachusetts.
Stuart, A.P.S. 1857-1859. Acadia, Nova Scotia.
Stuntz, George R. 1853-1855. Lancaster, Wisconsin.
Sturtevant, Julian Monson. 1856-1859. Jacksonville, Illinois.
Suckley, George. 1855-1857. New York, New York.
Sullivant, Joseph. 1853-1855. Columbus, Ohio.
Sullivant, William Starling. 1853-1860. Columbus, Ohio.
Sumner, Charles. 1849-1854. Boston, Massachusetts.
Sumner, George. 1854-1856. Boston, Massachusetts.
Sutherland, William. 1851--. Montreal, Canada.
Swan, Lansing B. 1854-1859. Rochester, New York.
Swann, Thomas. 1857 only. Baltimore, Maryland, L.C.
Sweeny, Peter Barr. 1856-1859. Buffalo, New York.
Swift, Paul. 1848-1851. Philadelphia, Pennsylvania.
Swift, William. 1848-1854. New York, New York.
Swinburne, John 1851-- (inactive 1855). Albany, New York.

Tabor, Azor. 1851-1859. Albany, New York.
Talcott, Andrew. 1853-1859. Cincinnati, Ohio. SFB.
Tallmadge, James. 1848-1853.* New York, New York.
Tappan, Benjamin. 1848-1851. Steubenville, Ohio.
Tappan, Henry Philip. 1856-1859. Ann Arbor, Michigan.
Tatlock, John. 1856-1860. Williamstown, Massachusetts.
Tatum, Joel H. 1856-1859. Baltimore, Maryland.
Taylor, George W. 1856-1858. Albany, New York.
Taylor, J. W. 1857-1859. Montreal, Canada.
Taylor, James. 1851-1854. Albany, New York.
Taylor, John W. 1856-1859. Wampsville, New York.
Taylor, Julius S. 1848-1859. Montgomery County, Ohio. SFB.
Taylor, Morse K. 1853-1859. Galesburg, Illinois. SFB.
Taylor, Morse R. 1853-1854. Jackson County, Michigan.
Taylor, Thomas M. 1857-1859. Montreal, Canada.
Taylor, W. I. 1858-1860. Worcester County, Maryland.
Taylor, William. 1857-1859. Montreal, Canada.
Taylor, William Johnson. 1851-1855. Philadelphia, Pennsylvania.
Tefft, Thomas Alexander. 1855-1858. New York, New York.
Tellkampf, Theodore A. 1848-1851. New York, New York.
Tenney, Benjamin J. 1851-1854. Kingston, New York.
Terlecki, Ignatius. 1853-1854. Paris, France.
Teschemacher, James Englebert. 1848-1853. *Boston, Massachusetts. SFB.
Teschermakst, E.D. 1848-1851. Schenectady, New York.
Tevis, Robert C. 1851-1859. Shelbyville, Kentucky.
Thayer, Henry White. 1848-1851. Boston, Massachusetts.
Thayer, S.W. 1855 only. Burlington, Vermont.
Thayer, Solomon. 1848-1853. Lubec, Maine.
Thickstun, T.F. 1857-1860. Meadville, Pennsylvania.
Thomas, David. 1853 only. Aurora, New York.

Thomas, Richard. 1857-1859. Montreal, Canada.
Thomas, William A. 1856--. Irvington, New York.
Thomas, William S. 1851-1854. Norwich, New York.
Thompson (Dr.) 1851-1853. Aurora, New York.
Thompson, Aaron R. 1848--. New York, New York.
Thompson, Alexander. 1848--(inactive 1851-1853). Aurora, New York.
Thompson, H.C. 1859--. Chapel Hill, North Carolina.
Thompson, J.W. 1855-1857. Wilmington, Delaware.
Thompson, John A. 1856-1859. Geneva, New York.
Thompson, John Edgar. 1848-1857. Philadelphia, Pennsylvania.
Thompson, Robert. 1853-1855. Columbus, Ohio.
Thompson, Waddy. 1850-1854. Greenville, South Carolina.
Thompson, Zadock. 1848--. Burlington, Vermont. SFB.
Thomson, John. 1857-1859. New York, New York.
Thoreau, Henry David. 1853 only. Concord, Massachusetts.
Thorn, James. 1856-1859. Troy, New York.
Thurber, George. 1848-1856. Providence, Rhode Island, LC.
Thurber, Isaac. 1855--. Providence, Rhode Island. LC.
Thurston, E.M. 1853-1855. Charleston, Maine. SFB.
Tiffany, Charles C. 1851-1854. Baltimore, Maryland. SFB.
Tiffany, Otis H. 1851-1854. Carlisle, Pennsylvania. LC. SFB.
Tilghman, Tench. 1858-1860. Baltimore, Maryland.
Tillinghast, Nicholas. 1853 only. Bridgewater, Massachusetts. SFB.
Tingley, Joseph. 1851-- (inactive 1856-1859). Greencastle, Indiana. SFB.
Tobey, Samuel Boyd. 1855-1857. Providence, Rhode Island. LC.
Tolderoy, James B. 1857-1860. Fredericton, New Brunswick.
Tolman, Albert. 1851-1854. Pittsfield, Massachusetts. SFB.
Torrey, Joseph. 1849-1859. Burlington, Vermont.
Totten, Silas. 1848-1851. Hartford, Connecticut.
Town, Salem. 1853-1857. Aurora, New York.
Townsend, Franklin. 1850-- (inactive 1855), Albany, New York. LC. SFB.
Townsend, Howard. 1856-1859. Albany, New York.
Townsend, John Kirk. 1848-1851.* Philadelphia, Pennsylvania.
Townsend, Robert. 1855-1860. Albany, New York.
Townshend, D. J. 1850-1854. John's Island, South Carolina. SFB.
Treadwell, C.P. 1857--. L'Original, Canada.
Treadwell, Daniel. 1851-1854. Cambridge, Massachusetts.
Treadwell, J. 1848-1854. Cambridge, Massachusetts.
Treadwell, Oliver Wetmore. 1856 only. Rockville, Maryland.
Trego, Charles B. 1848-1854. Philadelphia, Pennsylvania.
Trenholm, George Alfred. 1850-1854. Charleston, South Carolina.
Troost, Lewis. 1849-1854. Nashville, Tennessee.
Trudeau, Alexis. 1857-1859. Montreal, Canada.
True, Nathaniel T. 1848-1851. Monmouth, New Jersey.
Truesdell, Samuel. 1856-1859. New York, New York.
Trumbull, James Hammond. 1850-1857. Stonington, Connecticut.
Tucker, George. 1848-1851. Philadelphia, Pennsylvania.
Tuckerman, Edward. 1848-1853. Boston, Massachusetts.
Turner, Henry E. 1860--. Newport, Rhode Island.
Turner, William C. 1853-1855. Cleveland, Ohio. SFB.
Turner, William Wadden. 1848-1859 (inactive 1851). Washington, D.C. LC. SFB.
Tuthill, Franklin. 1854--. New York, New York.
Tuttle, David Kitchell. 1858-1860. Charlottesville, Virginia.
Tyler, Edward R. 1848 (deceased AAGN member). New Haven, Connecticut.
Tyler, Moses Coit. 1853-1855. Detroit, Michigan. SFB.
Tyler, P.B. 1859--. Springfield, Massachusetts.
Tyler, Ransom Hebbard. 1856--. Fulton, New York.
Tyler, Robert S. 1857-1859. Montreal, Canada.
Tyson, Isaac. 1853. Baltimore, Maryland. SFB.

Uhler, Philip Reese. 1858--. Baltimore, Maryland.
Ulffers, H. A. 1857 only. Springfield, Illinois.
Ulrici, Richard W. 1857 only. St. Louis, Missouri.
Upham, G.B. 1860--. Boston, Massachusetts.
Upham, Nathaniel Gookin. 1856--. Concord, New Hampshire.

Vail, Hugh D. 1854--. Haverford, Pennsylvania.
Vail, Stephen Montfort. 1856. Concord, New Hampshire.
Van Benschoten, James Cooke. 1858-1860. Chenango County, New York.
Van Cortlandt (Dr.). 1857 only. Ottawa, Canada.
Van Derpool, S. Oakley. 1855-1859. Albany, New York.

Van Durzee, William S. 1853-1857. Williamsville, New York. SFB.
Van Lennup, Henry John. 1848-1855. Constantinople, Turkey.
Van Pelt, William. 1853-- (inactive 1860). Williamsville, New York. SFB.
Van Tuyle, Henry. 1851-1854. Dayton, Ohio.
Vanuxem, Lardner. 1848 (deceased AAGN member). Bristol, Pennsylvania.
Van Vleck, John Monroe. 1855--. Middletown, Connecticut.
Van Vliet, Stewart. 1853-1854. Brownsville, Texas.
Venable, Charles Scott. 1850-1854. Hampden-Sidney, Virginia.
Verreau, Hospice-Anthelme J.B. 1856-1859. Montreal, Canada.
Volck, Adalbert John. 1858. Baltimore, Maryland.
Vost, George Leonard. 1855. New York, New York.

Wadsworth, Charles F. 1857-1860. Genesee, New York.
Wadsworth, James S. 1849--. Geneseo, New York.
Wagner, Tobias. 1854--. Philadelphia, Pennsylvania.
Wailes, Benjamin L.C. 1848-1854. Washington, D.C. SFB.
Walker, James Barr. 1853-1859. Mansfield, Ohio.
Walker, Joseph Reddeford. 1853-1855. Platte City, Missouri. SFB.
Walker, Joseph. 1856--. Oxford, New York.
Walker, Timothy, 1850-1856.* Cincinnati, Ohio. LC.
Wallace, C. 1850-1851. Charleston, South Carolina.
Wallbridge, T. C. 1857-1859. Belleville, Canada.
Walling, Henry F. 1855-1857. Providence, Rhode Island.
Walworth, Reuben H. 1856-1859. Saratoga, New York.
Ward, Henry Augustus. 1859--. Rochester, New York.
Ward, Mathew A. 1850-1851. Athens, Georgia. SFB.
Waring, Charles B. 1851-1854. New York, New York. SFB.
Waring, Samuel Copp. 1851-1854. New York, New York.
Warner, Edward. 1853. New Brighton, Pennsylvania.
Warren, Gouveneur Kemble. 1858--. United States Army.
Warren, Jonathan. 1848-1851. Boston, Massachusetts.
Warriner, Justin B. 1848-1851. Burlington, New Jersey.
Waterton, Charles. 1857-1859. Wakefield, England.
Watson, James Craig. 1859--. Ann Arbor, Michigan.
Wayland, Francis. 1855-1857. Providence, Rhode Island. LC.
Wayne, Benjamin. 1856-1858. New Orleans, Louisiana.
Weaver, George Sumner. 1848-1851. Cambridgeport, New York.
Webster, Horace B. 1848. (deceased AAGN member). Albany, New York.
Webster, John White. 1848-1850* Cambridge, Massachusetts.
Webster, Matthew H. 1848 (deceased AAGN member). Albany, New York.
Webster, William Franklin. 1855-1857. Providence, Rhode Island. LC.
Weddell, Horace, P. 1851-1854. Cleveland, Ohio. LC.
Wedderburn, A.J. 1848-1851. New Orleans, Louisiana.
Weed, Monroe. 1851-1859. Wyoming, New York. SFB.
Welch, John. 1856-1860. Newark, New Jersey.
Weld, Henry Thomas. 1848-1854. Mt. Savage, Maryland.
Wells, Samuel. 1848-1851. Northampton, Massachusetts.
Wells, Thomas. 1850-1858. New Haven, Connecticut. SFB.
Wentworth, Erastus. 1853-1855. Carlisle, Pennsylvania. SFB.
Wescott, Isaac. 1854-1856. New York, New York.
Wethered, Charles. 1858-1860. Baltimore, Maryland. SFB.
Wetherill, Dr. Charles Mayer. 1850-1854. Philadelphia, Pennsylvania.
Wetherill, John Price. 1848-1851. Philadelphia, Pennsylvania. LC. SFB.
Wetherill, Leander. 1849-1854. Rochester, New York. SFB.
Wetherill, Samuel. 1857-only. Bethleham, Pennsylvania.
Weyde, Wander. 1856 only. New York, New York.
Wheatland, Henry. 1848--. Salem, Massachusetts. SFB.
Wheatland, Richard H. 1859--. Salem, Massachusetts.
Wheatley, Charles Moore. 1848-- (inactive 1855-1856). New York, New York. SFB.
Wheeler, T.B. 1857--. Montreal, Canada.
Whepley, W.J.D. 1848-1851. New York, New York.
Whipple, A.B. 1856-1859. Nantucket, Massachusetts.
Whipple, J.E. 1856-1859. Lansingburgh, New York.
Whipple, Milton D. 1848-1851. Lowell, Massachusetts.
Whipple, Walter. 1853-1858. Adrian, Michigan. SFB.
Whitcomb, Joseph M. 1856-1858. Salem, New York.
White, Aaron. 1856-1859. Cazenovia, New York.
White, Charles. 1856--. Crawfordsville, Indiana.
White, E. B. 1850-1854. Charleston, South Carolina.
White, George. 1850-1854. Marietta, Georgia.
White, Henry H. 1860--. Harrodsburg, Kentucky.

White, Horace. 1856-1858. Chicago, Illinois.
White, Robert J.P. 1857-1859. Chambly, Canada.
Whiting, H. 1848-1854. Detroit, Michigan.
Whitman, William E. 1848-1854. Philadelphia, Pennsylvania.
Whitney, Asa. 1848-1860. Philadelphia, Pennsylvania.
Whitney, Eli. 1850-1854. New Haven, Connecticut.
Whitney, H.H. 1857--. Montreal, Canada. LC.
Whitney, John R. 1857-1859. Philadelphia, Pennsylvania.
Whitridge (Dr.). 1850-1851. Charleston, South Carolina.
Whittemore, Thomas J. 1849-1854. Cambridge, Massachusetts. SFB.
Whittich. 1850-1854. Pennfield, Georgia.
Whittlesey, Charles C. 1857--. St. Louis, Missouri. SFB.
Wightman, William May. 1850-1856. Charleston, South Carolina.
Wilcox, Lester. 1850-1854. Canajoharie, New York.
Wilder, Alexander. 1856-1859. New York, New York.
Wilder, Henry. 1848-1854. Lancaster, Massachusetts.
Wilder, Lyman. 1848-1859. Hoosick Falls, New York.
Wilgress, George. 1857-1859. Montreal, Canada.
Wilkes, Henry, 1857-1859. Montreal, Canada.
Willard, Samuel D. 1856-1859. Cayuga, New York.
Willey, George. 1853. Cleveland, Ohio.
Williams, Abram E. 1854 only. Albany, New York.
Williams, Abraham V. 1855-1857. New York, New York.
Williams, George Palmer. 1853-1855. Ann Arbor, Michigan. SFB.
Williams, Henry Willard. 1857--. Boston, Massachusetts.
Williams, L. D. 1851-1856. Meadville, Pennsylvania.
Williams, Matthew. 1859--. Syracuse, New York.
Williams, Matthew J. 1853. Columbia, South Carolina. SFB.
Williams, Moses B. 1848-1851. Boston, Massachusetts.
Williams, Peter O. 1851-1857. St. Lawrence County, New York. SFB.
Williams, S. K. 1851-1854. Canonsburg, Pennsylvania.
Williams, Samuel Wells. 1856-1859. Canton, China.
Williams, Thomas. 1858--. Portsmouth, Virginia.
Williams, William Fenwick. 1856-1858. Mosul, Turkey.
Williamson, James. 1857-1859. Kingston, Canada.
Williamson, Robert Stockton. 1858--. San Francisco, California.
Williman, A.B. 1850-1854. Charleston, South Carolina.
Wills, Frank. 1855-1857. New York, New York.
Wilson, George Francis. 1855-1856. Providence, Rhode Island.
Wilson, H.V. 1851-1854. Cleveland, Ohio. LC.
Wilson, John. 1853-1855. London, England.
Wilson, John. 1851-1854. Staunton, Virginia. SFB.
Wilson, John Z. 1851-1854. Albany, New York.
Wilson, Joseph. 1851-1854. Hampton-Sidney, Virginia. SFB.
Wilson, Samuel. 1850-1854. Charleston, South Carolina.
Wilson, W.C. 1858-1860. Carlisle, Pennsylvania.
Wilson, William. 1851-1854. Cincinnati, Ohio.
Wing, Joel A. 1850-1854. Albany, New York. LC. SFB.
Wing, Matthew G. 1850-1854. Albany, New York. SFB.
Winslow, Charles Frederick. 1856-1859. Newton, Massachusetts.
Winslow, Gordon. 1848-1854. Staten Island, New York.
Winslow, John Flack. 1856-1859. Troy, New York.
Winslow, Rufus K. 1851-1855. Cleveland, Ohio. LC.
Wolf, Elias. 1848-1851. New York, New York.
Wood, William. 1856-1859. Portland, Maine.
Woodall, John W. 1857-1859. Scarborough, England.
Woodbridge, George A. 1856--. Nashville, Tennessee.
Woodbridge, Jon E. 1851 only. Boston, Massachusetts. SFB.
Woodbury, Levi. 1848-1851.* Portsmouth, New Hampshire. SFB.
Woodbury, Peter T. 1853. New York, New York.
Woodhull, Maxwell. 1850-1855. United States Navy. LC.
Woodman, John Smith. 1857-1860. Hanover, New Hampshire.
Woodrow, James. 1853-1854. Milledgeville, Georgia. SFB.
Woodruff, H.W.B. 1857. New York, New York.
Woodruff, Luni. 1853 only. Ann Arbor, Michigan. SFB.
Wool, John Ellis. 1851-1854. Troy, New York. SFB.
Woollett, William L. 1851-1854. Albany, New York.
Woolsey, John Mumford. 1851-1854. Cleveland, Ohio. LC.
Woolsey, Theodore Dwight. 1850-1854. New Haven, Connecticut. LC.
Woolworth, Samuel Buell. 1856-1859. Albany, New York. LC.
Worcester, Joseph Emerson. 1849-1857. Cambridge, Massachusetts.

Wormley, Theodore George. 1853-1855. Columbus, Ohio. SFB.
Wragg, William. 1850-1854. Charleston, South Carolina.
Wright, A.D. 1848-1851. Brooklyn, New York. SFB.
Wright, Albert B. 1853-1855. Perrysburg, Ohio.
Wright, Arthur Williams. 1860--. New Haven, Connecticut.
Wright, Charles. 1856-1859. Wethersfield, Connecticut.
Wright, Charles W. 1851-1854. Cincinnati, Ohio. SFB.
Wright, Joseph J. B. 1853-1854. Carlisle Barracks, Pennsylvania. SFB.
Wright, John. 1848 (deceased AAGN member). Troy, New York.
Wright, Joseph C. 1857 only. Oswego, New York.
Wurtele, Louis C. 1857--. Lennoxville, Canada.
Wynne, James. 1860--. New York, New York.
Wynne, Thomas H. 1854--. Richmond, Virginia.

Yarnall, M. 1854-1856. United States Army.
Yates, Tiles F. 1848-1851. Albany, New York.
Yeadon, Richard. 1850-1854. Charleston, South Carolina.
Youmans, Edward Livingston. 1851--. Saratoga Springs, New York. SFB.
Young, Ira. 1848-1858.* Hanover, New Hampshire. SFB.
Young, J.A. 1853-1854. Camden, South Carolina. SFB.
Young, John. 1857-1859. Montreal, Canada.
Young, William H. 1858--. Baltimore, Maryland. LC.
Young, William L. 1851-1854. Vicksburg, Mississippi.

Zimmerman, C. 1850-1856 (inactive 1855). Columbia, South Carolina.

C: Non members

A group of men who apparently never solicited nor otherwise gained
AAAS membership became at least nominally involved in the annual meetings of
the organization. Such men may or may not have actually participated in the
meetings, because papers could be submitted by mail and read by the AAAS
secretary; committeemen might not fulfill any responsibility. The following
list includes, however, all persons who were named in the published Proceedings as participants without any attempt being made to verify their
activity. After each name is the person's residence (if known) and his type
of participation: LC for local committee, PP for a paper or report, and
simply Committee if some other assigned committee responsibility.

Allen, Philip. Providence, Rhode Island. LC.
Alvord, Benjamin. Washington, D.C. PP.
Ames, Samuel. Providence, Rhode Island. LC.
Andrews, John. Columbus, Ohio. Committee.
Andrews, Sherlock James. Cleveland, Ohio. Committee.
Angell, William G. Providence, Rhode Island. LC.
Arnold, Richard. Providence, Rhode Island. LC.
Atwater, George M. Springfield, Massachusetts. LC.

Badger, Oscar C. United States Navy. PP.
Barnard, Daniel Dewey. Albany, New York. LC.
Bartlett, D.L. Baltimore, Maryland. LC.
Barton, E.H. New Orleans, Louisiana. PP.
Beaujeu, P.D. Montreal, Canada. LC.
Benham, H.W. Washington, D.C. LC.
Bowles, Samuel. Springfield, Massachusetts. LC.
Bordley, James. Baltimore, Maryland. LC.
Bowditch, J. Ingersoll. Boston, Massachusetts. Committee.
Brent, John Carroll. Washington, D.C. LC.
Brown, George. Baltimore, Maryland. LC.
Brown, Henry. Cambridge, Massachusetts. PP.
Brown, Nicholas. Providence, Rhode Island. LC.
Brown, William W. Providence, Rhode Island. LC.
Brownell, A.C. Cleveland, Ohio. LC.
Bryan, Joseph. Washington, D.C. LC.
Bullock, William Peckham. Providence, Rhode Island. LC.

Chaffee, C.C. Springfield, Massachusetts. LC.
Chapin, Chester William. Springfield, Massachusetts. LC.
Chapman, Edward J. Canada. PP.

Chapman, Reuben Atwater. Springfield, Massachusetts. LC.
Chase, T. Massachusetts. PP.
Clark, Myron Holley. Albany, New York. LC.
Coffin, W.C.C. Committee.
Corcoran, William Wilson. Washington, D.C. LC.
Coxe, Richard Smith. Washington, D.C. LC.

Davis, Thomas. Providence, Rhode Island. LC.
De Haas, Willis. Committee.
Donan, A. A. Montreal, Canada. LC.
Dowden, C. PP.
Dundas, James Hepburn. Philadelphia, Pennsylvania.
Dunnell, Jacob. Providence, Rhode Island. LC.
Dunnell, Thomas Lyman. Providence, Rhode Island. LC.

Ferrier, James. Montreal, Canada. LC.
Foote, Eunice. PP.
Fowler, Joseph Smith. Davidson County, Tennessee. Committee.

Genth, Frederick Augustus. New York, New York. PP.
Giles, W. F. Baltimore, Maryland. LC.
Gorham, John. Providence, Rhode Island. LC.
Gray, Andrew B. Mexican Boundary Survey. PP.

Habich, G. Roxbury, Massachusetts. PP.
Hale, Nathan. Boston, Massachusetts. LC. SFB.
Halton, L.H. Montreal, Canada. LC.
Hardeland, L.P. PP.
Hartshorn, Joseph Charles. Providence, Rhode Island. LC.
Hawley, Gideon. Albany, New York. LC.
Headley, Joel Taylor. Albany, New York. LC.
Hill, Silas. Washington, D.C. LC.
Hoffman, S.O. Baltimore, Maryland.
Hooker, Josiah. Springfield, Massachusetts. LC.
Hoppin, Thomas Frederick. Providence, Rhode Island. LC.
Hoppin, William Warner. Providence, Rhode Island. LC.
Howard, Frederick. Washington, D.C. LC.
Hughes, W.E. PP.
Hunt, C.B. United States Navy. PP.
Hunt, G.H. Baltimore, Maryland. LC.

Ives, Robert H. Providence, Rhode Island. LC.

James, John H. Urbana, Ohio. Committee.

Kane, John Kintzing. Philadelphia, Pennsylvania. Committee.
Kane, Paul. Toronto, Canada. PP.
Kendall, C.D. PP.
Kendall, H.L. Providence, Rhode Island. LC.
Kennedy, Joseph C.G. Washington, D.C. LC.
King, John L. Springfield, Massachusetts. LC.
Kingsbury, John. Providence, Rhode Island. LC.
Kingsley, H.L. Cleveland, Ohio. LC.
Knight, John. New Haven, Connecticut. LC.
Kohl, John George. Washington, D.C. PP.

LaFontaine, Louis Hypolite. Montreal, Canada. LC.
Lane, Jonathan H. Washington, D.C. PP.
Lea, Matthew Carey. Philadelphia, Pennsylvania. PP.
Lewis, James. Mohawk, New York. PP.
Lowell, John Amory. Boston, Massachusetts. LC.

Manton, Walter. Providence, Rhode Island. LC.
Maury, J.W. Washington, D.C. LC.
Mayer, Alfred M. Baltimore, Maryland. PP.
Mayer, Charles F. Baltimore, Maryland. LC.
Maynadier (Capt.). Washington, D.C. LC.
Medary, Samuel. Columbus, Ohio. Committee.
Meredith, J.F. Baltimore, Maryland. LC.
Meredith, Jonathan. Baltimore, Maryland. LC.
Mohun, Francis. Washington, D.C. LC.

Moore, Francis. Texas. PP.
Murchison, Roderick Impey. Scotland. PP.

Neilson, J. Crawford. Baltimore, Maryland. LC.
Nicholson, Alfred O.P. Washington, D.C. LC.

Oakes, John. PP.

Paterson, John. Albany, New York. PP.
Perry, Oliver Henry. Cleveland, Ohio. LC.
Phelps, Ansel. Springfield, Massachusetts. LC.
Phillips, James. Chapel Hill, North Carolina. Committees.
Poey, Andres. Havana, Cuba. PP.
Pyne (Rev.). Washington, D.C. LC.

Salter, John William. Edinburgh, Scotland. PP.
Sangston, Lawrence. Baltimore, Maryland. LC.
Schubert, M.E. Cambridge, Massachusetts.
Seaton, William Winston. Washington, D.C. LC.
Secchi, Angelo. Washington, D.C. PP.
Smith, Stephen H. Providence, Rhode Island. LC.
Smyth, William. Brunswick, Maine. Committee.
Sprague, William Buell. Albany, New York. LC.
Stanley, Anthony D. New Haven, Connecticut. Committees.
Stead, Thomas J. Providence, Rhode Island. LC.
Stone, Amasa. Cleveland, Ohio. LC.
Stuart, R.S. Baltimore, Maryland. LC.
Sunderland, B. Washington, D.C. LC.

Tayloe, Banjamin Ogle. Washington, D.C. LC.
TenEyck, Philip. Albany, New York. PP.
Thompson, J.M. Springfield, Massachusetts. LC.
Tiffany, Francis. Springfield, Massachusetts. LC.
Trimble, Allen. Highland County, Ohio. Committee.
Tully, William. Springfield, Massachusetts. LC.

Van Rensselaer, Stephen. Albany, New York. LC.
Vansant, Joshua. Baltimore, Maryland. LC.
Van Winkle, Peter Godwin. Baltimore, Maryland. LC.

Waesche, George F. R. Baltimore, Maryland. LC.
Walker, George. Springfield, Massachusetts. LC.
Warder, George A. Baltimore, Maryland. LC.
Waterman, Richard. Providence, Rhode Island. LC.
Waterman, Rufus. Providence, Rhode Island. LC.
Weed, Thurlow. Albany, New York. LC.
William, C.P. PP.
Wurdeman, William. Washington, D.C. PP.
Wyckoff, Isaac Newton. Albany, New York. LC.